Transitioning to Microsoft Power Platform

An Excel User Guide to Building Integrated Cloud Applications in Power BI, Power Apps, and Power Automate

David Ding

Apress®

Transitioning to Microsoft Power Platform: An Excel User Guide to Building Integrated Cloud Applications in Power BI, Power Apps, and Power Automate

David Ding
Sydney, NSW, Australia

ISBN-13 (pbk): 978-1-4842-9238-9
ISBN-13 (electronic): 978-1-4842-9239-6
https://doi.org/10.1007/978-1-4842-9239-6

Managing Director, Apress Media LLC: Welmoed Spahr
Acquisitions Editor: Joan Murray
Development Editor: Laura Berendson
Editorial Assistant: Gryffin Winkler
Copyeditor: Kim Burton

Cover image designed by kjpargeter on Freepik (www.freepik.com)

Distributed to the book trade worldwide by Springer Science+Business Media New York, 233 Spring Street, 6th Floor, New York, NY 10013. Phone 1-800-SPRINGER, fax (201) 348-4505, e-mail orders-ny@springer-sbm.com, or visit www.springeronline.com. Apress Media, LLC is a California LLC and the sole member (owner) is Springer Science + Business Media Finance Inc (SSBM Finance Inc). SSBM Finance Inc is a **Delaware** corporation.

For information on translations, please e-mail booktranslations@springernature.com; for reprint, paperback, or audio rights, please e-mail bookpermissions@springernature.com.

Apress titles may be purchased in bulk for academic, corporate, or promotional use. eBook versions and licenses are also available for most titles. For more information, reference our Print and eBook Bulk Sales web page at http://www.apress.com/bulk-sales.

Any source code or other supplementary material referenced by the author in this book is available to readers on GitHub.

Printed on acid-free paper

To my wife, Yuan Yao, for her endless love and support

To my boys, Martin and Harry, for making me care more about the future

To Dr. Peter Critchley for his friendship, wisdom, and guidance

To Dmitry Osipenko for providing excellent feedback on this book

To Joan Murray for believing in me on my first book idea

Table of Contents

About the Author

David Ding is the director and lead consultant of SDInnovation, a consulting company that provides customized decision support and training services to organizations across multiple industries. David is a certified Power BI developer with a master's degree in data science. Previously, David held multiple senior business and technical positions. This book is part of his mission to help everyone to get better with data.

About the Technical Reviewer

Fabio Claudio Ferracchiati is a senior consultant and analyst/developer using Microsoft technologies. He works for Bluarancio (`www.bluarancio.com`). He is a Microsoft Certified Solution Developer for .NET, a Microsoft Certified Application Developer for .NET, a Microsoft Certified Professional, and a prolific author and technical reviewer. Over the past ten years, he's written articles for Italian and international magazines and coauthored more than ten books on a variety of computer topics.

Introduction

The self-service BI (business intelligence) movement in recent years has vastly improved the capability of analysts in finance and front-line businesses. The level of entry for Power BI is so low that everyone who cares about data can start using it in a few days. Microsoft Power Platform has further empowered everyone to build business applications and automation solutions.

What Is This Book About?

This book empowers you to build better data solutions, respond faster to stakeholder requests, and contribute to a better data culture in your areas of choice. You learn the basics of building interactive reports using Power BI, building applications using Power Apps, building process automation flows using Power Automate, and finally, integrating all three tools to build a comprehensive data-driven decision solution.

Who Is This Book For?

This book is for all Excel users with a good business understanding and looking for ways to do things better and faster.

This book is for business analysts and reporting specialists who have watched some YouTube videos on Power BI and are humble enough to understand that a 30-minute video plus trial-and-error strategy has certain limitations.

This book is for business and team leaders passionate about improving decisions with available data but may be frustrated with the collaboration process with some technical teams.

This book is also for university students willing to learn about one of the most popular data platforms in 2023.

How Is This Book Structured?

The book is structured to address some of the most common business challenges analysts face in their day-to-day job. It discusses business challenges and the methods for solving them. In key chapters of the book, you are given some mini-hackathon-style challenges as an effective way to solidify the learning.

The book covers 11 chapters.

Chapter 1 quickly gets you started with building your first Power BI report. Starting from a simple Power BI licensing guide to help you to understand the common license rules and some limitations. You then work on translating an existing Excel report into Power BI.

Chapter 2 introduces you to the three core components of Power BI: Power Query, data models, and DAX. You use different components to solve some of this chapter's most common reporting challenges.

Chapter 3 helps you understand the best practices in collaboration and automation. Power BI Service (cloud server) is another key advantage for Power BI over its many competitors. This chapter focuses on how Power BI Service can help you to facilitate collaboration and automation.

Chapter 4 provides a guide on visualizing data for self-service BI solutions. This chapter starts with various visualization options and why you should choose some charts over others. It introduces the concept of data explanation vs. exploration and how to achieve both in Power BI. It ends with some very practical report design concepts.

Chapter 5 provides an effective framework for collaboration between multiple developers or teams of developers. This chapter starts with governance. You learn about the utilization-driven approach to improve build efficiency.

Chapter 6 provides an entry-level guide for data manipulation in SQL databases for reporting purposes. You start with creating a cloud database for practice purposes. You then learn to construct the most common SQL queries.

Chapter 7 is a comprehensive guide on the SharePoint list for Power Platform data storage. You start with creating and editing a SharePoint list. You then focus on access and security considerations of the list. You also learn about the different ways of connecting to Power BI.

Chapter 8 introduces creating simple applications with Power Apps. You learn about Power Apps licensing arrangements. You first create an app using the default Apps template. You then create an app from a blank canvas.

Chapter 9 helps you to understand the fundamentals of Power Apps. You learn about many important concepts, including variables and delegation. You then improve the basic app using several techniques, including the color palette, user interface features, dynamic calculation, and more on user experience.

Chapter 10 helps you to understand the basics of cloud workflows. You start with Power Automate licensing. You then learn about the basic concepts of Cloud Flow. You build three common flows: email notification, creating a backup, and creating approval workflows.

Chapter 11 introduces you to the Power Platform integrated solution architecture. You learn how to build reporting, back-end automation, and front-end user applications into a single solution.

CHAPTER 1

Power BI First Report

Microsoft defines Power BI as follows: "Connect to and visualize any data using the unified, scalable platform for self-service and enterprise business intelligence (BI) that's easy to use and helps you gain deeper data insight."

Whereas this makes sense, no one else would describe Power BI that way. Over the past seven years, I have worked with many new Power BI users, including many from a technology background. Some of them see Power BI as a glorified Excel, while others mainly focus on the visualization aspect of the tool. While these are not wrong, they are just the tip of the iceberg. As you can see in Figure 1-1, most people only see the reporting or dashboard side of Power BI. People don't see the back-end connections, transformation, modeling, and cloud services. Arguably the power of Power BI has more to do with what people don't see than what they see.

D. Ding, *Transitioning to Microsoft Power Platform*,
https://doi.org/10.1007/978-1-4842-9239-6_1

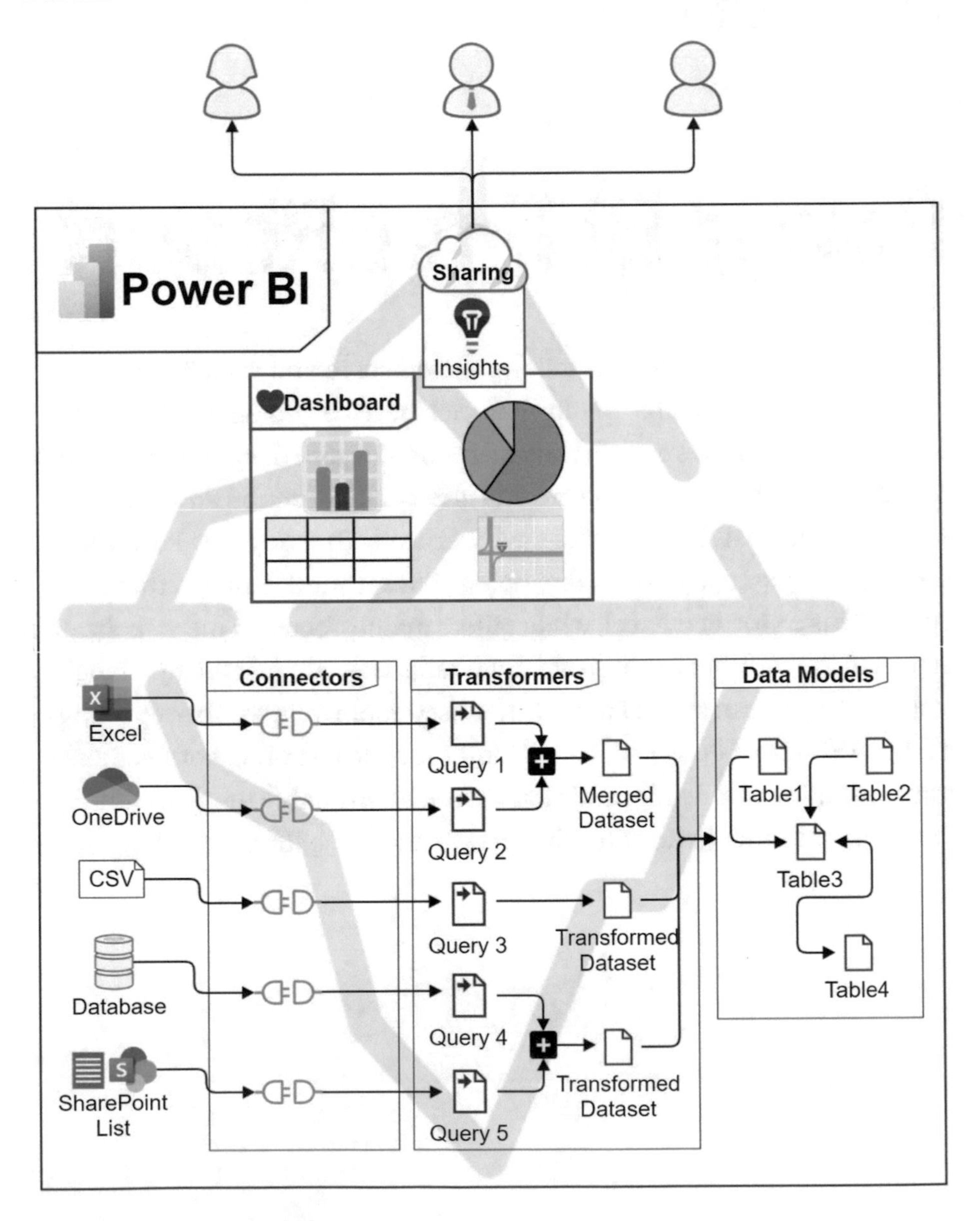

Figure 1-1. *Power BI process view*

The first three chapters tackle some practical challenges linked to different business scenarios. Each challenge focuses on a specific knowledge area, including data source connection, data transformation, data model, various interactive visuals, sharing, and automation. You can access the dataset and report files to follow the development steps.

Note Let's start with a basic example and progress quickly into more practical and in-depth scenarios. Check out Microsoft's tour if you want more time with the basics; visit `https://docs.microsoft.com/en-us/power-bi/fundamentals/desktop-getting-started`.

This first chapter focuses on developing the first report. Before you get into the development, you first gain some knowledge about Power BI subscriptions and costs.

Business Scenario

When it comes to real-world challenges, understanding the *why* is often more important than knowing the *how*. This is also true for learning; it is much easier to absorb the *how* if you understand the importance of the content. This book revolves around Kim's career development at AWM Bicycle Company Limited.

AWM is a US-based bike manufacturing business that produces bikes, bike parts, clothes, and accessories. The business faces strong overseas competition, local labor shortages, and high operational costs. After careful market analysis, leaders in AWM decided to consolidate its product lines to produce fewer items that better meet the evolving needs of most cyclists. Figure 1-2 shows AWM's organizational structure.

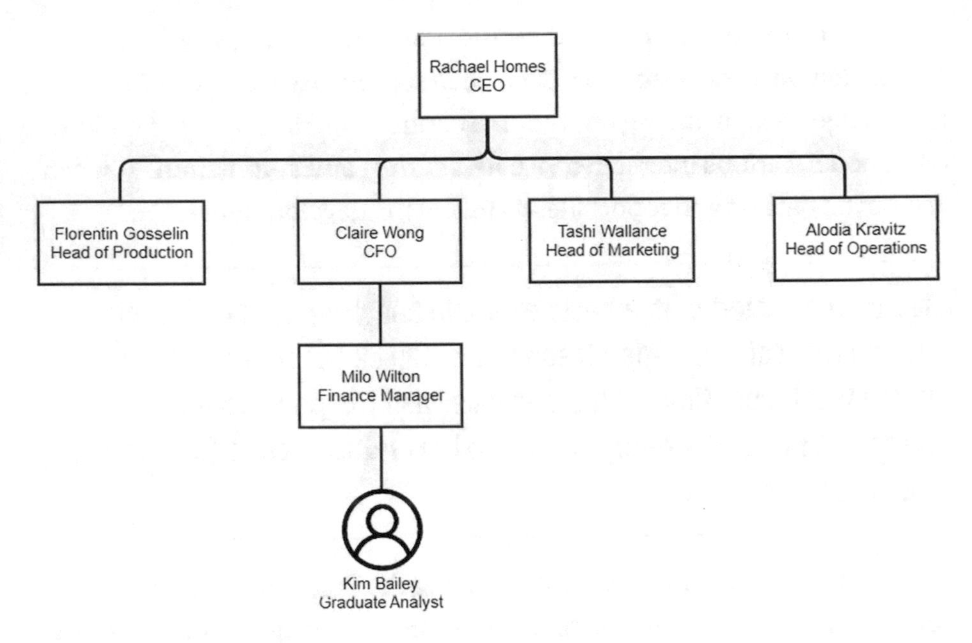

Figure 1-2. *AWM organizational structure*

Kim is a new university graduate with a degree in accounting. She is passionate about data analytics and has the drive to make better decisions. Kim joined AWM two months ago, reporting to Milo, the finance manager under Claire's CFO team. John and Claire were two financial partners that reported to Milo.

After joining AWM for three months, Kim has learned about what the company is doing and her role within the financial reporting team. Instead of just giving her the administration tasks, Milo, the finance manager, has decided to balance her exposure with small, ad hoc analysis projects. Microsoft Excel is the only tool for data analysis; the visualization is done in Excel and PowerPoint.

Power BI Software Licensing

Software licensing and subscriptions are confusing but important aspects of any tool. Understanding some basic rules help you to determine if the tool is fit for purpose and cost-effective for your organization.

Kim told Milo that she taught herself how to use Microsoft Power BI during her last semester at the university. To her pleasant surprise, Milo showed a lot of interest in the tool. However, he was concerned about the associated cost and asked her to investigate before proceeding.

Kim went to the Power BI pricing site at `https://powerbi.microsoft.com/en-us/pricing/` to read about the types of licenses and also the difference between the license types. She summarized the situation in Table 1-1.

Table 1-1. *Power BI Licensing Summary (Sep-2022)*

License	Cost	Use
Power BI Desktop	Free	Desktop can be used to develop reports; this is also the full version.
Power BI Pro	US$10 per user per month (60-day free trial)	Pro is required for every user sharing and viewing reports.
Premium Per User	US$20 per user per month (60-day free trial)	Expensive options that unlock additional features are not required.
Premium Capacity	$5,000 per capacity per month (60-day free trial)	More capacity features and sharing capabilities.

In addition, she discovered different types of "workspaces" for sharing the reports, as shown in Table 1-2.

Table 1-2. *Power BI Workspaces*

Workspace Type	Required License	Description
My Workspace	No license required	Reports can be published here only for the publisher to see. It can't be shared without a Pro license.
Normal Workspace	Power BI Pro	Reports can be published here with excellent sharing capabilities.
Premium Workspace	Premium	There are more capacity features and sharing capabilities.

Generally speaking, this licensing arrangement makes it easy and low-cost at the start. The process for acquiring a new license is also very simple. Premium capacity provides another option for organizations to cap their spending.

The following are a few additional licensing rules to note.

- **The professional license applies to people**. There is no "viewer license." The Pro license allows developers and viewers to access reports in workspaces. A professional license is required to publish reports. Except for personal workspace
- **Premium capacity applies to workspaces**. One premium capacity can be assigned to multiple workspaces. It gives all users access to read reports. Workspaces with Premium Capacity also enable premium functionalities. This includes the DevOps function and paginated reports.

Figure 1-3 shows the relationship between users, developers, and Power BI workspaces.

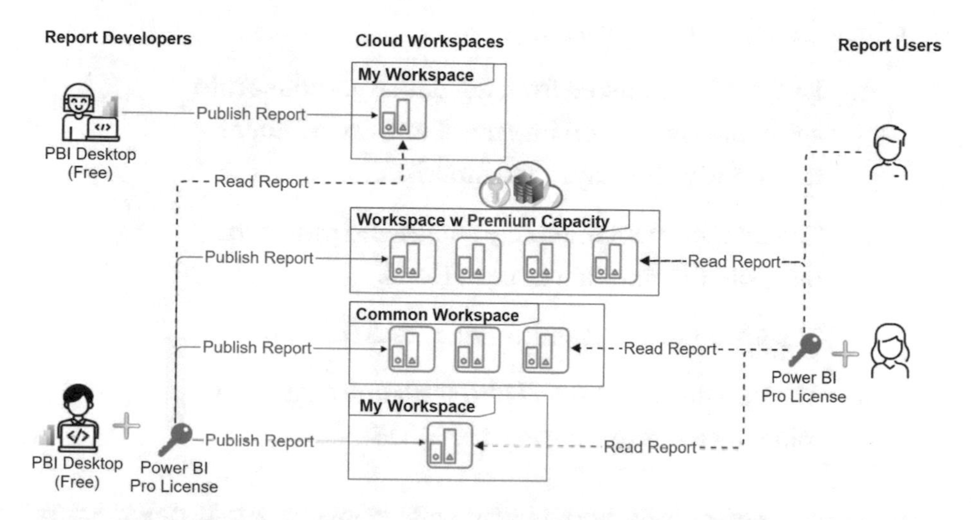

Figure 1-3. User and developer workspaces licenses

Build Your First Power BI Report

Now that you understand the basics of Power BI licensing and cost, it is time to build your first Power BI report.

Data and Requirements

In this task, let's help Kim analyze the product sales data in Power BI. Milo showed Kim the current Excel dashboard in Figure 1-4. He's happy with the layout, especially the interactive aspect of the data filters. He used this for the monthly senior management team updates to show month-on-month sales performance. A copy of the report is in the file01 Excel report.xlsx exercise file.

The technology team generated the data sitting behind the chart. It is in the file02 monthly product sales data.csv exercise file. The following describes the monthly process.

1. Raw data is extracted from the data warehouse into a CSV file and placed into the share-drive folder by the 2nd working day of the month.

2. The finance team manually copies and pastes the data into the raw data table in Excel.

3. The pivot table and charts are refreshed.

4. The updated file is saved into the same share-drive folder by the fifth working day (WD5).

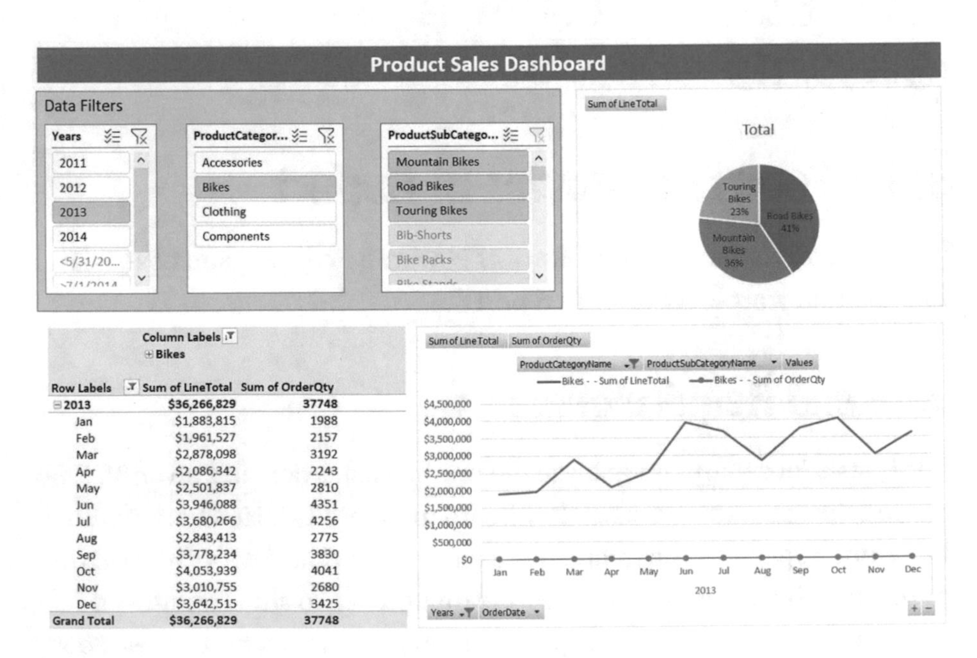

	Column Labels	
	Bikes	
Row Labels	Sum of LineTotal	Sum of OrderQty
2013	$36,266,829	37748
Jan	$1,883,815	1988
Feb	$1,961,527	2157
Mar	$2,878,098	3192
Apr	$2,086,342	2243
May	$2,501,837	2810
Jun	$3,946,088	4351
Jul	$3,680,266	4256
Aug	$2,843,413	2775
Sep	$3,778,234	3830
Oct	$4,053,939	4041
Nov	$3,010,755	2680
Dec	$3,642,515	3425
Grand Total	$36,266,829	37748

Figure 1-4. *Excel report on sales revenue*

Install Power BI Desktop

Your first task is to install Power BI Desktop. There are two installation options.

- Download from the Power BI site at `https://powerbi.microsoft.com/en-us/downloads/` and execute the installation file.
- Download from Microsoft Store (recommended).

Microsoft Store method is preferred because it includes an auto-update feature. It is important to note that Microsoft rolls out new features, improvements, and bug fixes monthly. Instead of downloading and installing each month, it is easier to let Power BI update itself when a new release is available.

Note It is possible to have two Power BI installed on the same machine using the two different methods. This can result in a lot of confusion. It is recommended to keep the installation by removing both and reinstalling from the Microsoft Store.

Report Building

Replicating Excel reports is a good place to start with Power BI report. There are two main reasons.

- Familiarity with data
- A separate report for data reconciliation purposes

Once Power BI is installed, you can build desktop reports. Figure 1-5 shows that connecting to some data is the first step.

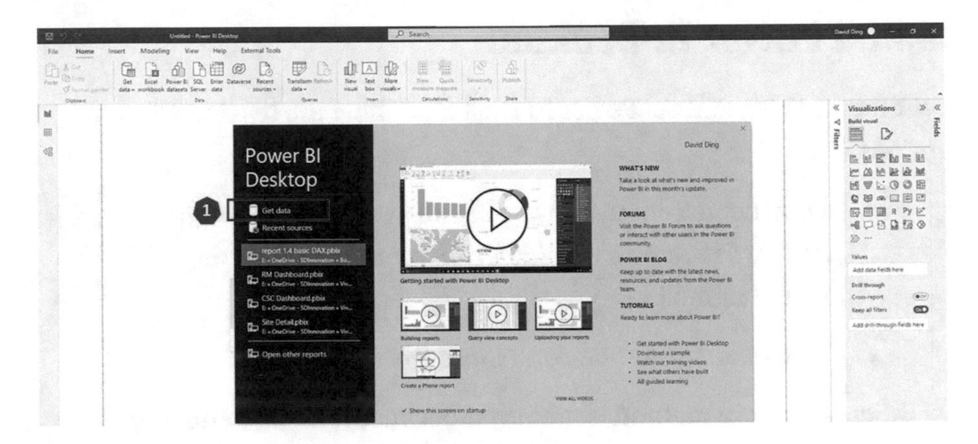

***Figure 1-5.** Connect to data from Power BI Desktop*

You can then load the data, as shown in the following steps and Figure 1-6.

1. Select **Text/CSV**.
2. Choose the file from the local folder.
3. Select **Load**.

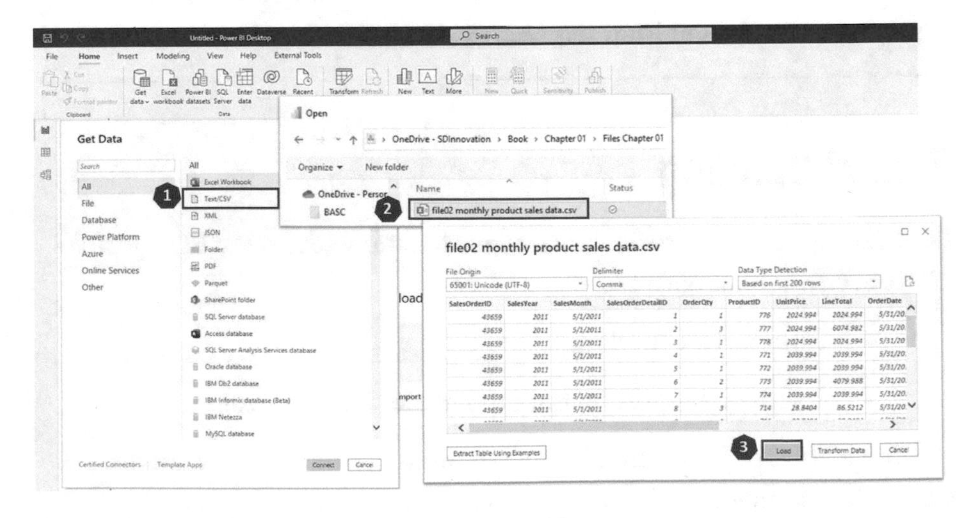

***Figure 1-6.** Steps to connect to a CSV file*

Now that the data has been loaded into Power BI, note the different functional areas of Power BI Desktop, as described next and illustrated in Figure 1-7.

1. **Report canvas** is where visuals are created and managed.
2. The **ribbon** displays common operations for reports and visuals.
3. **Chart options** is the default visual selection panel.
4. **Data fields** can be dragged to the formatting pane.
5. **Visual format** is for data fields and custom formatting.
6. **Filters** is the default area for adding filters to the visuals and all pages.
7. **Page** is where you select or add pages.
8. Other views include navigating to report, data, and model views.

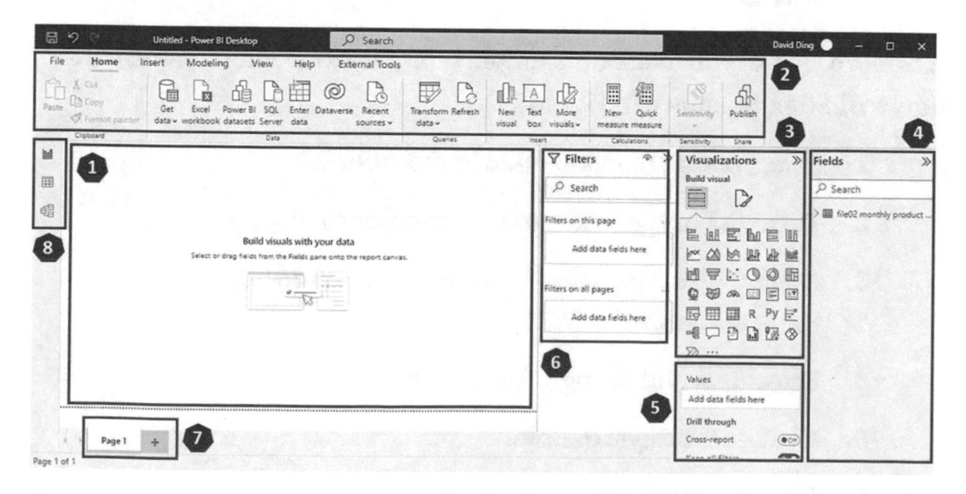

Figure 1-7. Power BI Desktop function areas

When replicating the Excel report, matching the visuals shown in Figure 1-8 to Power BI visuals is a good idea.

The visuals include (1) text box, (2) slicers, (3) pie chart, (4) matrix, and (5) line chart. They are added one by one into the dashboard.

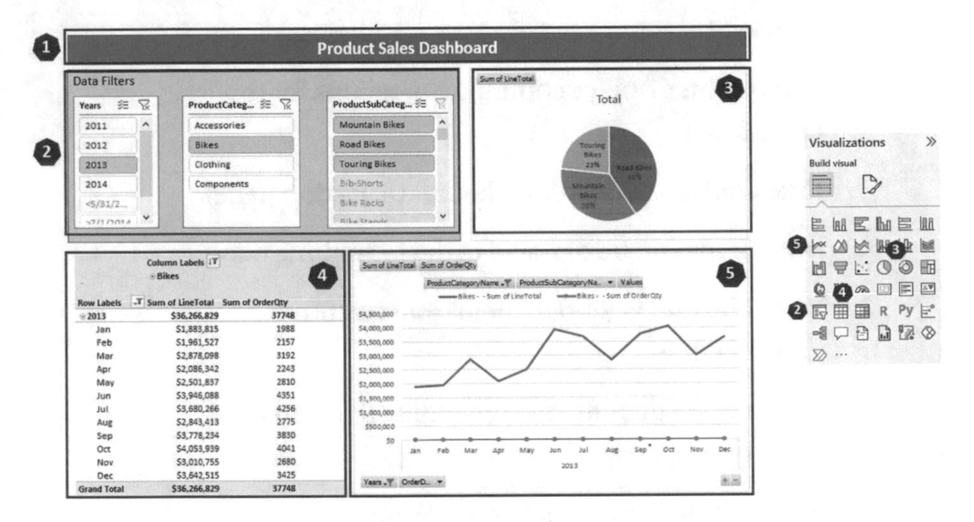

Figure 1-8. Excel report visual in Power BI

Add a Text Box

A text box allows you to create a simple title. The following steps explain how to do this, as shown in Figure 1-9.

1. Select Insert, then **Text box** in the ribbon.
2. Turn on Background and set the color to blue.
3. Add the **Produce Sales Dashboard** text and adjust the text box size.
4. Select text and change Font Size to 28.
5. Change font style to Bold.
6. Change text alignment to Center.
7. Save the change to Replace Excel.

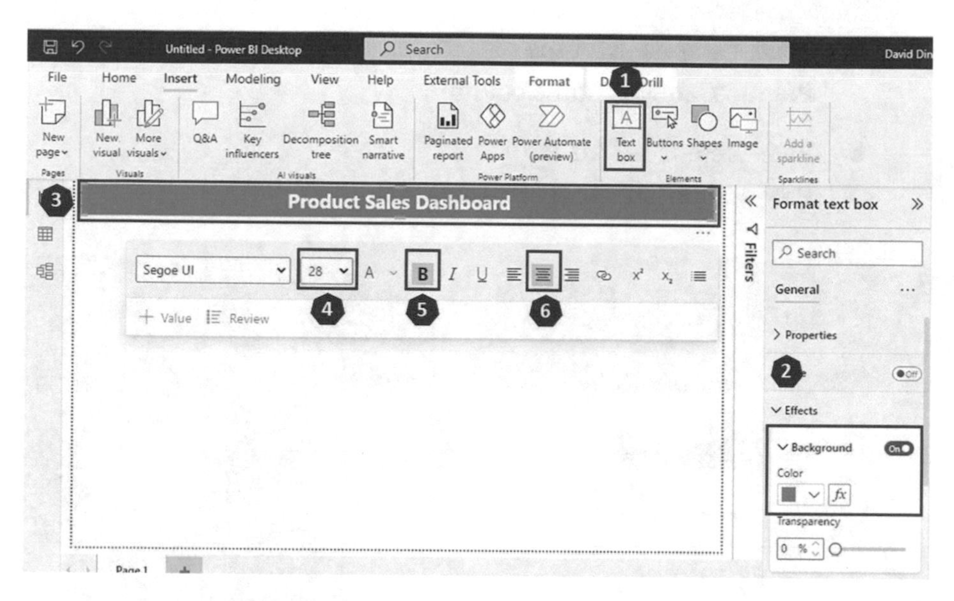

Figure 1-9. *Adding a text box*

Add Slicers

You can add the slicers using the following steps, as illustrated in Figure 1-10.

Tip Whenever you add a new element to the report canvas, you should click a blank area so that no other elements are selected.

---

1. In Visualizations, choose the slicer.
2. Drag and drop Sales/Year into the data field of the slicer visual.
3. From the Slicer drop-down, choose List.

4. Repeat steps 1–3 for the ProductCategoryName column.

5. Repeat steps 1–3 for the ProductSubCategoryName column.

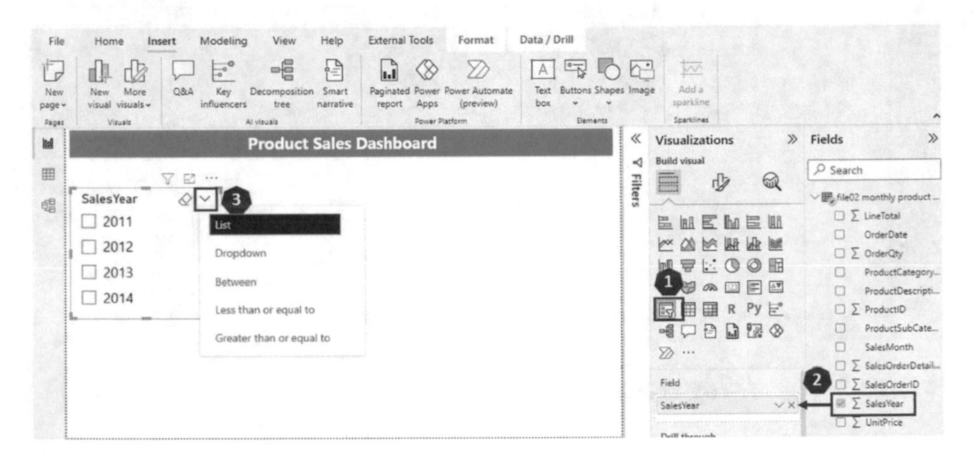

Figure 1-10. Adding a slicer

Tip The slicer drop-down box has a few options. The options depend on the column type. You can try the different drop-down options.

Add a Pie Chart

Add the pie chart. The following steps explain how to do it, as illustrated in Figure 1-11.

1. Click a blank area of the canvas (apply this for all future steps).

2. In Visualizations, choose the pie chart.

3. Drag and drop LineTotal into the pie chart values.

4. Drag and drop ProductSubcategoryName into the pie chart legend.

5. Select the format visual.

6. In the Label Content drop-down, choose Category, Percentage of Total.

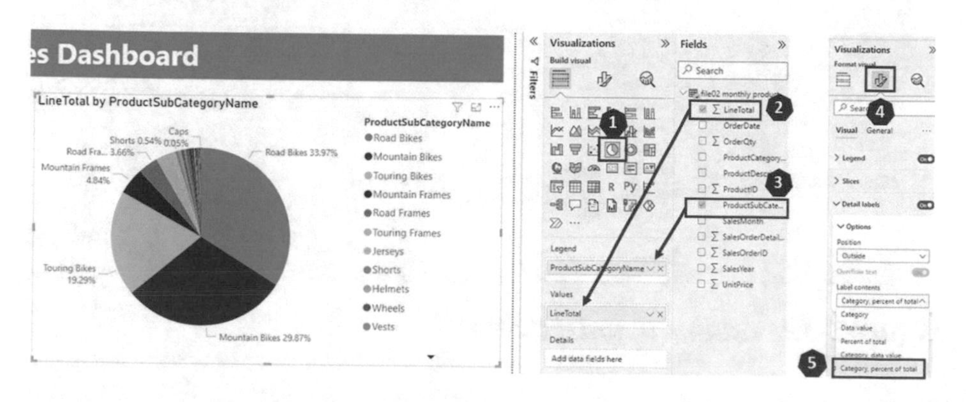

Figure 1-11. *Adding a pie chart*

Add a Matrix Table

The following steps add the matrix table, as illustrated in Figure 1-12.

1. In Visualizations, choose the matrix table.

2. Drag and drop LineTotal into matrix values.

3. Drag and drop ProductCategoryName into the matrix columns.

4. Drag and drop SalesMonth into the matrix rows.

5. Select SalesMonth.

6. Change column format to yyyy-mm.

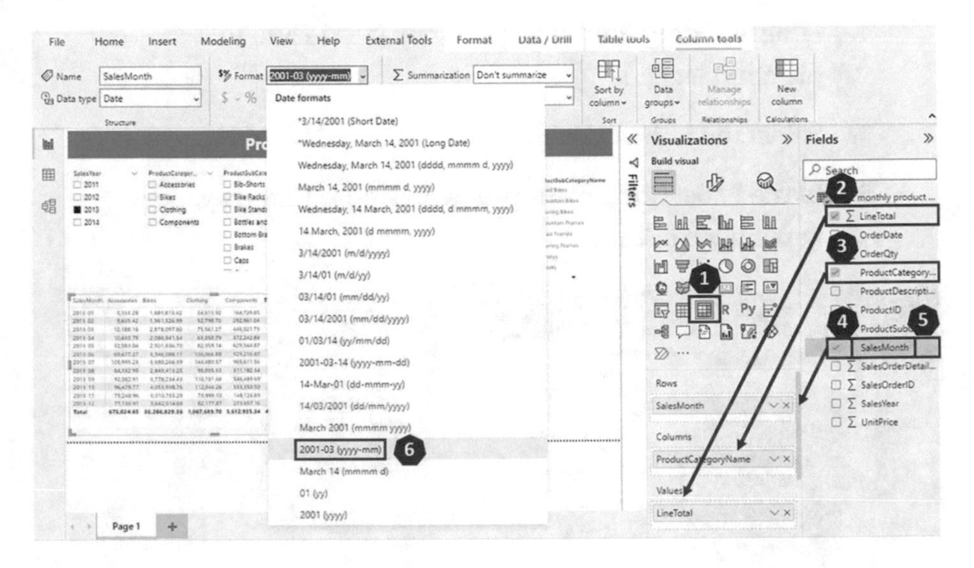

Figure 1-12. Adding a matrix table

Add a Line Chart

The following steps explain how to add the line chart, as illustrated in Figure 1-13.

1. In Visualizations, choose Line Chart.
2. Drag and drop LineTotal to the line chart Y-axis.
3. Drag and drop ProductCategoryName to the line chart legend.
4. Drag and drop SalesMonth to the X-axis.
5. Adjust the position and size of the different visuals.
6. Apply the year filter to 2013 to see the 2013 view.

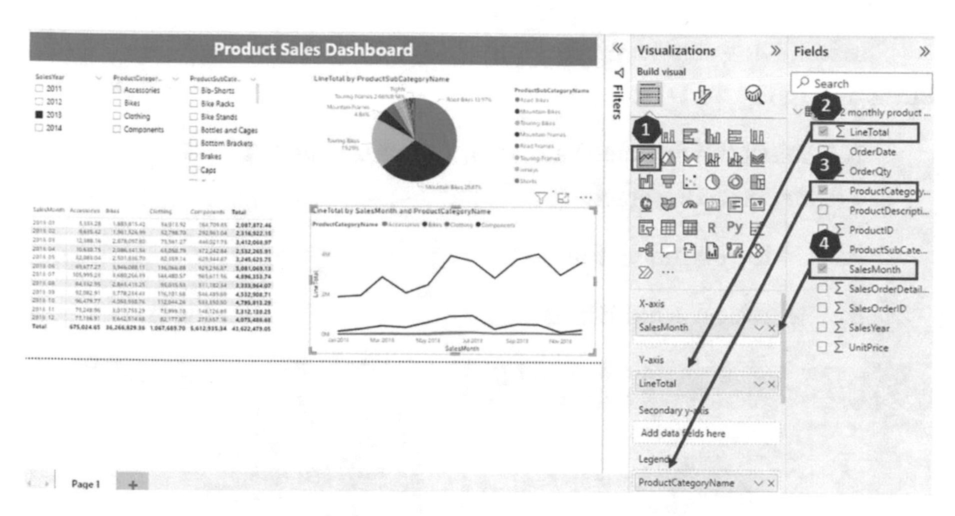

__Figure 1-13.__ Adding a line chart

Caution Power BI recognizes the SalesMonth column as a date column. Sometimes Power BI may recognize the SalesMonth column as a text column. This causes issues when you replicate the matrix and line chart processes. Changing a column type is covered in the Power Query session, so you can ignore the issues for now.

Congratulations, you have now replicated all elements from the Excel report. One of the main strengths of Power BI over Excel is its dynamic and interactive nature. You can click different sessions in the visual and see how it causes other parts of the report to change.

Clean up the Report

Tidying up the report is a critical step when developing Power BI reports. This may involve experimenting with different visuals, checking unexpected interactions, and making adjustments.

The following steps replace a line chart with a stacked column chart, as illustrated in Figure 1-14.

1. Select the line chart and change the visual to a stacked column chart.

2. Go to **Format visual.**

3. Switch on **Total labels.**

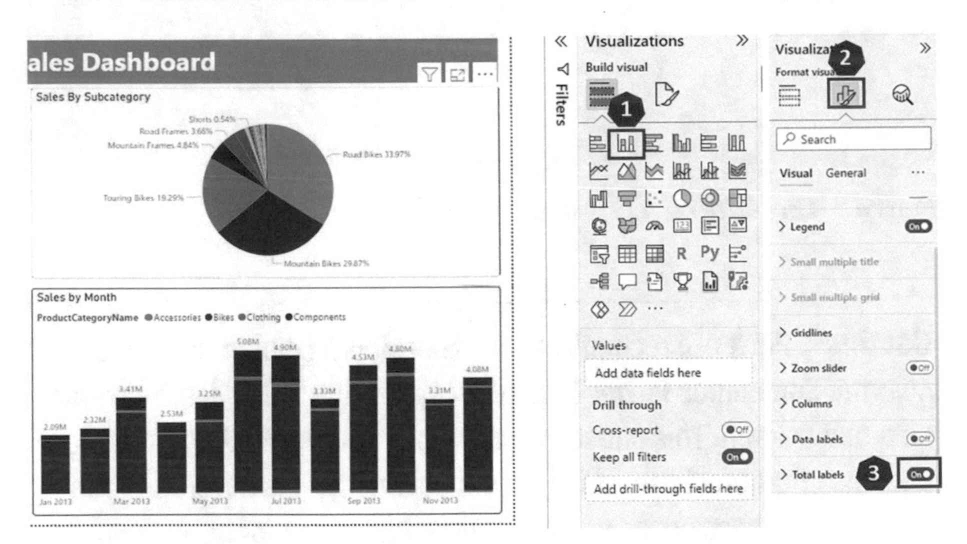

***Figure 1-14.** Change line chart to stacked column chart*

There are three interaction options: Filter, Highlight, and No interaction. If you pick a value in SalesMonth in the matrix table, the pie chart and the stacked bar chart are highlighted. As you can see in Figure 1-15, this interaction works well for the stacked bar chart but not so well for the pie chart.

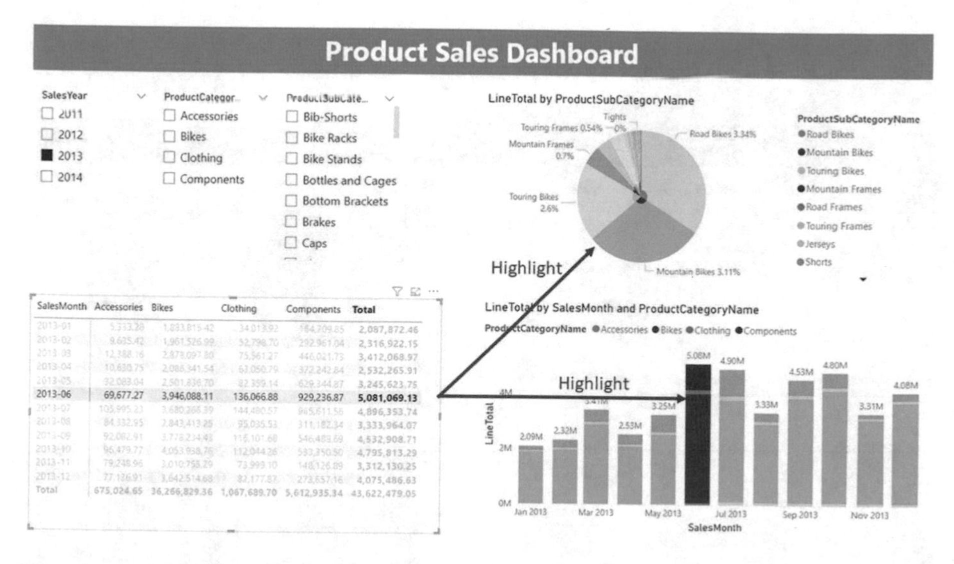

Figure 1-15. *The default Power BI interaction is Highlight*

As illustrated in Figure 1-16, the following steps explain how to change the interaction from a matrix table to a pie chart. In this case, you want to filter the value in a pie chart. In this context, *filter* means that when a month is chosen in the matrix table, the pie chart only shows the value and breakdown for the month.

1. Select the matrix table.
2. Choose **Format** and select **Edit interactions.**
3. Change the interaction to Filter.

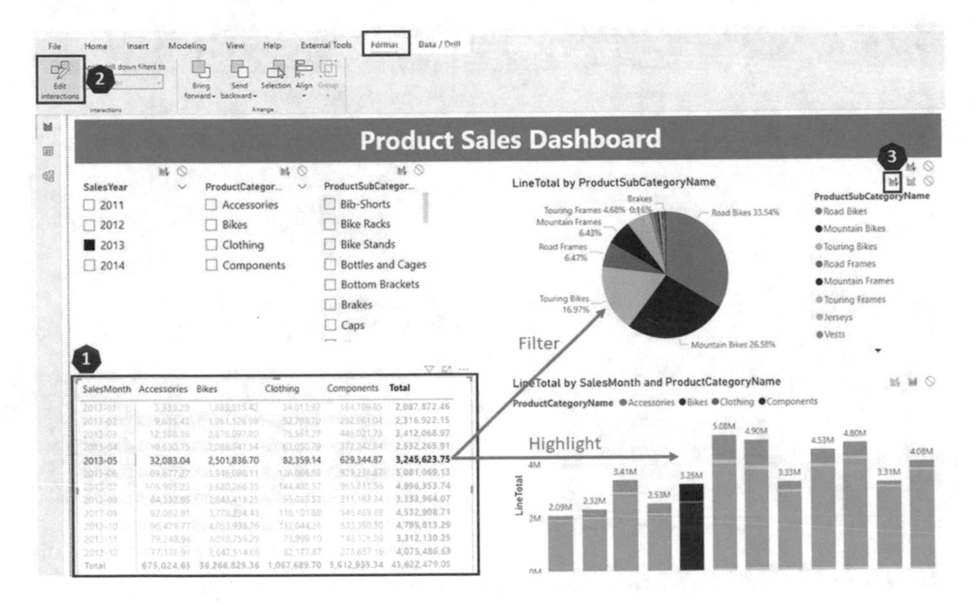

Figure 1-16. *Change interaction to Filter*

Simplification is one important way of improving the dashboard. The steps illustrated in Figure 1-17 remove unnecessary elements of the dashboard to improve the overall experience.

1. Simplify the pie chart title.
2. Simplify the stack column chart title.
3. Remove the pie chart legend.
4. Remove the Y-axis.
5. Remove the X-axis title.
6. Add borders to the visuals.

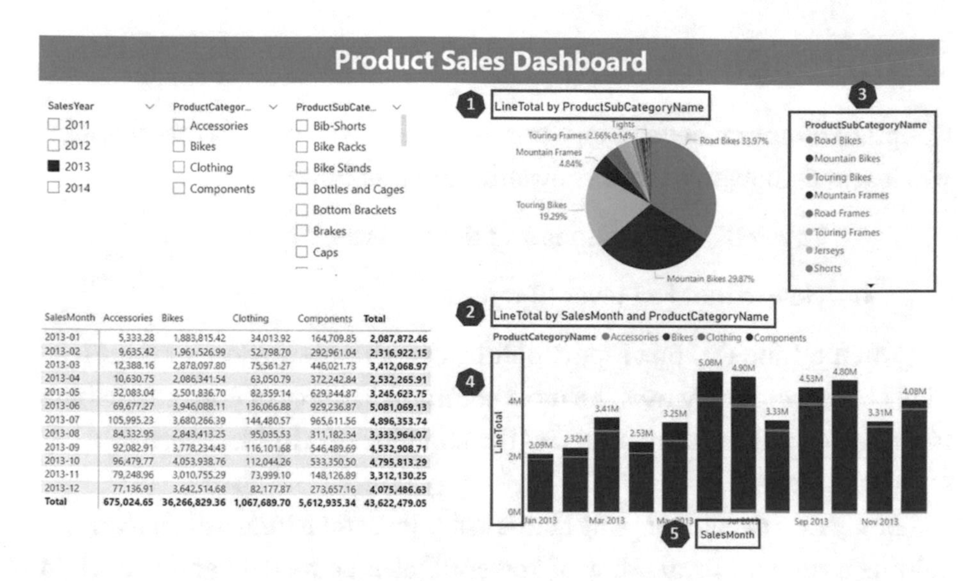

***Figure 1-17.** Areas of simplification*

These simple changes make the report look more professional. The change can be seen in Figure 1-18.

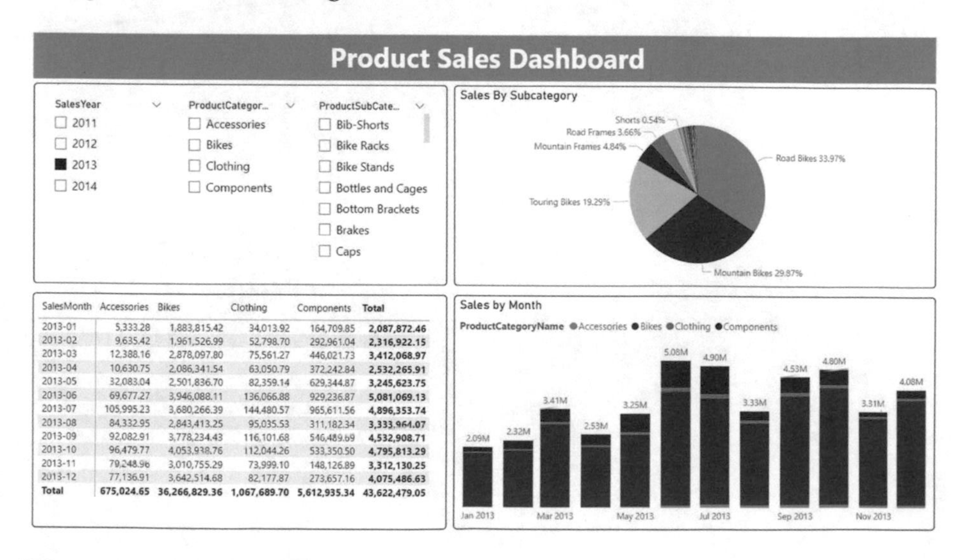

***Figure 1-18.** Simplified report*

Conclusion

Congratulations on developing your first Power BI report! In this chapter, you learned about two main knowledge areas in Power BI.

- Power BI subscriptions and the associated costs
- How to build a Power BI report

You may find this first report-building experience to be very intuitive. This is because it only uses the most common features of Power BI, which have been optimized, and much of the complexity is hidden from the report developers.

Many new developers stay at this entry level for a long time because gaining a good understanding of Power BI takes some deliberate effort. In the next chapter, you learn about the core components of Power BI.

CHAPTER 2

Power BI Core Components

This chapter uncovers the three key components of Power BI Desktop: Power Query, data models, and Data Analysis Express (DAX). Each component performs a critical aspect in the data analysis. Many reports developer overly focus on DAX and neglects the importance of Power Query and data models. This often results in overcomplicated and unpredictable DAX. This chapter explores the following.

- **Power Query** for loading and transforming data tables
- **Data models** for creating relationships between data tables
- **DAX** to create measures, columns, and tables in the report

Build a Report with Power Query

Prior to the introduction of Power Query, data automation in Excel was typically done through macros. Macros are recorded in the Visual Basic for Applications (VBA) programming language. Macros store a set of ordered VBA instructions to perform the different tasks required for data automation.

D. Ding, *Transitioning to Microsoft Power Platform*,
https://doi.org/10.1007/978-1-4842-9239-6_2

Compared to VBA, Power Query M is much more intuitive and user-friendly. It also provides a set of very practical data transformation functions to help simplifying common data transformation tasks. The Power Query M language was first introduced in Excel 2010. It is a more data centric style of programming language. Power Query was packaged into Power BI to import, transform and combine datasets. Power Query is one of the main reasons why Power BI is ahead of many of its competitors.

In this session, you learn how Power Query can automate complex data processes.

Business Scenario

After the initial success of the Sales Dashboard migration in Power BI. Milo has asked Kim to push the report one step further by including the product cost data. This dataset is useful for understanding both the *cost* and *margin* calculations.

The product cost data is in the file03productcostdata.csv exercise file.

Data Processing Automation

Before you continue with the new file, let's take a step back and check out the data refresh process. If you open the last file from Chapter 1 and click the refresh button in Figure 2-1, the dataset is reloaded into Power BI.

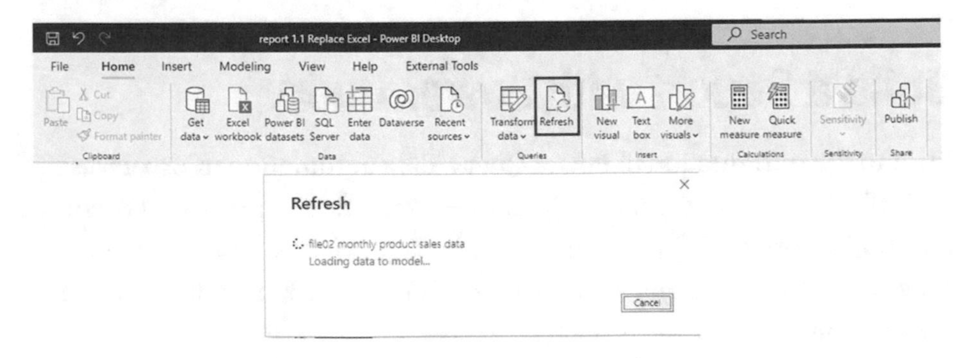

Figure 2-1. *Refresh data in Power BI*

The following steps, as shown in Figure 2-2, provide information to help you understand the data loading process.

1. Right-click the Sales table in Fields.

2. Select Edit query.

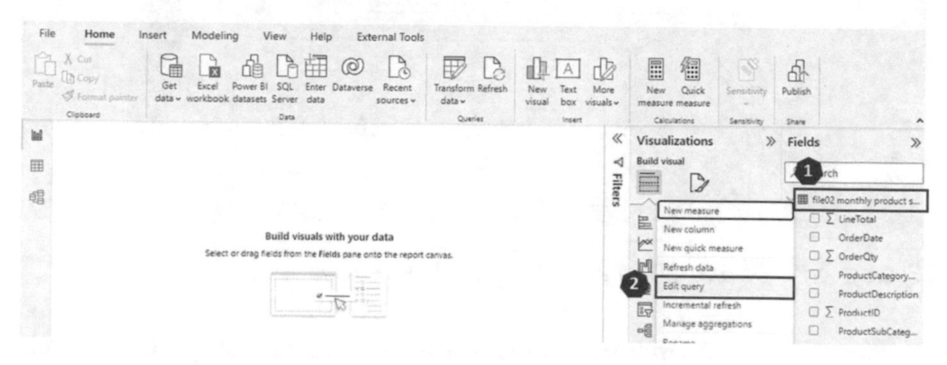

Figure 2-2. *Edit query*

This action takes you to the Power Query Editor, as shown in Figure 2-3. There are a few different function areas in the editor.

1. The **ribbon** displays common tasks.

2. **Queries** lists stored queries.

3. **Table preview** displays the first 1,000 records in the query.

4. **APPLIED STEPS** query executes step by step following the function order.

5. The **formula bar** is the Power Query function executed in each step.

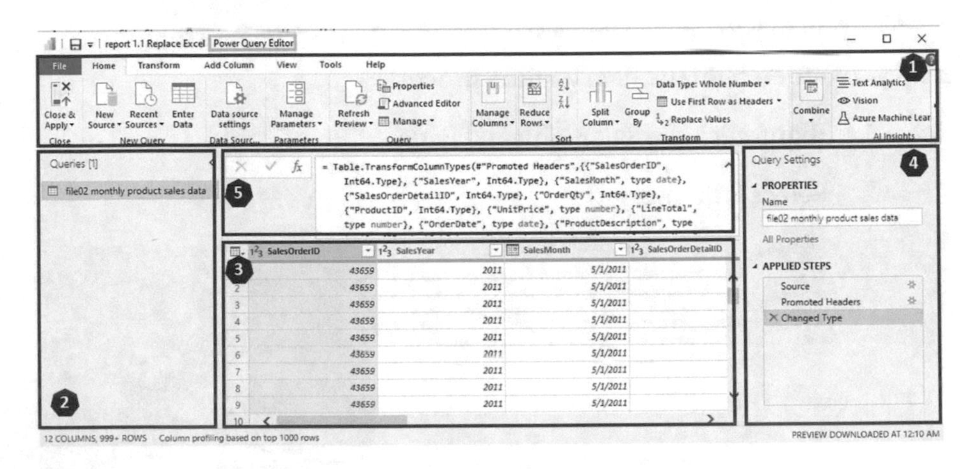

***Figure 2-3.** Power Query Editor function areas*

Figure 2-4 shows that there are three transformation steps. If you click through each step, you can see the formula.

1. **Source** = *Csv.Document*(File.Contents("E:\folder...\monthly product sales data.csv"),[Delimiter=",", Columns=12, Encoding=65001, QuoteStyle=QuoteStyle.None])

2. **Promoted Headers** = *Table.PromoteHeaders*(Source, [PromoteAllScalars=true])

3. **Changed Type** = *Table.TransformColumnTypes*(#"Promoted Headers",{{"OrderDate", type date}, {"LineTotal", type number}, {"UnitPrice", type number}, {"OrderQty", Int64.Type}})

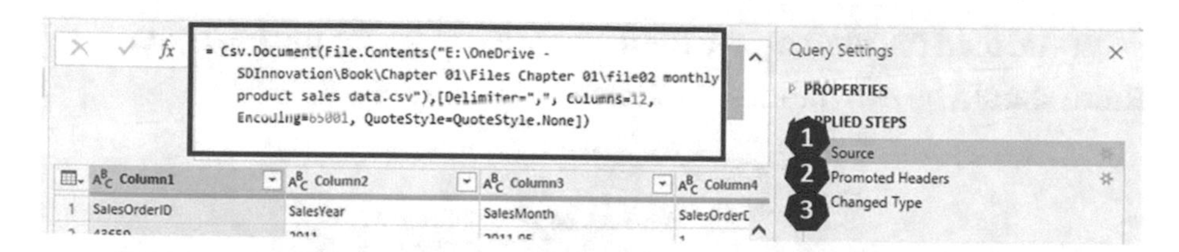

Figure 2-4. *Power Query transformation steps*

While the formula may look daunting initially, it is easy to understand. By focusing on the function names, you know that the steps are loading a CSV file, using the first row as the header, and changing the column type.

Create a New Query

Now that you have a basic understanding of Power Query, create a new query. The following steps, as shown in Figure 2-5, explain how to connect to the product data.

1. In the Power Query Editor, select New Source → Text/CSV.

2. Choose the product cost data file.

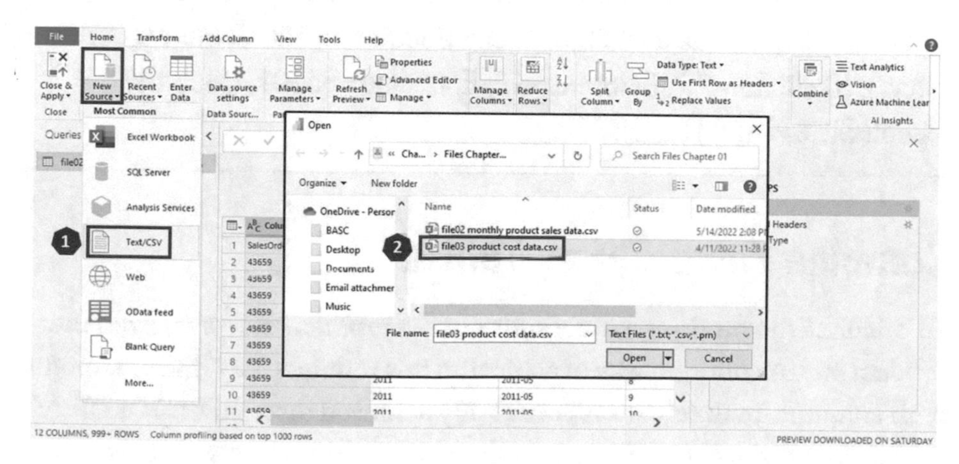

Figure 2-5. *Create a new query*

If you load the file, you see the same three steps for the new query. There should be two queries, as shown in Figure 2-6.

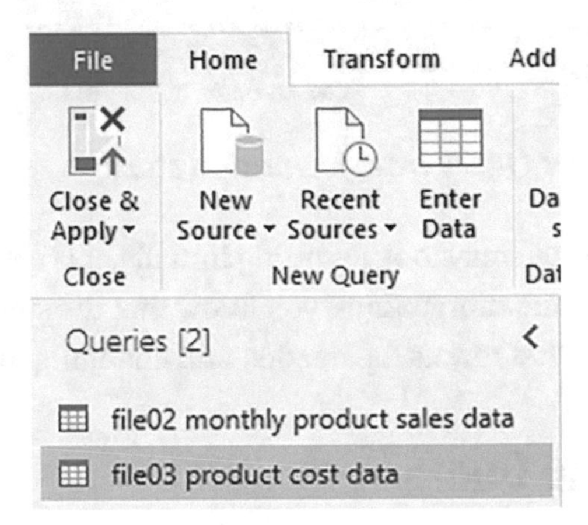

Figure 2-6. *A new query has been added*

Tip Sometimes, when you add a new column in the *source file* and refresh the report, the new column is not loaded. This is often because of the Columns=x input parameter from the load function. This parameter tells Power BI to only load the first 12 columns. If you remove this, it loads all columns. Csv.Document(File.Contents("data.csv"), [Delimiter=",", ~~Columns=12~~, Encoding=65001, QuoteStyle = QuoteStyle.None])

Combine Power Query Tables

To calculate the profit, you must subtract the *product sales price* from the *product cost*. A common way of achieving this requires the table to contain both columns. In Excel, this is commonly achieved through VLOOKUP. In Power Query, you can achieve a similar result by *merging* two tables.

Tables are merged via a common *key column* in both tables. For example, the left-hand table contains product sales records, including a Product Key column, and the right-hand table contains product details, also including a Product Key column. If I want to bring a column (e.g., Product Category) from the Product Detail table into the Product Sales Record table, I can use the Key column to merge the two tables in Power Query.

It is important to note that the two tables often do not contain the same number of keys; for example, the Product Sales Record table may contain less or more of the Customer Keys than the Customer Detail table. Also, the Product Sales Record table may contain duplication of the Product keys (the same product being sold to different customers at different times), but the Product Key should be unique in the Product Detail table.

As shown in Figure 2-7, all the keys from two tables can be fitted into the following three scenarios (or areas in the graph).

1. The key in the left table is not in the right table.
2. The key is in both tables.
3. The key in the right table is not in the left table.

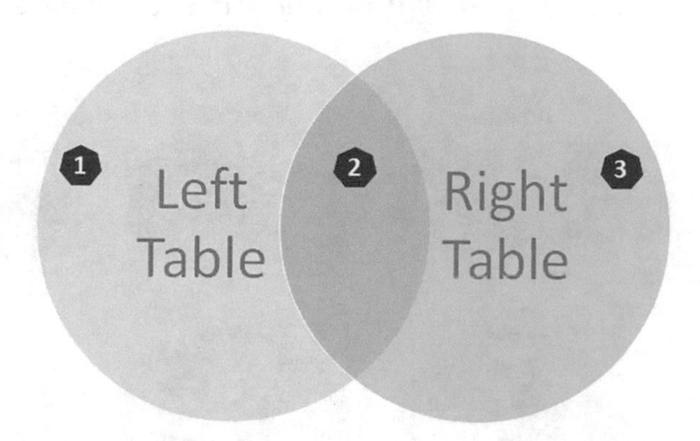

Figure 2-7. *Left and right tables joined to create three areas*

This joint can also result in three special cases, as shown in Figure 2-8.

1. There are no matching keys between the two tables.
2. The keys are a complete match (rarely happens).
3. One of the tables contains all the keys of the other table and more.

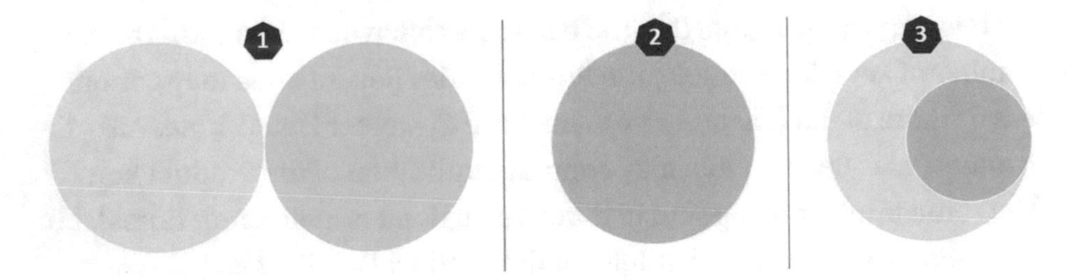

Figure 2-8. *Special scenarios in two tables join*

Power BI supports the six different join types. The different join types are represented by the selected red areas in Figure 2-9. Even though it is referred to as *joint types*, it is really the output type. For example, the Left Outer joint type produces the output for keys that matches the left-hand table; the Full Outer joint type produces the output for all keys from both table.

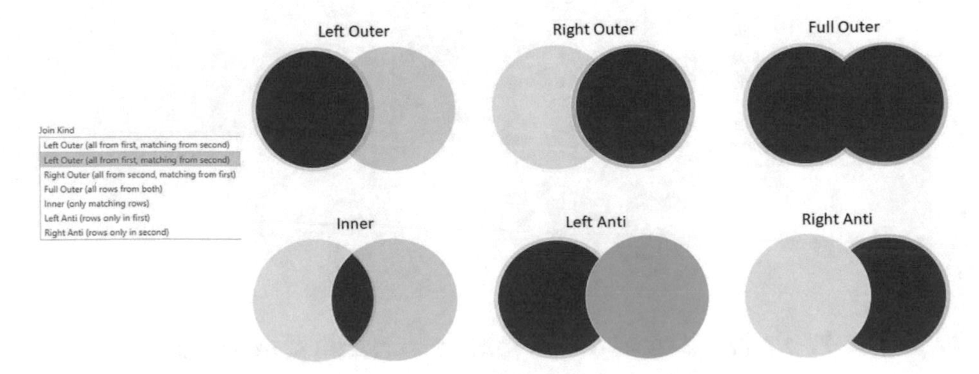

Figure 2-9. *Join types in Power Query*

To continue with the business scenario, you can use Left Join to achieve the requirement. The following steps explain how to do this, as shown in Figure 2-10.

1. Select the product sales query (left).
2. Choose **Merge Queries** from the Home ribbon.
3. Choose the product cost query (right).
4. Choose **ProductID** as the key on the left table.
5. Choose **ProductID** as the key on the right table.
6. Check that Left Outer is used in Join Kind.

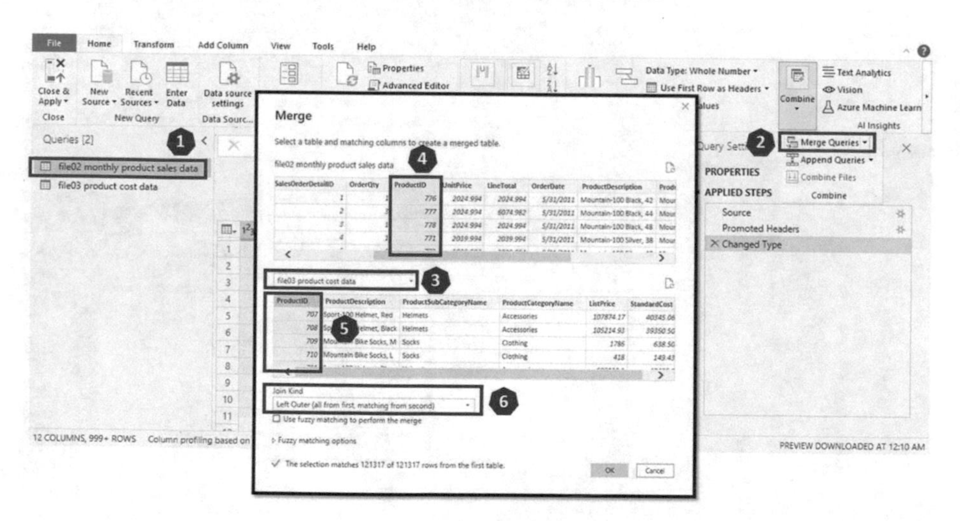

***Figure 2-10.** Merge two queries*

Caution The OK button may be grayed out because of the error message. One rule for the join key is that the key columns must be the same data type.

A new column with the word *Table* in each cell is created after the merge. This table column allows you to pick the columns within the table. The following steps explain how, as shown in Figure 2-11.

1. Select the Expand symbol from the column.
2. Select the StandardCost and ListPrice required fields.
3. Deselect **Use original column name as prefix** (optional).

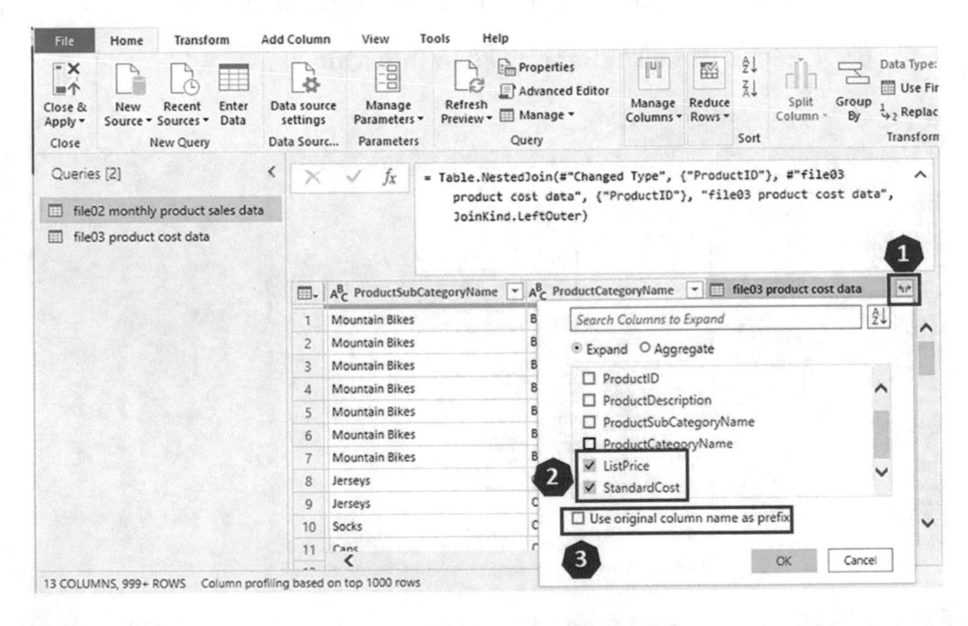

Figure 2-11. *Add columns from the product table*

Tip If you find this session confusing, it is best to practice this a few times in different scenarios. This is a critical concept for data analysis. If you have some data using VLOOKUP( ), try to replace it with Left Join. For the reference table, try to create some duplications to see the impact on Left Join.

Add a New Column

You might have noticed that the merge created additional steps in the query. Next, you need to create a new LineProfit column. It is OrderQty multiplied by the difference between UnitPrice and StandardCost. It can be created using the following steps, as shown in Figure 2-12.

1. Choose **Custom Column** from Add Column Ribbon.
2. Enter the new column name as LineProfit.
3. Enter the calculation formula as ([UnitPrice] - [StandardCost]) * [OrderQty].
4. Repeat steps 1–3 for LineCost = [StandardCost] * [OrderQty].

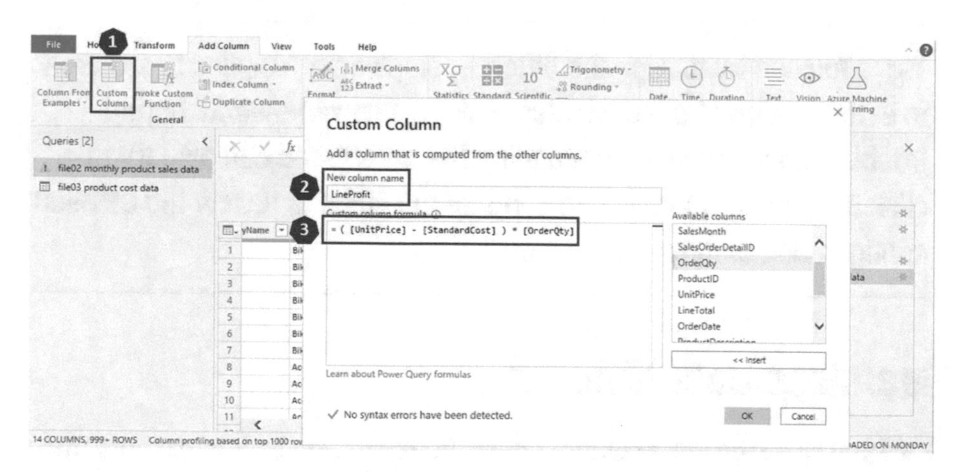

***Figure 2-12.** Create a new column*

Lastly, you need to change the column types of the new columns into decimal numbers, as shown in Figure 2-13.

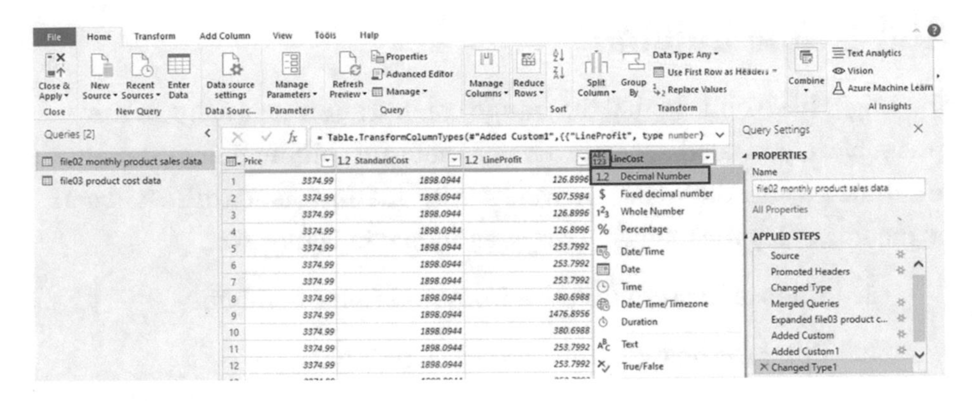

***Figure 2-13.** Changing column type*

Once the new columns have been added, the new table is ready to be loaded into Power BI by selecting Close & Apply in the Home ribbon.

Tip You can try to click the different steps in the Applied Steps area to see the output from each step. This can be very helpful during troubleshooting or in understanding the output. You can also try to click the settings icon next to some of the steps to review the detailed configuration of the selected step.

Seamless Data Integration

Once you close the Power Query Editor, Power BI proceeds to execute the queries. The merged table is now loaded into Power BI for front-end visuals.

You normally move on to data models and visuals, but it is important to appreciate the seamless method for data integration in Power BI. You can easily switch between backend data processing and front-end data models and visualizations. This data integration capability saves a lot of time during the development process and is a critical part of report automation.

Add a New Report Page

After merging the new pricing columns and calculating the profit column, it is time to add the new report page. This new page combines trends in revenue, costs, and profits.

The following steps, shown in Figure 2-14, create initial content on the new page by copying and pasting from the previous page.

1. Click the New Page button.
2. Rename the page **Product Summary** by double-clicking the page name.
3. Copy and paste the heading, filter, and matrix table from the original page.
4. Rename the title.

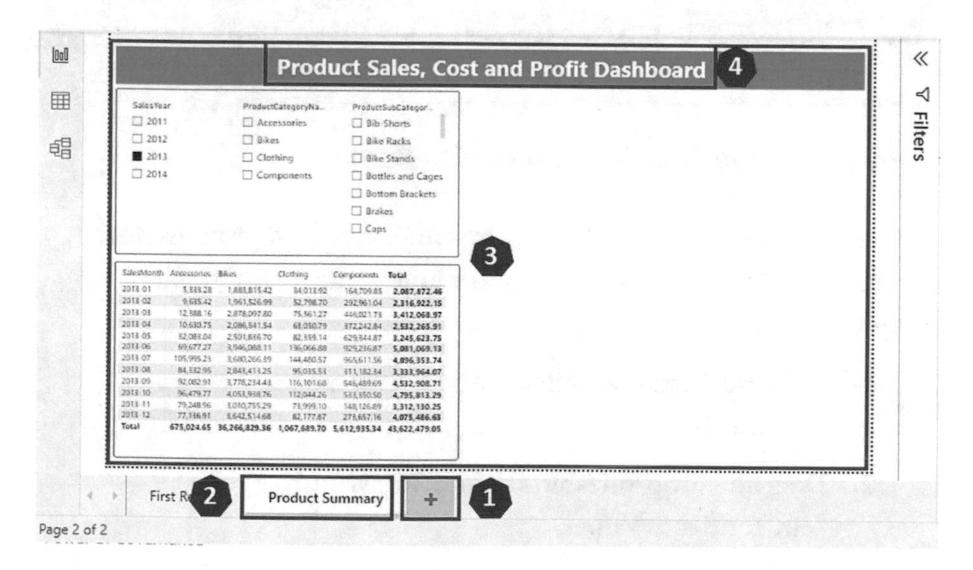

Figure 2-14. *Adding a new report page*

The next step is to change the matrix table to display cost and profit. This is done by doing the following, as shown in Figure 2-15.

1. Remove the data field in Columns.
2. Add **LineCost** and **LineProfit** to Values.

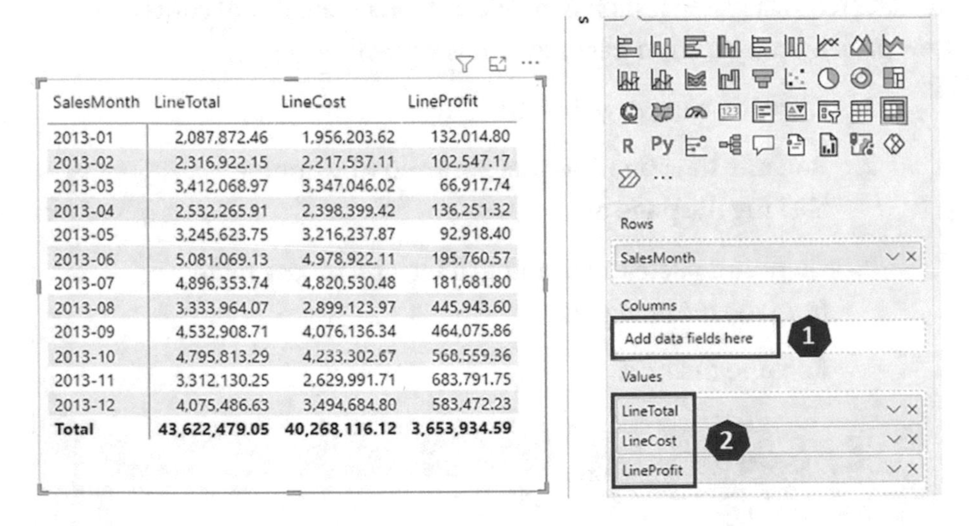

SalesMonth	LineTotal	LineCost	LineProfit
2013-01	2,087,872.46	1,956,203.62	132,014.80
2013-02	2,316,922.15	2,217,537.11	102,547.17
2013-03	3,412,068.97	3,347,046.02	66,917.74
2013-04	2,532,265.91	2,398,399.42	136,251.32
2013-05	3,245,623.75	3,216,237.87	92,918.40
2013-06	5,081,069.13	4,978,922.11	195,760.57
2013-07	4,896,353.74	4,820,530.48	181,681.80
2013-08	3,333,964.07	2,899,123.97	445,943.60
2013-09	4,532,908.71	4,076,136.34	464,075.86
2013-10	4,795,813.29	4,233,302.67	568,559.36
2013-11	3,312,130.25	2,629,991.71	683,791.75
2013-12	4,075,486.63	3,494,684.80	583,472.23
Total	**43,622,479.05**	**40,268,116.12**	**3,653,934.59**

Figure 2-15. *Add Line Cost and Profit columns*

You can now add charts showing revenue, cost, and profit trends. Figure 2-16 and the following steps show how to create a simplified column chart.

1. From the Visual selection, choose a stacked column chart.
2. Drag and drop **SalesMonth** into X-axis and **LineTotal** into Values.
3. Remove the title from the X-axis.
4. Remove the Y-axis.

5. Change the chart title.

6. Add a data label.

7. Add a visual border.

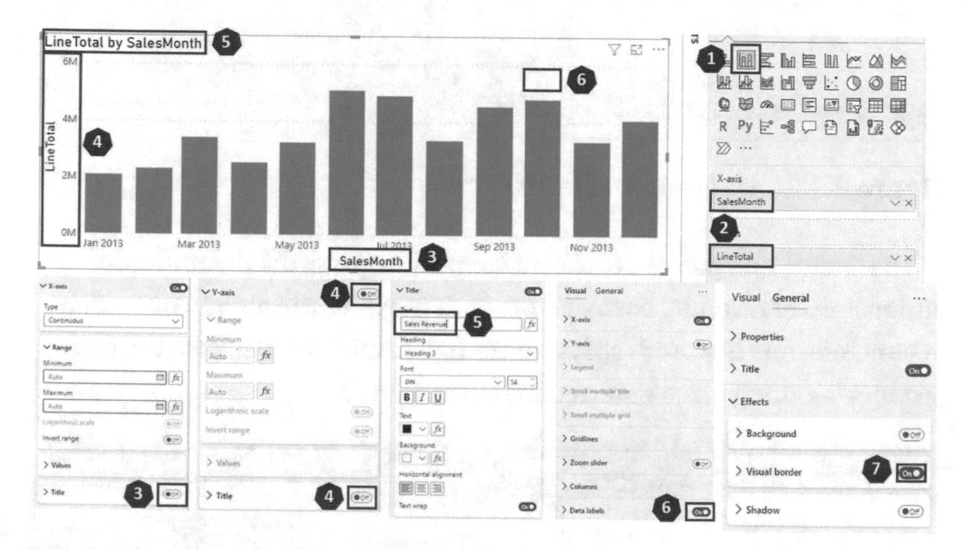

Figure 2-16. *Add and configure column chart*

You can then copy and paste this visual for LineCost and LineProfit. Use the following steps, as shown in Figure 2-17.

1. Replace LineTotal with LineCost.

2. Change the column color to yellow.

3. Repeat steps 1 and 2 for LineProfit (change the color to purple).

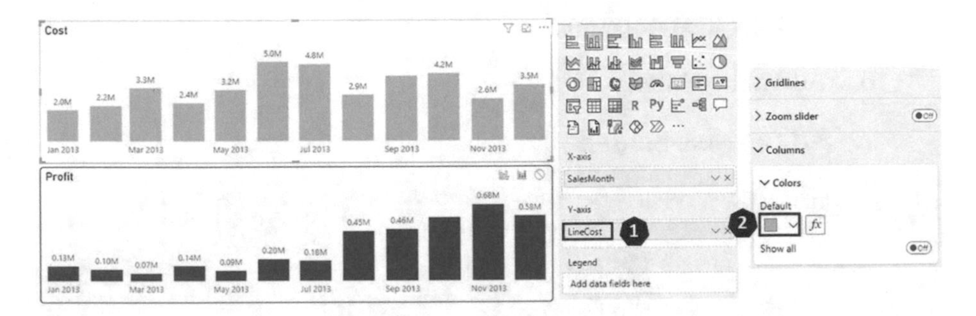

Figure 2-17. Add column charts

As shown in Figure 2-18, the new report page covers the multiple dimensions of revenue, cost, and profit. It provides the ability for leaders to conclude that the profit spike from August 2013 was because the new product—touring bikes—started to return a profit.

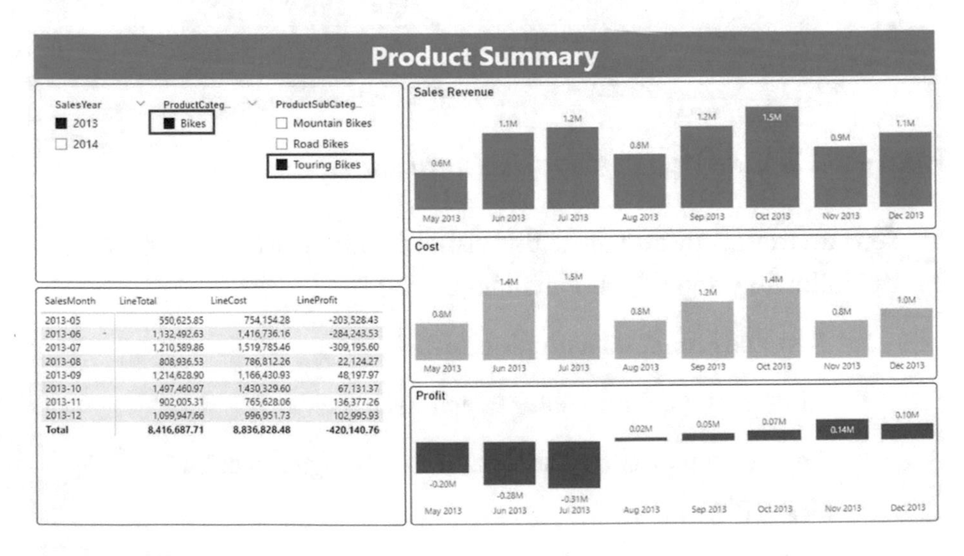

Figure 2-18. New report showing revenue, cost, and profit

Power Query Summary

After reading this session, you should have gained some basic understanding of Power Query. You have used Power Query to connect to data sources, create new calculated columns, and merge with other tables.

Building a Report with a Data Model

A data model in Power BI is a *collection of relationships* between multiple tables. The tables are connected via column relationships. The example in Figure 2-19 is a basic data model created by connecting the monthly sales table with a date table.

The monthly product sales table is typically called a *fact table* because it contains facts about sales. The date table is referred to as a *dimension or reference table*. Other examples of dimension tables could be customer details and product details tables.

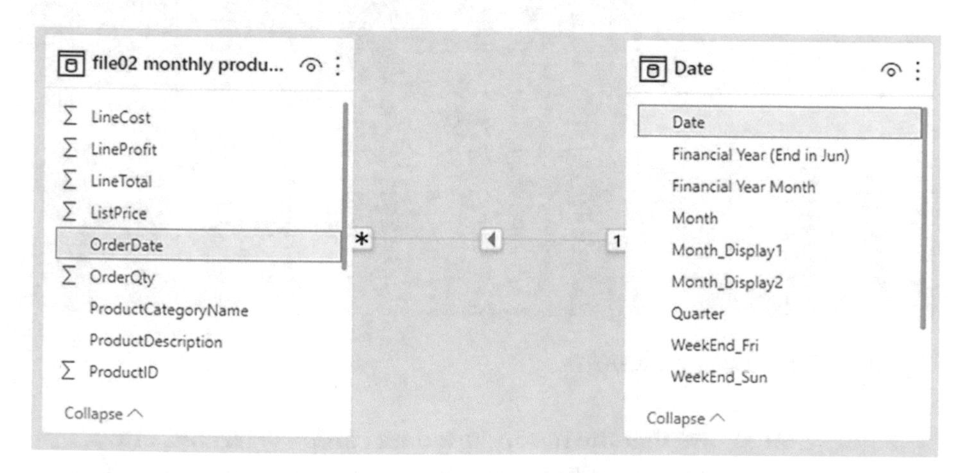

Figure 2-19. *Basic data model*

Why a Data Model?

Although you can implement data logic using Power Query, using a data model has three main benefits.

- Minimizes data redundancy
- Facilitates data reuse
- Filters across multiple fact tables

Minimize Data Redundancy

In the previous exercise, the Product Sales Data table contains many redundant values relating to products. As shown in Figure 2-20, the column in the blue boxes shows the five columns that contain a lot of redundant data.

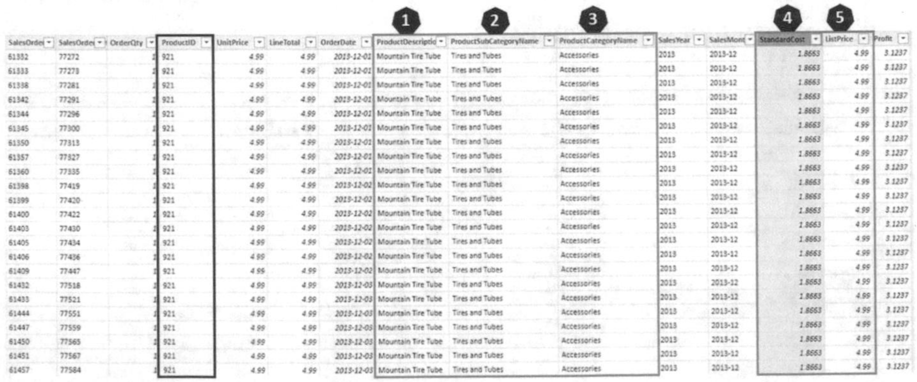

SalesOrde	SalesOrde	OrderQty	ProductID	UnitPrice	LineTotal	OrderDate	ProductDescriptio (1)	ProductSubCategoryName (2)	ProductCategoryName (3)	SalesYear	SalesMont	StandardCost (4)	ListPrice (5)	Profit
61332	77272	1	921	4.99	4.99	2013-12-01	Mountain Tire Tube	Tires and Tubes	Accessories	2013	2013-12	1.8663	4.99	3.1237
61333	77273	1	921	4.99	4.99	2013-12-01	Mountain Tire Tube	Tires and Tubes	Accessories	2013	2013-12	1.8663	4.99	3.1237
61338	77281	1	921	4.99	4.99	2013-12-01	Mountain Tire Tube	Tires and Tubes	Accessories	2013	2013-12	1.8663	4.99	3.1237
61342	77291	1	921	4.99	4.99	2013-12-01	Mountain Tire Tube	Tires and Tubes	Accessories	2013	2013-12	1.8663	4.99	3.1237
61344	77296	1	921	4.99	4.99	2013-12-01	Mountain Tire Tube	Tires and Tubes	Accessories	2013	2013-12	1.8663	4.99	3.1237
61345	77300	1	921	4.99	4.99	2013-12-01	Mountain Tire Tube	Tires and Tubes	Accessories	2013	2013-12	1.8663	4.99	3.1237
61350	77313	1	921	4.99	4.99	2013-12-01	Mountain Tire Tube	Tires and Tubes	Accessories	2013	2013-12	1.8663	4.99	3.1237
61357	77327	1	921	4.99	4.99	2013-12-01	Mountain Tire Tube	Tires and Tubes	Accessories	2013	2013-12	1.8663	4.99	3.1237
61360	77335	1	921	4.99	4.99	2013-12-01	Mountain Tire Tube	Tires and Tubes	Accessories	2013	2013-12	1.8663	4.99	3.1237
61398	77419	1	921	4.99	4.99	2013-12-02	Mountain Tire Tube	Tires and Tubes	Accessories	2013	2013-12	1.8663	4.99	3.1237
61399	77420	1	921	4.99	4.99	2013-12-02	Mountain Tire Tube	Tires and Tubes	Accessories	2013	2013-12	1.8663	4.99	3.1237
61400	77422	1	921	4.99	4.99	2013-12-02	Mountain Tire Tube	Tires and Tubes	Accessories	2013	2013-12	1.8663	4.99	3.1237
61403	77430	1	921	4.99	4.99	2013-12-02	Mountain Tire Tube	Tires and Tubes	Accessories	2013	2013-12	1.8663	4.99	3.1237
61405	77434	1	921	4.99	4.99	2013-12-02	Mountain Tire Tube	Tires and Tubes	Accessories	2013	2013-12	1.8663	4.99	3.1237
61406	77436	1	921	4.99	4.99	2013-12-02	Mountain Tire Tube	Tires and Tubes	Accessories	2013	2013-12	1.8663	4.99	3.1237
61409	77447	1	921	4.99	4.99	2013-12-02	Mountain Tire Tube	Tires and Tubes	Accessories	2013	2013-12	1.8663	4.99	3.1237
61432	77518	1	921	4.99	4.99	2013-12-03	Mountain Tire Tube	Tires and Tubes	Accessories	2013	2013-12	1.8663	4.99	3.1237
61433	77521	1	921	4.99	4.99	2013-12-03	Mountain Tire Tube	Tires and Tubes	Accessories	2013	2013-12	1.8663	4.99	3.1237
61444	77551	1	921	4.99	4.99	2013-12-03	Mountain Tire Tube	Tires and Tubes	Accessories	2013	2013-12	1.8663	4.99	3.1237
61447	77559	1	921	4.99	4.99	2013-12-03	Mountain Tire Tube	Tires and Tubes	Accessories	2013	2013-12	1.8663	4.99	3.1237
61450	77565	1	921	4.99	4.99	2013-12-03	Mountain Tire Tube	Tires and Tubes	Accessories	2013	2013-12	1.8663	4.99	3.1237
61451	77567	1	921	4.99	4.99	2013-12-03	Mountain Tire Tube	Tires and Tubes	Accessories	2013	2013-12	1.8663	4.99	3.1237
61457	77584	1	921	4.99	4.99	2013-12-03	Mountain Tire Tube	Tires and Tubes	Accessories	2013	2013-12	1.8663	4.99	3.1237

Figure 2-20. *Data redundancy*

Figure 2-21 shows that the redundant data can be extracted into a separate product table with a link to the Sales table. This simple method can sometimes reduce data storage by 60% compared to the single flat file approach.

SalesOrderID	SalesOrderDetailID	OrderQty	ProductID	UnitPrice	LineTotal	OrderDate	SalesYear	SalesMonth
64680	87276	1	921	4.99	4.99	2014-01-21	2014	2014-01
64663	87230	1	921	4.99	4.99	2014-01-20	2014	2014-01
64653	87200	1	921	4.99	4.99	2014-01-20	2014	2014-01
64644	87172	1	921	4.99	4.99	2014-01-20	2014	2014-01
64643	87168	1	921	4.99	4.99	2014-01-20	2014	2014-01
64632	87145	1	921	4.99	4.99	2014-01-20	2014	2014-01
64630	87142	1	921	4.99	4.99	2014-01-20	2014	2014-01
64629	87139	1	921	4.99	4.99	2014-01-20	2014	2014-01
64627	87132	1	921	4.99	4.99	2014-01-20	2014	2014-01
64622	87117	1	921	4.99	4.99	2014-01-20	2014	2014-01
64615	87104	1	921	4.99	4.99	2014-01-20	2014	2014-01
64614	87102	1	921	4.99	4.99	2014-01-20	2014	2014-01

ProductID	ProductDescription	ProductSubCategoryName	ProductCategoryName	ListPrice	StandardCost
921	Mountain Tire Tube	Tires and Tubes	Accessories	15444.05	5776.1985

Figure 2-21. *Use reference table to reduce data redundancy*

Facilitate Data Reuse

Take organizational structure as an example. In a large organization, the historical org structure table can be surprisingly complicated, with different reports using slightly different structure tables. This can result in many hours arguing about which figure is correct.

One solution to this problem is for the different teams to adopt a standardized org structure table for all reporting. As shown in Figure 2-22, the org structure dimension table is imported and linked using a specific approach across all reports regardless of the fact table being used.

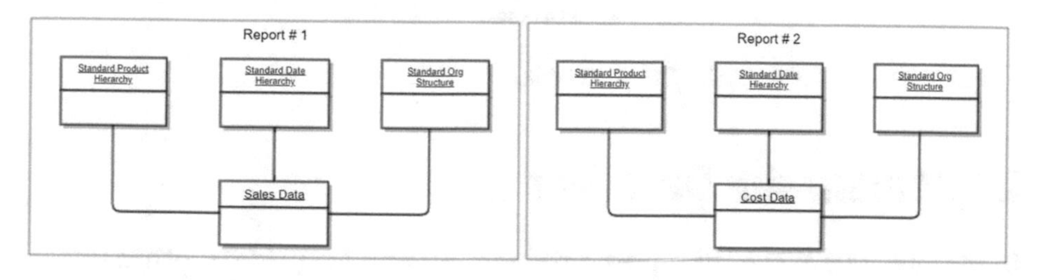

Figure 2-22. *Reuse standard reference tables*

Filter Across Multiple Fact Tables

It is common for leaders to demand summary reports that include multiple key performance indicators (KPI). Because many reports are designed in parallel, it does not allow the users to see multiple KPIs on a single page.

Figure 2-23 shows one way of achieving this using a data model. This data model lets users see sales and cost data for the same product, period, and business units.

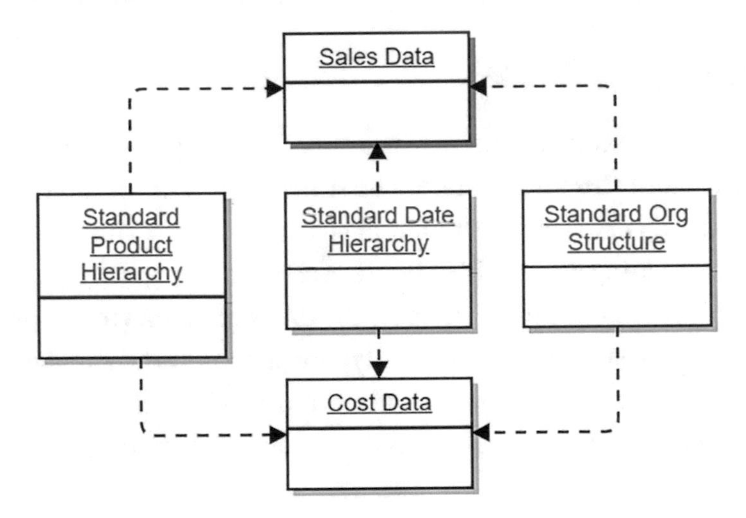

Figure 2-23. *Common filter tables*

Implement the Data Model

In the previous Power BI report, product cost and list price information were brought in through merge (Left Join) using Power Query. There was only a single table in this report. You can achieve the same result using a data model.

Create a Basic Data Model

The Product Cost table was merged into the Monthly Sales table in the previous session. Instead of merging the two tables, you can achieve the same result using a simple data model between the two tables.

You can now reverse this merging step by removing the steps highlighted in Figure 2-24 in the Power Query Editor.

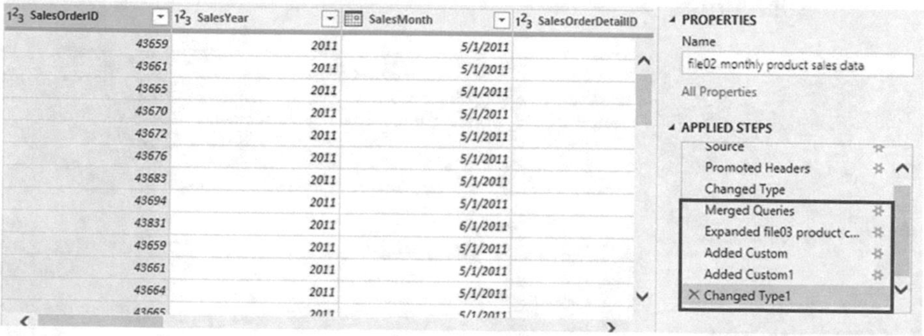

Figure 2-24. Steps to be removed in Power Query Editor

Once the merge steps are removed, create a basic data model by establishing a relationship between the tables. The following steps explain how to do it, as shown in Figure 2-25.

1. Go to Models view.
2. Drag and drop the **ProductID** field from the monthly Product Sales table into the ProductID field in the Product Cost table.
3. A relationship is established automatically from step 2.
4. Note that this area can be used to edit the relationship.
5. Note that Cardinality is automatically determined.

6. Note that the relationship status can be Active (default) or Inactive.

7. Note that Filter Direction options can be Single or Both. The direction is also indicated in step 3.

8. Note that when a filter direction is set to Both, there is also the option to set a security filter in both directions.

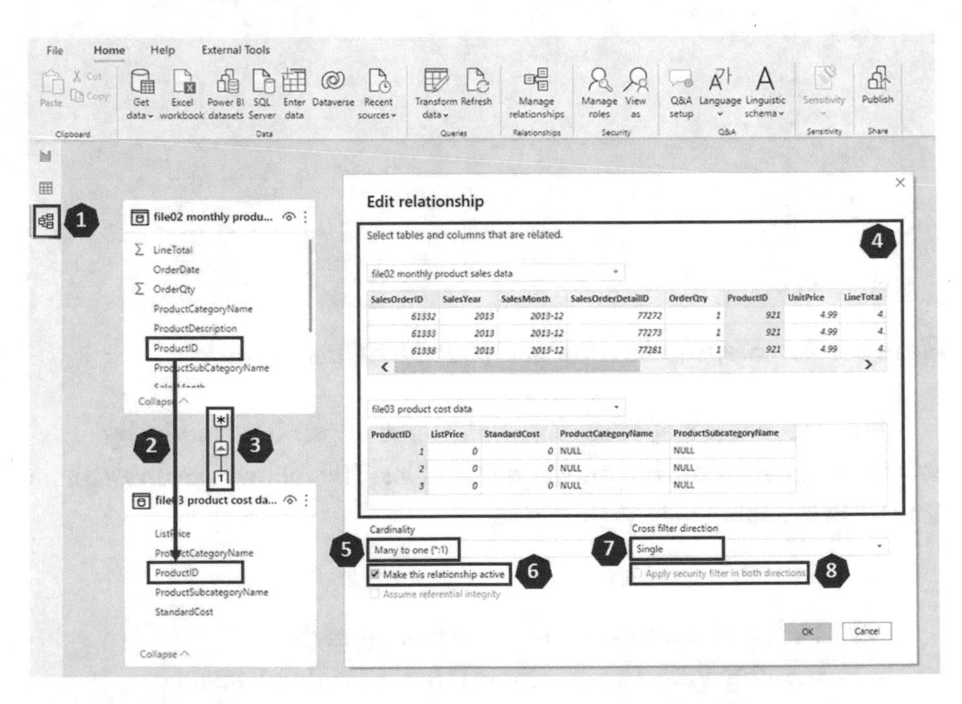

***Figure 2-25.** Creating a relationship between two tables*

Caution There can only be one active relationship between two tables. Avoid many-to-many relationships at all costs.

If you go back to the combined report, you see the errors shown in Figure 2-26. These errors are caused by the LineCost and LineProfit columns being removed in earlier steps.

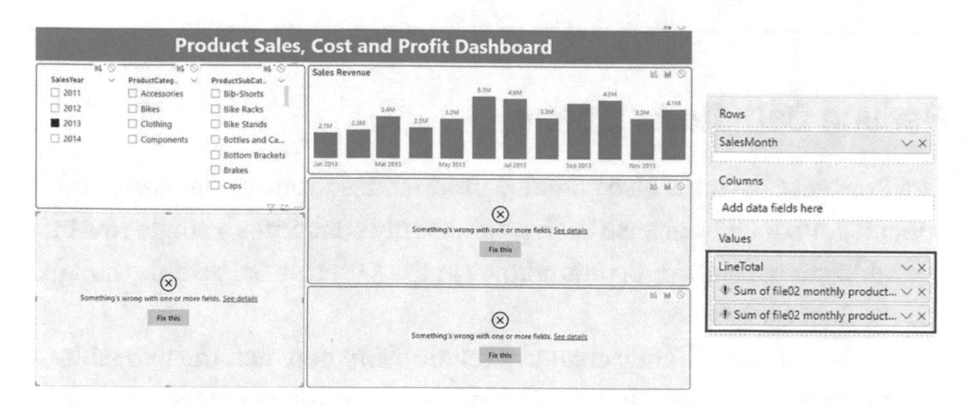

Figure 2-26. *Report errors*

The following steps explain how you can create two DAX columns of the same name to replace them. Once the new columns are created, the visual errors should be resolved.

1. Right-click the Monthly Sales table and choose **New Column**.
2. Enter the column formula: LineCost = [OrderQty] * RELATED('file03 product cost data'[StandardCost]).
3. Create another LineProfit column: LineProfit = LineTotal - LineCost.

Note The RELATED function is like a simplified VLOOKUP function in Excel. It can bring values from another related table. The related function only works in many-to-one or one-to-one relationships.

It does not work if there is more than one possible value (i.e., trying to create a column in both tables in a many-to-one relationship); one of them is not allowed.

Reduce Data Redundancy

The Product Cost table also contains product description, category, and subcategory detail. Because each product only consumes a single row in the table, there is less data redundancy in the Monthly Sales table than in the Product Cost table.

Since you have already created a relationship between the two tables, you can remove the product description category and subcategory tables from the Monthly Sales table. The following steps and Figure 2-27 explain how.

1. Hold the Ctrl key to select the three columns.

2. From the Home ribbon, select Remove Columns.

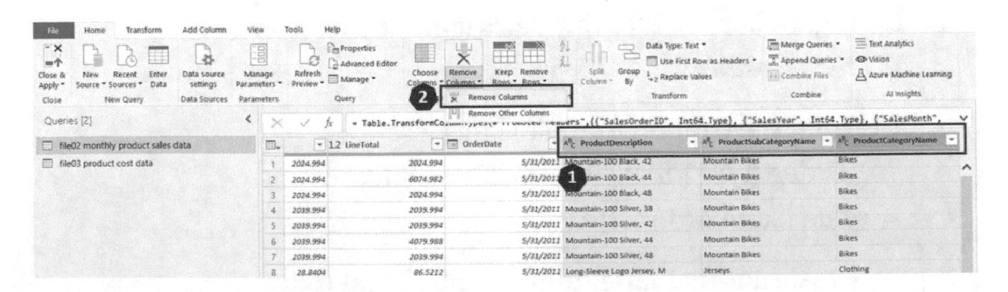

Figure 2-27. *Remove the column in Power Query*

Once the changed data is loaded into Power BI, it creates new errors in the slicer visuals. The following steps explain how to resolve the errors, as shown in Figure 2-28.

1. Select the first slicer.

2. Drag and drop the ProductCategory field from the Product Cost table into the Slicer field.

3. Repeat steps 1 and 2 for the second slicer with the ProductSubCategory field.

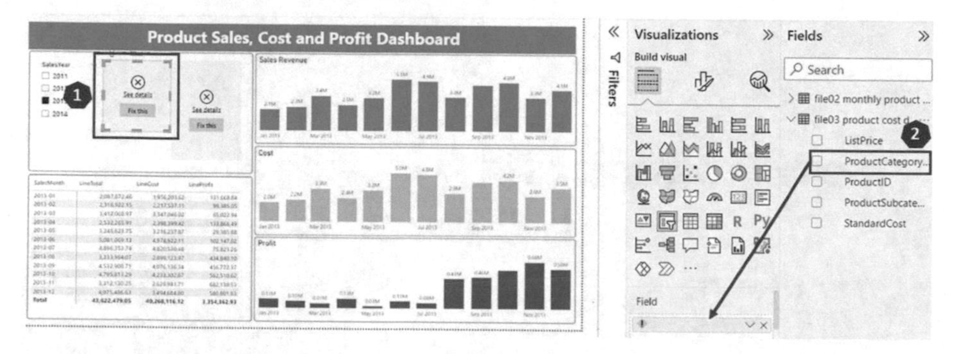

Figure 2-28. *Resolve errors in slicer*

In this case, there is no need to bring the information into the table as new columns. You can use data fields as is for visuals. The RELATED function is only required for specific calculation purposes requiring *row context*. You understand more about context in the DAX session.

Tip Using a data model to reduce data redundancy can significantly reduce file size for large datasets. You may consider a big flat table for smaller datasets to simplify the data logic. When developing Power BI reports, no method is better than the other. It's about choosing the best implementation for the specific requirements.

Customized Calendar Table

When a data column is classified as a date column in Power Query, Power BI generates an auto-date hierarchy (table). This is indicated by the data hierarchy in Figure 2-29. There are options for year, quarter, month, and day.

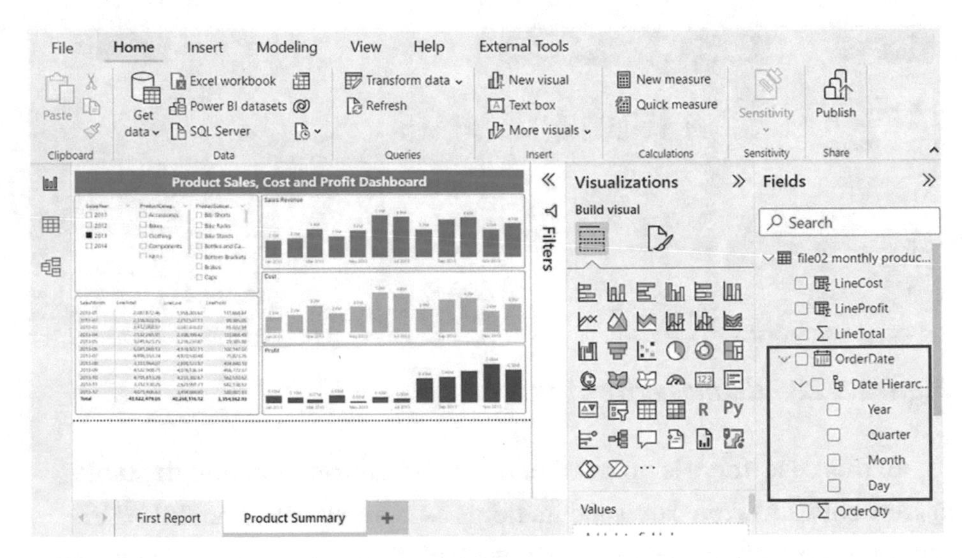

Figure 2-29. Autodate hierarchy

When developing Power BI reports, there is often a need to use different date displays and groupings. Some of this may not be supported in Power BI by default. For example, Power BI may not support some specific date formats, such as Jan-22 (or Jan/2022). Maybe you need to display weekly grouping instead of monthly, which is also not supported in the default date hierarchy. One way to get around this issue would be to use an Excel calendar table, as shown in Figure 2-30.

Date	Month	Year	Quarter	Financial Year (End in Jun)	Financial Year Month	WeekEnd_Sun	WeekEnd_Fri	Month_Display1	Month_Display2	Year Quarter	YearMonthInt
1/1/2011	1	2011	1	2011	7	12/26/2010	12/31/2010	Jan-11	Jan/2011	2011 Q1	201101
1/2/2011	1	2011	1	2011	7	12/26/2010	12/31/2010	Jan-11	Jan/2011	2011 Q1	201101
1/3/2011	1	2011	1	2011	7	1/2/2011	12/31/2010	Jan-11	Jan/2011	2011 Q1	201101
1/4/2011	1	2011	1	2011	7	1/2/2011	12/31/2010	Jan-11	Jan/2011	2011 Q1	201101
1/5/2011	1	2011	1	2011	7	1/2/2011	12/31/2010	Jan-11	Jan/2011	2011 Q1	201101
1/6/2011	1	2011	1	2011	7	1/2/2011	12/31/2010	Jan-11	Jan/2011	2011 Q1	201101
1/7/2011	1	2011	1	2011	7	1/2/2011	12/31/2010	Jan-11	Jan/2011	2011 Q1	201101
1/8/2011	1	2011	1	2011	7	1/2/2011	1/7/2011	Jan-11	Jan/2011	2011 Q1	201101
1/9/2011	1	2011	1	2011	7	1/2/2011	1/7/2011	Jan-11	Jan/2011	2011 Q1	201101
1/10/2011	1	2011	1	2011	7	1/9/2011	1/7/2011	Jan-11	Jan/2011	2011 Q1	201101
1/11/2011	1	2011	1	2011	7	1/9/2011	1/7/2011	Jan-11	Jan/2011	2011 Q1	201101
1/12/2011	1	2011	1	2011	7	1/9/2011	1/7/2011	Jan-11	Jan/2011	2011 Q1	201101
1/13/2011	1	2011	1	2011	7	1/9/2011	1/7/2011	Jan-11	Jan/2011	2011 Q1	201101
1/14/2011	1	2011	1	2011	7	1/9/2011	1/7/2011	Jan-11	Jan/2011	2011 Q1	201101
1/15/2011	1	2011	1	2011	7	1/9/2011	1/14/2011	Jan-11	Jan/2011	2011 Q1	201101

Figure 2-30. *Excel calendar table*

Although this approach has several downsides, it is an easy way to introduce users to the idea of having a calendar table. You see a better way to achieve this later in the DAX session.

Note The YearMonthInt column and similar integer columns can be used to set display order. This column type must follow the year-month-date order with no gaps (e.g., 20220510).

The following columns must be set to the right column type when loading the new table into Power Query, as shown in Figure 2-31.

1. Month_Display1 as text
2. Month_Display2 as text
3. YearMonthInt as a whole number

	ABC Month_Display1	ABC Month_Display2	ABC Year Quarter	123 YearMonthInt
1/2010	Jan-11	Jan/2011	2011 Q1	201101
1/2010	Jan-11	Jan/2011	2011 Q1	201101
1/2010	Jan-11	Jan/2011	2011 Q1	201101
1/2010	Jan-11	Jan/2011	2011 Q1	201101
1/2010	Jan-11	Jan/2011	2011 Q1	201101
1/2010	Jan-11	Jan/2011	2011 Q1	201101
1/2010	Jan-11	Jan/2011	2011 Q1	201101
7/2011	Jan-11	Jan/2011	2011 Q1	201101

Figure 2-31. *Calendar column types*

Once you have loaded the file04calendartable.xlsx file, you can connect the Date column from the Calendar table with the Order Date in the Monthly Product Sales table, as shown in Figure 2-32.

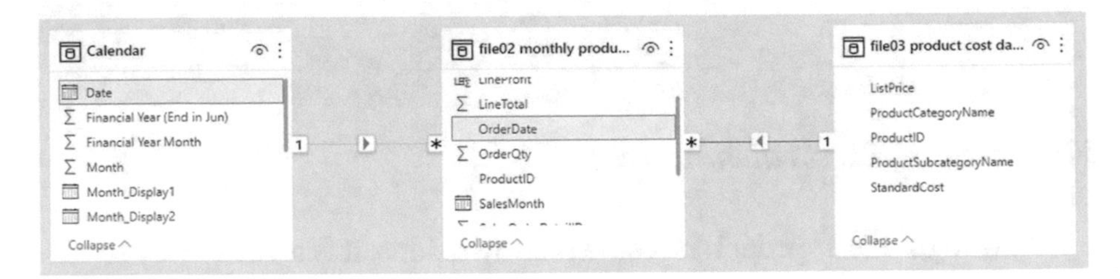

Figure 2-32. *Connect calendar table to a data model*

Instead of using the sales month (2013-02) format in the report dashboard, you can now use some different date format, such as Month Display 1 (mmm-yy). While alphabetical order is the default order for text columns, this can be changed. To do that, use the following steps, as shown in Figure 2-33.

1. Choose the **Month_Display1** column from the Calendar table.

2. In the **Column tools** ribbon, choose **Sort by column** and select **YearMonthInt**.

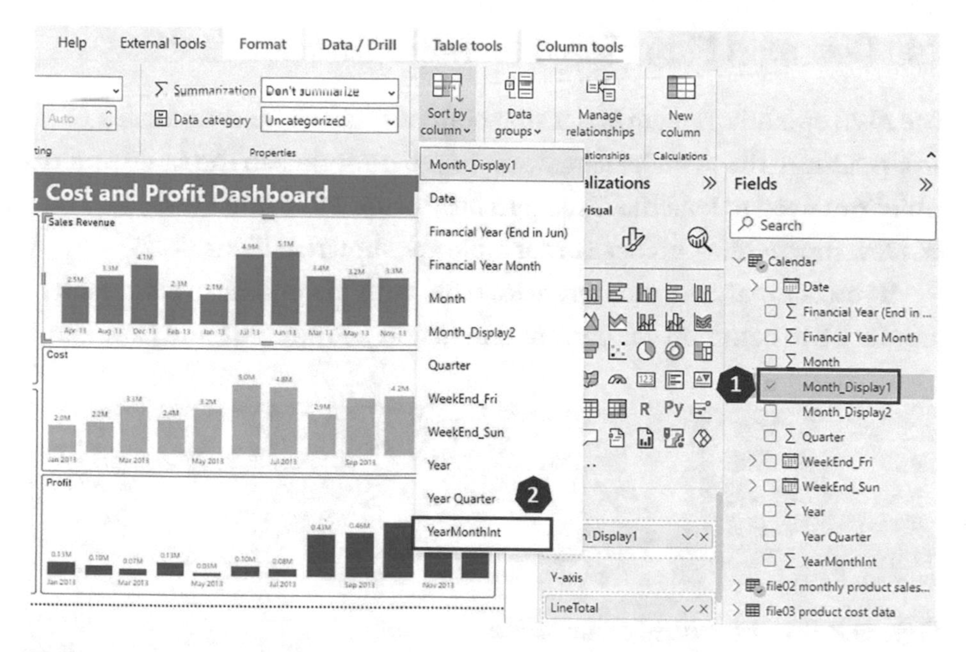

Figure 2-33. *Set sort order for columns*

Some of the visuals may be displayed by the values in descending order, as shown in Figure 2-34. You may consider changing this to monthly chronological order. The following steps explain how to do it.

1. Select the visual and choose more options.
2. Under **Sort axis**, choose Month_Display1 and Sort ascending.

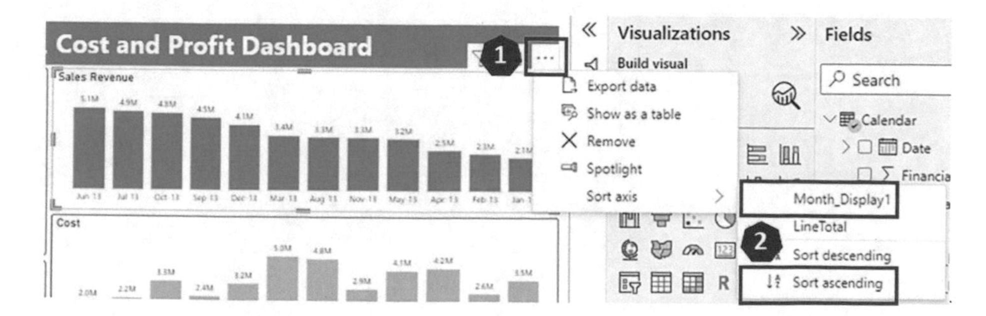

Figure 2-34. *Change display order*

Use Common Filter Table on Multiple Fact Tables

The AWS operations team is tracking both purchase orders and sales orders. One of the team leaders wants to see both stats in the same product table. You need to load the file05 purchase order table into Power BI. The first few rows of the purchase order table are shown in Figure 2-35.

If you compare the monthly sales table with the purchase order table, there are two common columns or dimensions: ProductID and DueDate.

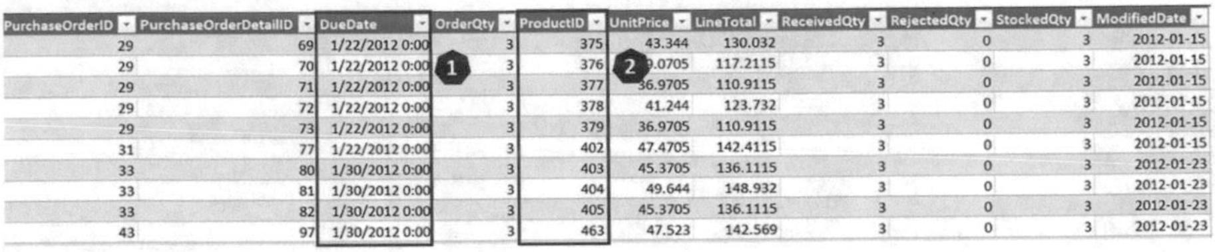

PurchaseOrderID	PurchaseOrderDetailID	DueDate	OrderQty	ProductID	UnitPrice	LineTotal	ReceivedQty	RejectedQty	StockedQty	ModifiedDate
29	69	1/22/2012 0:00	3	375	43.344	130.032	3	0	3	2012-01-15
29	70	1/22/2012 0:00	3	376	[illegible].0705	117.2115	3	0	3	2012-01-15
29	71	1/22/2012 0:00	3	377	36.9705	110.9115	3	0	3	2012-01-15
29	72	1/22/2012 0:00	3	378	41.244	123.732	3	0	3	2012-01-15
29	73	1/22/2012 0:00	3	379	36.9705	110.9115	3	0	3	2012-01-15
31	77	1/22/2012 0:00	3	402	47.4705	142.4115	3	0	3	2012-01-15
33	80	1/30/2012 0:00	3	403	45.3705	136.1115	3	0	3	2012-01-23
33	81	1/30/2012 0:00	3	404	49.644	148.932	3	0	3	2012-01-23
33	82	1/30/2012 0:00	3	405	45.3705	136.1115	3	0	3	2012-01-23
43	97	1/30/2012 0:00	3	463	47.523	142.569	3	0	3	2012-01-23

Figure 2-35. *Purchase order table*

One of a data model's most powerful use cases is that the common dimensions from different fact tables can be connected via common filter tables.

Once the table has been loaded via Power Query, you can connect the additional table following the relationships, as shown in Figure 2-36.

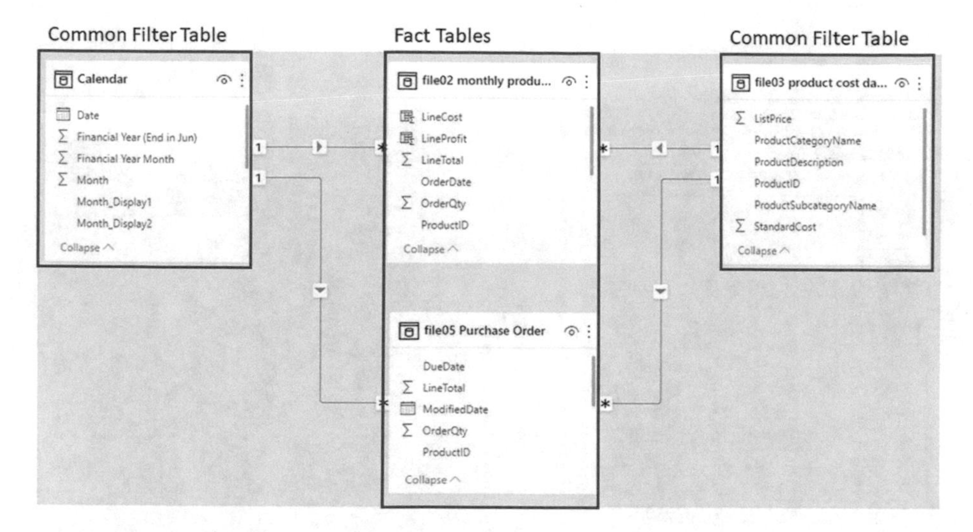

Figure 2-36. *Data model with multiple fact tables and common filter tables*

Because both the monthly sales table and the purchase order table contain the LineTotal column. To differentiate the two, you must rename them **Sales LineTotal** and **PO LineTotal**.

Once you have completed the data connection, you can create the required table. The following steps and Figure 2-37 explain how to do it.

1. Select the table visual.
2. Drag and drop Monthly_Display1 into Columns.
3. Drag and drop Sales LineTotal from monthly sales into Columns.
4. Drag and drop PO LineTotal from Purchase Order into Columns.

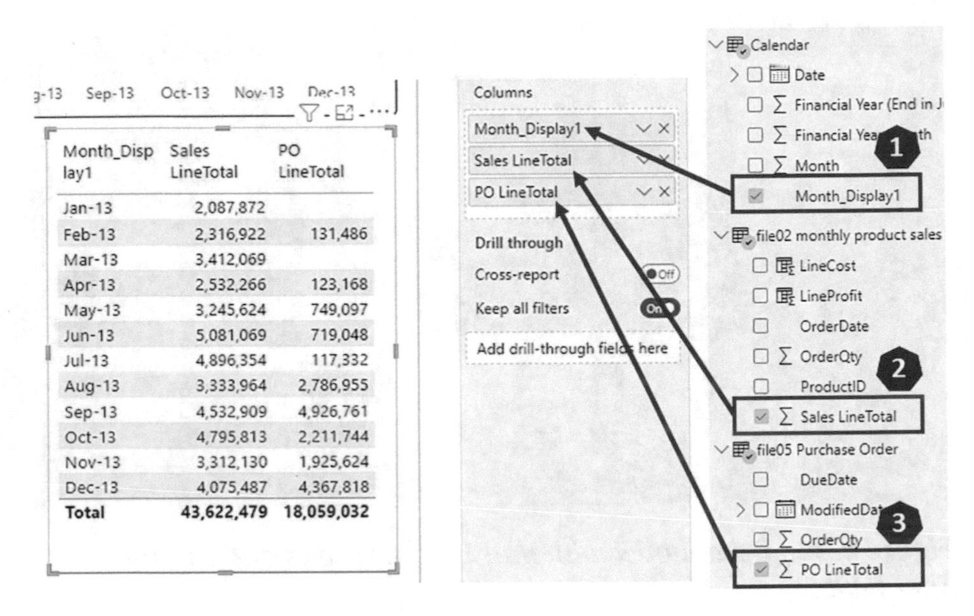

Month_Display1	Sales LineTotal	PO LineTotal
Jan-13	2,087,872	
Feb-13	2,316,922	131,486
Mar-13	3,412,069	
Apr-13	2,532,266	123,168
May-13	3,245,624	749,097
Jun-13	5,081,069	719,048
Jul-13	4,896,354	117,332
Aug-13	3,333,964	2,786,955
Sep-13	4,532,909	4,926,761
Oct-13	4,795,813	2,211,744
Nov-13	3,312,130	1,925,624
Dec-13	4,075,487	4,367,818
Total	**43,622,479**	**18,059,032**

***Figure 2-37.** Create a table visual from multiple fact tables*

You can use the same fields to create a clustered column chart, as shown in Figure 2-38.

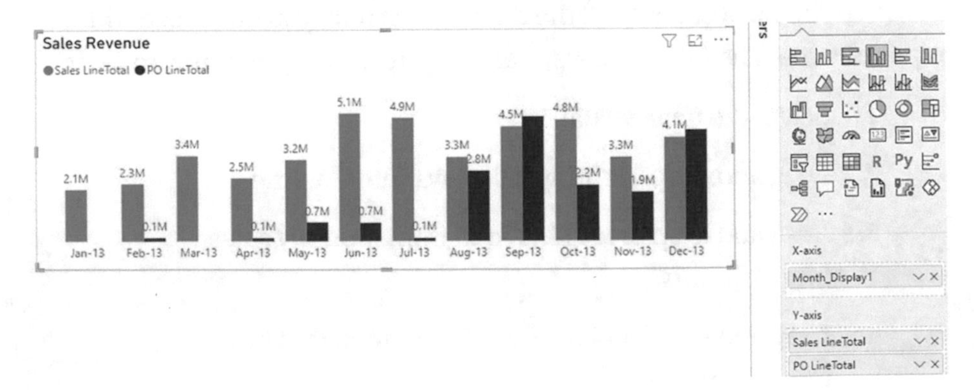

***Figure 2-38.** Create a clustered column chart*

You can also create a matrix table visual to add details to the report. The following steps, shown in Figure 2-39, explain how.

1. Choose a matrix table from the visual selection.
2. Fill in the visual fields for Rows, Columns, and Values.

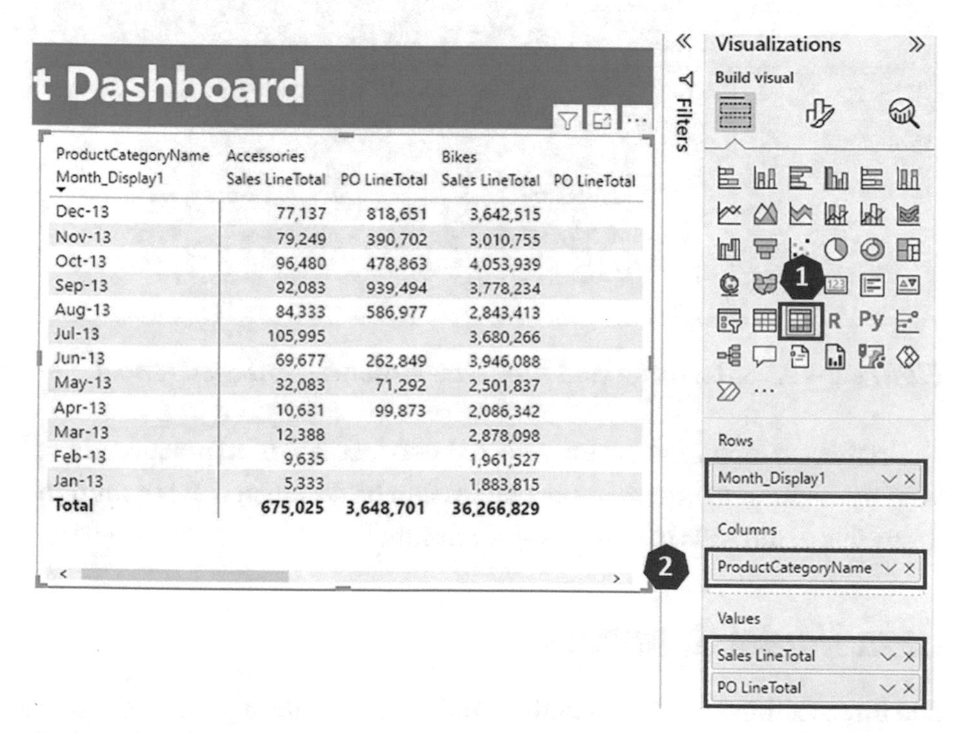

Figure 2-39. *Create a new matrix table visual*

The combined report may look like the report layout in Figure 2-40.

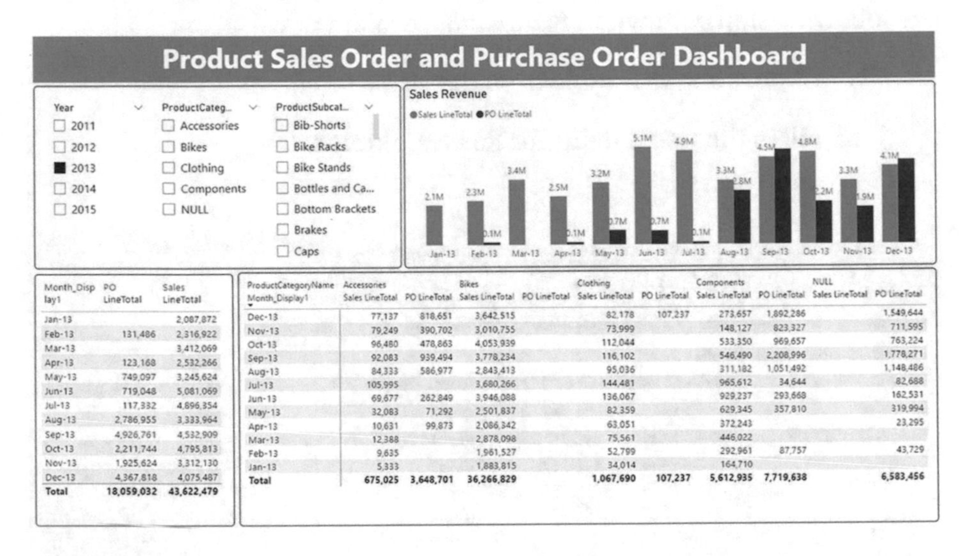

Figure 2-40. *Sales order and purchase order a combined report*

In this example, you have connected two fact tables. In practice, there is no upper limit to how many fact tables can be connected. It is common to see five to ten KPIs in a single report or table.

Data Model Summary

The three examples help you understand three common purposes of using a data model.

- Reduce data redundancy.
- Encourage data reuse (e.g., same Calendar table on multiple reports).
- Use common filter tables to filter multiple fact tables.

Data models can also be used for many other purposes, including simplifying DAX, allowing inactive relationships, and configuring row-level security.

DAX: Data Analysis Expression

DAX is a critical skill set in Power BI because it is close to the front-end visual. It is the most researched topic in Power BI. As shown in Figure 2-41, according to Google, searches for DAX are about three times higher than Power Query.

It is also the reason this DAX session is placed after Power Query and data models. When it comes to Power BI, you can implement many logics in either Power Query M or DAX. But a balanced approach between Power Query, data models, and DAX may work out to be much simpler and more efficient than trying to problem-solve in DAX.

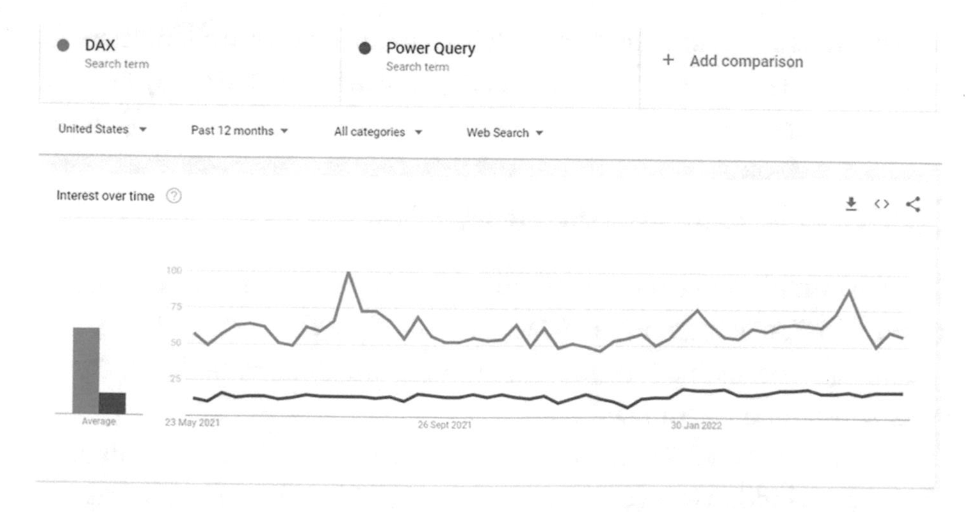

***Figure 2-41.** Google search trend shows DAX more popular than Power Query*

DAX is one of the main reasons why Power BI is easy to begin but difficult to master. As you explore in more detail in this session, DAX can be used to create measures, columns, and tables. It also has many "tricks" to achieve complex logic with simple DAX function(s).

This session takes the following step-by-step approach to dive into DAX.

- Basic aggregation calculations
- Filter context
- Row context
- Create a calendar table
- Calculate() function
- Date calculation
- Use of variables in DAX
- Organize DAX measures

The functions and examples contained in the following chapter are only designed to allow you to get a head start with DAX. This is by no means comprehensive.

Create a Basic Measure in DAX

The most simple and common use of DAX is the measures based on basic aggregation functions, such as SUM(), COUNT(), AVERAGE(), MIN(), and MAX(). You should recognize these table functions from Excel. They behave in similar ways in DAX.

The reports created in previous examples used the SUM() function. The usage is hidden from the developer to simplify the creation process. For example, in our previous example, when the Sales TotalLine and PO TotalLine are used in the visuals using the SUM() function. This is evident if you do the following, as shown in Figure 2-42.

1. Expand the drop-down next to [Sales Total] in the Values field.

2. Notice the different aggregation options and that Sum is selected.

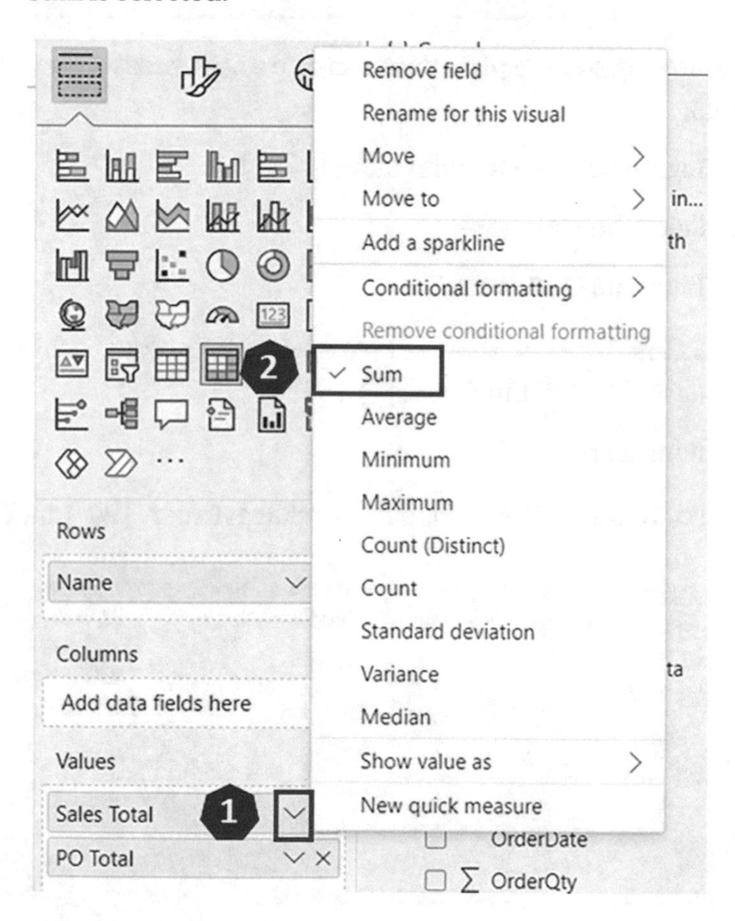

Figure 2-42. *Hidden use of DAX*

Instead of using the columns, you can replicate this by creating a measure with the SUM() function.

> **Note** DAX measures do not belong to any table but must be created inside a table. You will learn how to organize measures later; for now, all measures are created in the Calendar table.

The following steps explain how to create a DAX measure, as shown in Figure 2-43.

1. Right-click the Calendar table.
2. Select **New measure**.
3. Enter the DAX formula.

   ```
   Sales Total = SUM( 'file02 monthly product sales
   data'[Sales LineTotal] )
   ```

4. Repeat steps 1–3 for

   ```
   PO Total = SUM( 'file05 Purchase Order'[PO LineTotal] )
   ```

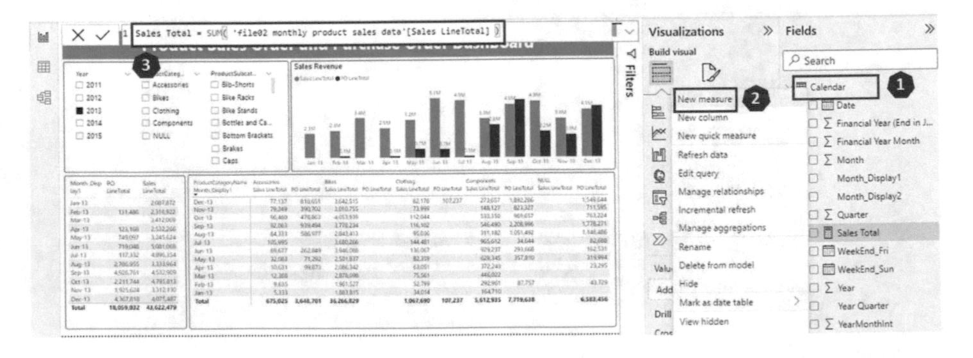

***Figure 2-43.** Create a new DAX measure*

With the new DAX measure, you can check out the Power BI fields relating to DAX measures shown in Figure 2-44.

1. Note the calculator logo indicating that Sales Total is a measure.
2. Note the Measure formula.
3. Note the Measure tools, including Structure, Format, and Properties.

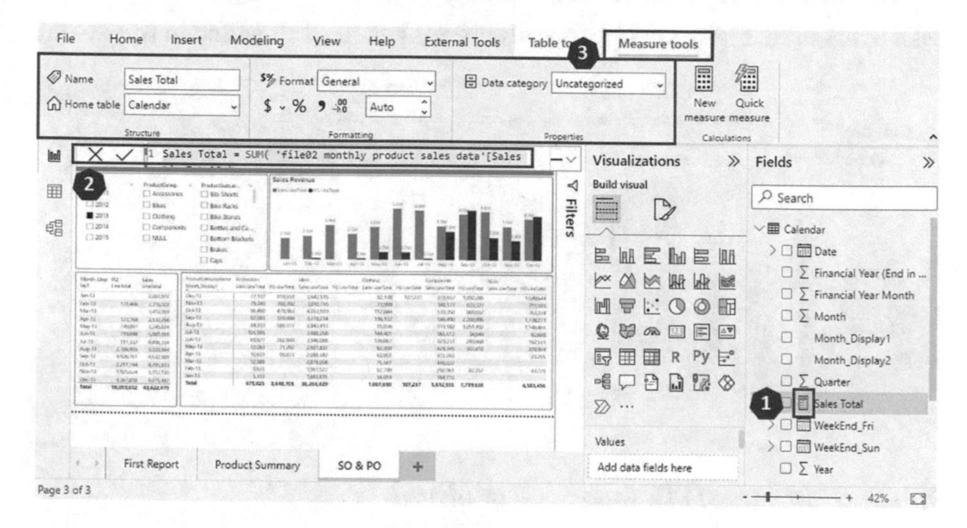

Figure 2-44. *DAX measure-related fields*

Now you can dive deeper into the DAX measure formula. Creating a total sales measure consists of the following three parts, as illustrated in Figure 2-45.

1. Sales *Total* is the measure name.
2. SUM() is the function.
3. Column Sales LineTotal is the function parameter.

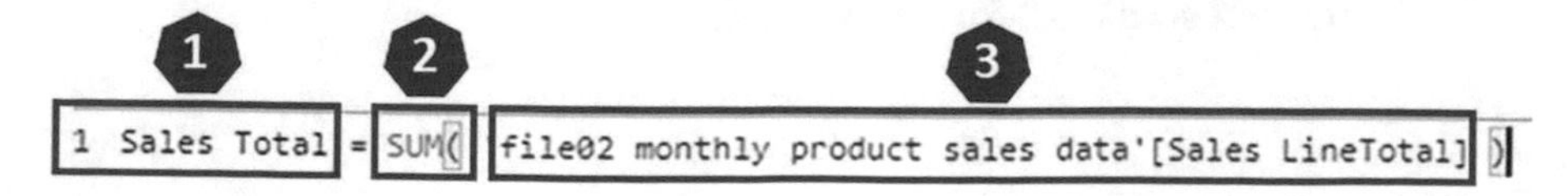

Figure 2-45. *DAX measures formula breakdown*

According to the Microsoft DAX documentation, the SUM function only accepts one column as its parameter (docs.microsoft.com/en-us/dax/sum-function-dax). Some functions, such as CALCULATE(expression, [Filter 1] ...), accept multiple parameters. The DAX measure also allows a variable declaration to construct more complex measures.

You can now replace the values in the clustered column visual with DAX measures. Figure 2-46 shows that it makes no difference in the visual.

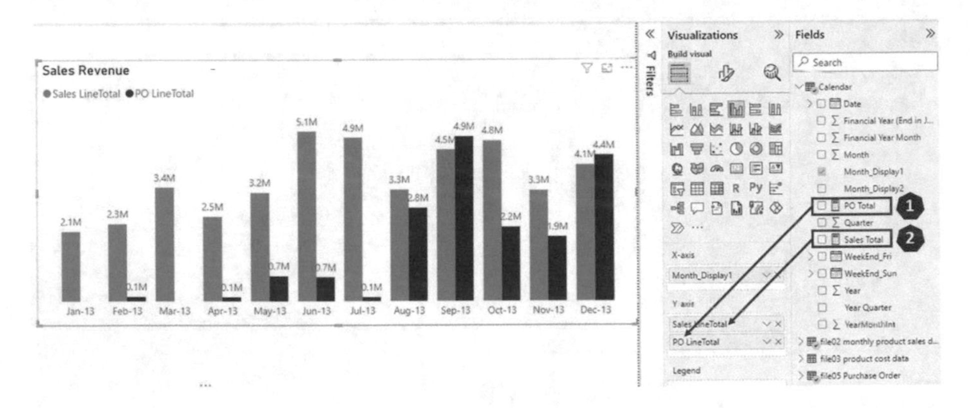

Figure 2-46. *Use DAX measures in visual*

When displaying the DAX measure in tables, it is a common requirement to format the measures differently. The following steps show how to format the PO total with a $ prefix and zero decimal places, as shown in Figure 2-47.

1. Select PO Total.
2. Select the $ symbol.
3. Set the decimal place to 0.
4. Repeat steps 1–3 for Sales Total.

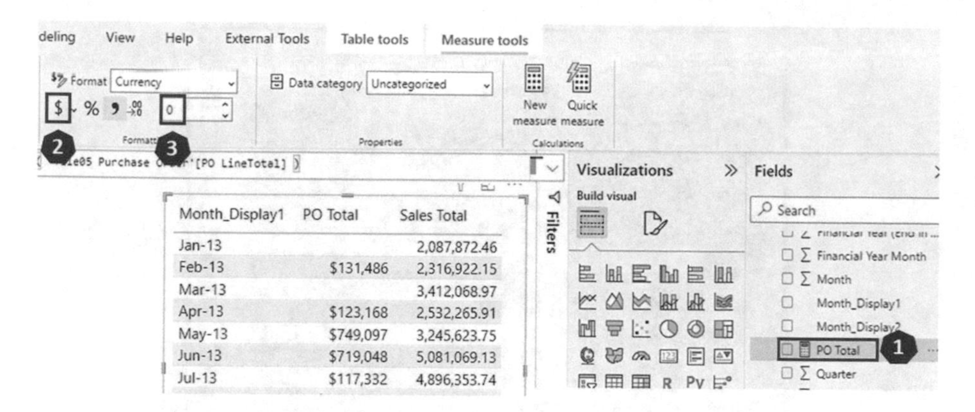

Figure 2-47. *Format DAX measure*

Filter Context in DAX

Once you understand the basics of DAX, the next critical thing to understand is that DAX is all about context. DAX has two types: *row context* and *filter context*.

Once you understand DAX context, you can answer a silly but important question. How can Power BI use the same DAX measure to display different values for each product description? See the example shown in Figure 2-48.

ProductDescription	Sales Total
All-Purpose Bike Stand	21,624.00
AWC Logo Cap	23,582.96
Bike Wash - Dissolver	11,611.21
Cable Lock	4,700.52
Chain	6,451.01
Classic Vest, L	5,854.70
Classic Vest, M	61,255.31
Classic Vest, S	103,570.71
Fender Set - Mountain	23,914.24
Total	**43,622,479.05**

Figure 2-48. *Sales Total Measure display*

The answer is the *filter context.*

Figure 2-48 shows the Sales Total measure calculated multiple times (separate calculation for each product description), once for each row. If there are 100 product names in the table, the sales total is calculated 100 times. Because each row produces a different filter context, 100 different results are displayed in the Sales Total column.

This behavior is further explained in Figure 2-49. Calculate the displayed value for the row in Product A. The following steps explain how to do it.

1. The filter context filters the Monthly Sales table for product A.

2. The sales total DAX measure sums the Sales LineTotal column.

3. The calculated total from step 2 is displayed in the table next to Product A.

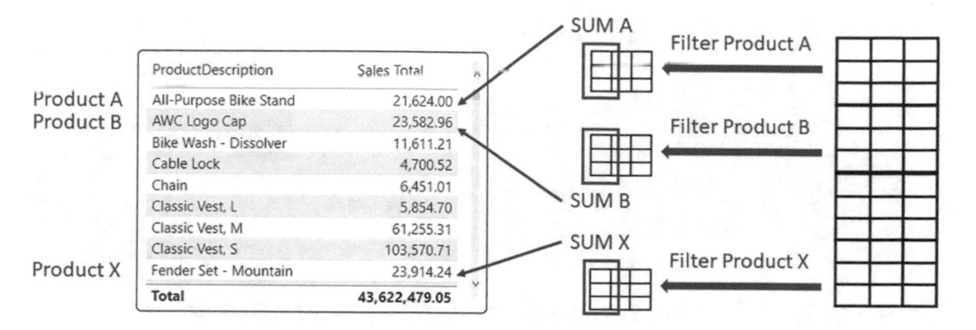

***Figure 2-49.** The filter context in action*

When there are multiple columns in the table, this creates multiple filter contexts. Figure 2-50 shows that the product and month filters are applied to the calculation when the filter context is combined.

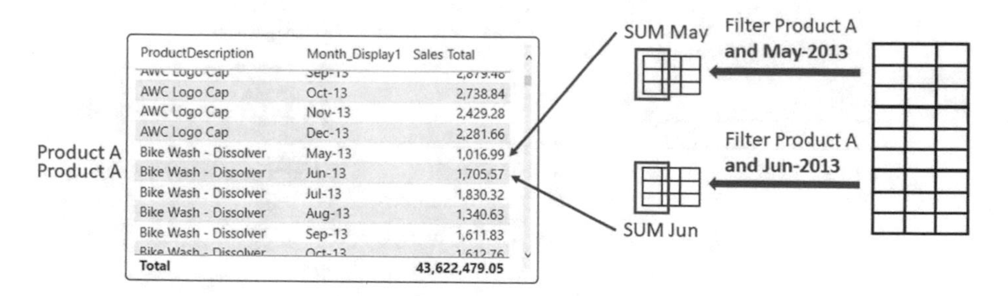

***Figure 2-50.** Multiple filter contexts*

Now you understand how each value of the Sales Total column is calculated. How is the total value calculated at the bottom of the table?

It is a common mistake for developers to believe that the total value of $43,622,479 is calculated by adding up all the individual values in the Sales Total column. The total value is calculated *without* a filter context, as shown in Figure 2-51.

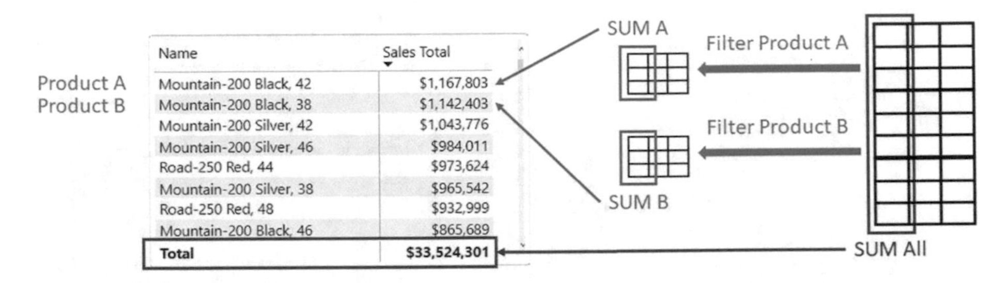

***Figure 2-51.** The total is calculated with no filter context*

The previous examples showed the filter context within the same visual. It is important to note that the report could have additional filter context. Figure 2-52 shows that the selections from the slicer and year introduce additional filter context.

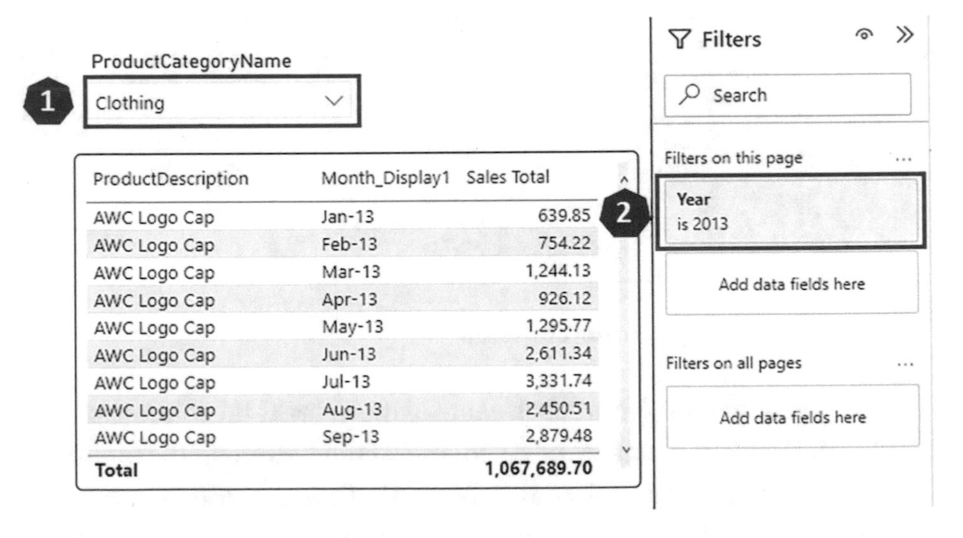

***Figure 2-52.** Additional filter contexts*

As you can see, any Power BI report page could have layers of filter contexts. When the filter contexts change, all measures displayed in the report are re-evaluated, causing the report figures to change on the fly.

You also find out later that the filter context can be "modified" using the CALCULATE() DAX function. This allows developers to use a filter context to its full advantage.

Row Context in DAX

Row Context was typically triggered by the iterator functions in DAX, such as SUMX(), MINX(), and AVERAGEX(). It is a relatively complex concept for new Power BI developers to understand. Even if you are not 100% clear about row context after reading, you may continue with the book but wish to revisit this session later.

The following are a few key concepts to understand row context.

- Iteration means going through a table one row at a time until it reaches the end.
- Each row in the table has a different row context.
- Row context evaluates a function or expression and produces an output value.
- A DAX iterator specifies the function to be applied for the outputs from each row in step 3 (e.g., SUMX() sums all output values). AVERAGEX() averages all the output values.

SUMX() and SUM() functions often produce the same result. For example, using the following DAX formula, you can create a new measure called Sales TotalX using the SUMX() instead of the SUM() function.

```
Sales TotalX =
  SUMX( 'file02 monthly product sales data', [Sales
  LineTotal] )
```

Compare this to the following original Sales Total DAX formula.

```
Sales Total =
  SUM( 'file02 monthly product sales data'[Sales LineTotal] )
```

Even though the results are identical, as shown in Figure 2-53, SUMX() and SUM() functions are fundamentally different in that SUMX() utilizes row context while SUM() does not.

ProductDescription	Sales Total	Sales TotalX
AWC Logo Cap	23,582.96	23,582.96
Classic Vest, L	5,854.70	5,854.70
Classic Vest, M	61,255.31	61,255.31
Classic Vest, S	103,570.71	103,570.71
Full-Finger Gloves, L	20,760.91	20,760.91
Full-Finger Gloves, M	14,929.82	14,929.82
Full-Finger Gloves, S	3,588.16	3,588.16
Half-Finger Gloves, L	12,649.81	12,649.81
Half-Finger Gloves, M	31,678.24	31,678.24
Total	**1,067,689.70**	**1,067,689.70**

Figure 2-53. *SUM and SUMX produce the same result*

Iterator functions are much more powerful than the prior example has demonstrated. There is a better example of using an iterator for calculating the total cost. Previously, the total cost was calculated following a two-step process.

First, create a DAX column in the Monthly Sales table.

```
Line Cost = '
  monthly product sales data'[OrderQty]
  * RELATED( 'Production Cost'[StandardCost] )
```

Then, create a DAX measure.

```
Total Cost = SUM( 'monthly product sales data'[Line Cost] )
```

With row-level context, this can be calculated using an iterator in a single step.

```
Cost TotalX =
  SUMX(
    'file02 monthly product sales data',
    // below expression to be completed for each row \\
     [OrderQty]*RELATED('file03 product cost
     data'[StandardCost])
  )
```

Row-level context and iterators are common optimization techniques in Power BI. They are more efficient and simplify the data logic.

Note Because there are no for-loop functions in DAX, iterators can replace for-loop logic.

Create a Calendar Table Using DAX

The DAX functions in the previous examples all returned single values (e.g., SUM(), SUMX()). Many DAX functions can also create tables (e.g., Filter(), Summarise(), and AddColumns()). One of the most common DAX tables is the Calendar table from the previous session. You can create the same Calendar table. The following steps and Figure 2-54 explain how.

1. Select **New table** in the Modeling ribbon.
2. Enter the following DAX formula.

```
DAX Calendar =
ADDCOLUMNS(
  // create a [Date] column with start and end date
  CALENDAR(DATE(2011, 1, 1), DATE(2015, 12, 31 )),
```

```
        // create additional columns based on the [Date]
        column above
        "Calendar Year", YEAR( [Date] ),
        "Month_Display1", FORMAT( [Date], "mmm-yy" ),
        "Month_Display2", FORMAT( [Date], "mmm/yyyy" ),
        "YearMonthInt", FORMAT( [Date], "YYYYMM" )
    )
```

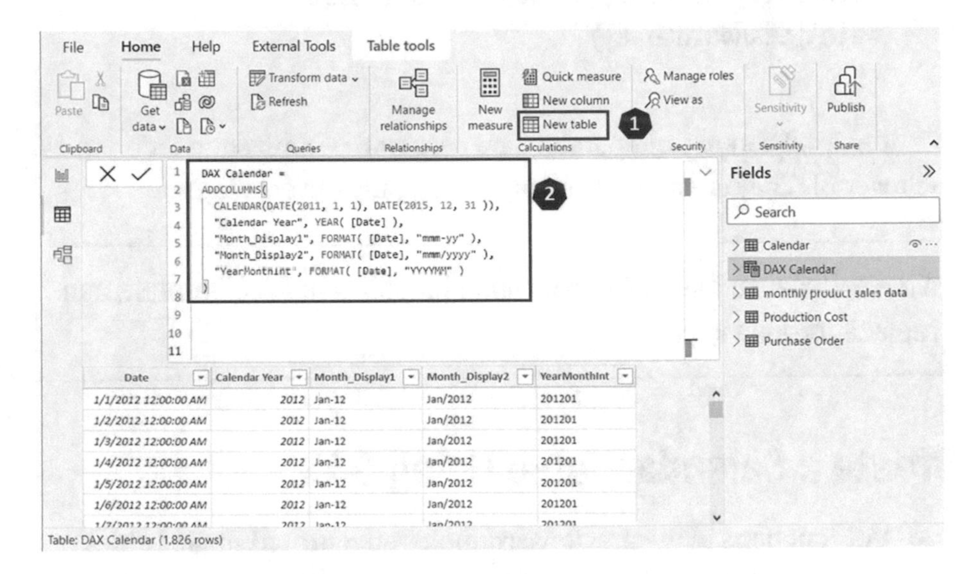

Figure 2-54. *Create a new DAX calendar table*

This DAX function may look more complex than the ones you have created previously. When exploring complex DAX formulas, you often need to follow the order of *inside out*, then *top to bottom*. You may also consider the functions at two levels: *summary* and *detail*. The DAX Calendar formula can be viewed in the following order.

- Functions that provide a high-level summary
  - **CALENDAR()** generates a table with a column named Date that contains a date range with specified start and end dates.
  - **ADDCOLUMNS()** adds additional calculated columns based on the Date column in the Filtered table from step B.
- Detail functions
  - **DATE()** creates a date value to allow comparison with the [Date] column in the Filter() function
  - **YEAR()** extracts the calendar year values from the [Date] column.
  - **FORMAT()** is a very flexible function that translates [Date] column into specified text formats, such as mmm-yy (Jan-12) or mmm/yyyy (Jan/2012).

One key benefit of using this DAX Calendar approach is that the date range can be determined according to the fact table. This way, there is no need to keep updating a big data table in Excel. Figure 2-55 shows how to achieve a dynamic date range.

```
DAX Calendar =
ADDCOLUMNS(
    CALENDAR(
        // DATE(2011, 1, 1), date(2015, 12, 31)
        DATE( YEAR(MIN('file02 monthly product sales data'[OrderDate])), 1, 1 ),
        DATE( YEAR(MAX('file02 monthly product sales data'[OrderDate])), 12, 31 )
    ),
    "Calendar Year", YEAR( [Date] ),
    "Month_Display1", FORMAT( [Date], "mmm-yy" ),
    "Month_Display2", FORMAT( [Date], "mmm/yyyy" ),
    "YearMonthInt", FORMAT( [Date], "YYYYMM" )
)
```

Figure 2-55. *Dynamic DAX calendar*

Note Some developers may be hesitant to introduce DAX tables because they worry this may slow down the report. For small tables with simple logic like the Calendar table, this is not going to have any noticeable impact. It may be useful to understand the execution sequence in Power BI, as shown in Figure 2-56.

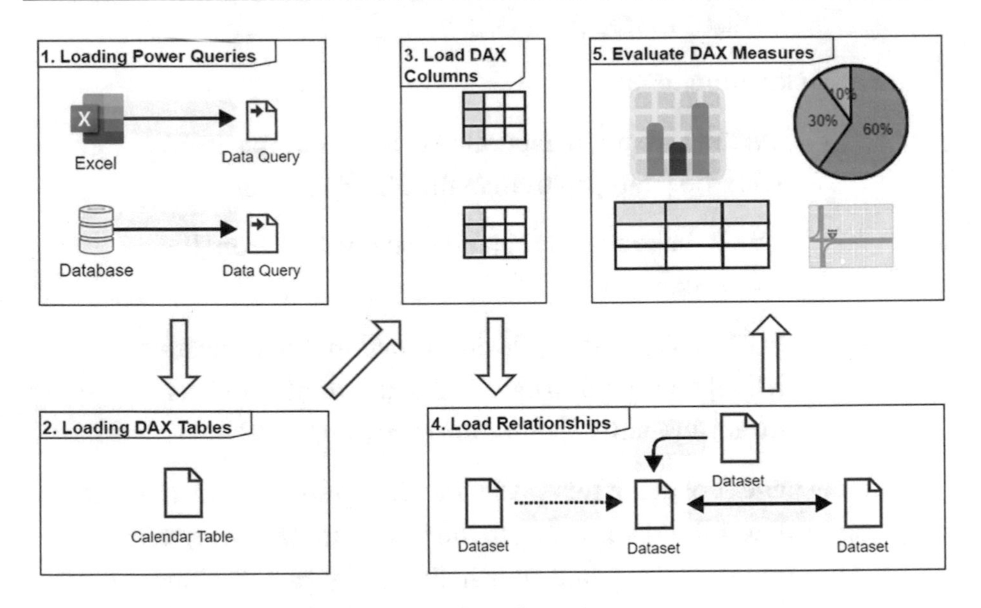

Figure 2-56. *Power BI execution order*

The Calculate Function in DAX

At this point, you should understand the basics of DAX and the filter context in Power BI. It is enough for exploring the very important CALCULATE() function.

The CALCULATE() function accepts two types of arguments: expressions and filters.

CALCULATE(<expression> , <filter1> , <filter2> ...)

Most of the time, CALCULATE() changes the filter context. The following are the three main ways.

- Ignore the filter context.
- Add a filter context.
- Update the filter context.

Ignore the Filter Context

The following example shows a common reason for ignoring filter context: to calculate the total percentage.

As shown in Figure 2-57, you must divide the row value by the Total value to calculate the percentage. This logic conflicts with the filter context, which prevents the total from being calculated. The Calculate() function is the only way to ignore the filter context with the RemoveFilter() function.

First, create a Fixed Total measure.

```
Fixed Total =
CALCULATE(
  [Sales Total], // re-use predefined measure
  // ignore filter context for category name \\
  REMOVEFILTERS('file03 product cost
  data'[ProductCategoryName]) )
```

Then, create a Percentage of Total measure.

```
Percentage of Total = [Sales Total] / [Fixed Total]
```

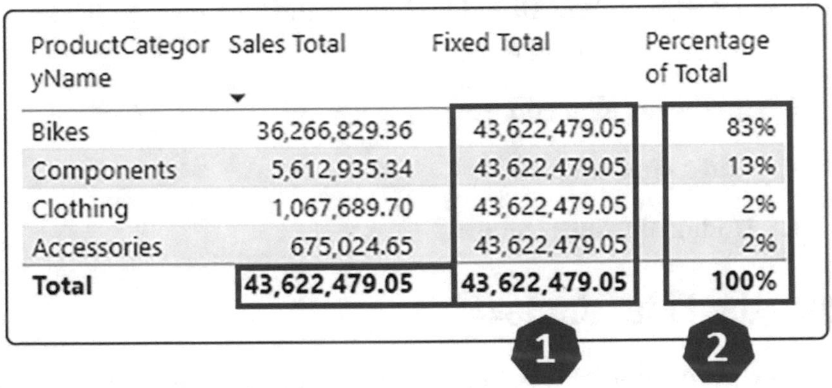

ProductCategoryName	Sales Total	Fixed Total	Percentage of Total
Bikes	36,266,829.36	43,622,479.05	83%
Components	5,612,935.34	43,622,479.05	13%
Clothing	1,067,689.70	43,622,479.05	2%
Accessories	675,024.65	43,622,479.05	2%
Total	**43,622,479.05**	**43,622,479.05**	**100%**

Figure 2-57. *Calculate percentage of total*

RemoveFilter() can apply to a column or a table. When it is applied to the table, it ignores all filter contexts applied to the table.

Note RemoveFilter( ) is the same as the All( ) function. You should use RemoveFilter( ) for columns and All( ) for tables to improve readability.

RemoveFilters(), All(), AllExcept(), and many other functions are referred to as *filter modifiers* in the Calculate() function because they can modify the filter context.

Add a Filter Context

The opposite of ignoring filter context is to introduce a new filter context. For example, the business may consider that any line item total less than $200 is considered a low-value item. The following steps create the columns shown in Figure 2-58.

First, create a Low Value Sales Total measure.

```
Low Value Sales Total =
CALCULATF(
    [Sales Total],
    // adding new filter context below \\
    'file02 monthly product sales data'[Sales LineTotal] < 200
)
```

1. Then, create a Low Value Sales % measure.

```
Low Value Sales % = DIVIDE([Low Values Sales Total],[Sales Total])
```

ProductCategory Name	Sales Total	Low Values Sales Total	Low Value Sales %
Bikes	$36,266,829		
Components	$5,612,935	$326,325	6%
Clothing	$1,067,690	$529,315	50%
Accessories	$675,025	$484,358	72%
Total	**$43,622,479**	**$1,339,997**	**3%**

1 2

Figure 2-58. *Add a new filter context*

Tip DIVIDE( ) function is used in step 2 instead of the / operator. This is because DIVIDE( ) is much better at error handling (e.g., when denominator = 0, it returns blank( ) instead of error).

Update the Filter Context

In a different example, you might want to update the filter context to only display sales for bikes, as shown in Figure 2-59. This can be achieved with the following DAX formula.

```
Bikes Sales Total =
CALCULATE(
    [Sales Total],
    // update filter context \\
    'file03 product cost data'[ProductCategoryName] = "Bikes"
)
```

ProductCategoryName	Sales Total	Bikes Sales Total
Bikes	$36,266,829	$36,266,829
Components	$5,612,935	$36,266,829
Clothing	$1,067,690	$36,266,829
Accessories	$675,025	$36,266,829
Total	**$43,622,479**	**$36,266,829**

Figure 2-59. *Change filter context*

The formula in CALCULATE() is applied the same way as adding a new filter context, but this time it is updating the existing filter context. This is because each row in the table shown in Figure 2-59 already contains a filter context for the same column (i.e., ProductCategoryName). Instead of adding a new filter context of `ProductCategoryName = "Bikes"`, the Calculate() function replaces any existing filter context on the column. As a result, all rows display the total using filter context for Bikes.

One interesting function that prevents any update in the original filter context is KeepFilters(). Figure 2 60 shows that the total value disappears when the KeepFilter() function is applied. This is because the KeepFilter() function forces the original filter context on ProductCategoryName to be kept instead of removed. For example, the filter context for the second row becomes ProductCategoryName = "Bikes" AND ProductCategoryName = "Components". Since this is not possible, the total is blank.

ProductCategoryName	Sales Total	Bikes Sales Total
Bikes	$36,266,829	$36,266,829
Components	$5,612,935	
Clothing	$1,067,690	
Accessories	$675,025	
Total	**$43,622,479**	**$36,266,829**

Figure 2-60. *KeepFilters()*

Date Intelligence with CALCULATE

The examples you created have used several different filter modifiers inside the Calculate() function. DAX also has many filter modifiers for date-related calculations.

DatesYTD() is a filter modifier that is very useful in practice. It can be used to calculate the cumulative total. You can create an analysis using the following DAX code, as shown in Figure 2-61.

```
Sales YTD =
CALCULATE(
    [Sales Total],
    DATESYTD('DAX Calendar'[Date])
)
```

Month_Display1	Sales Total	Sales YTD
Jan-13	$2,087,872	$2,087,872
Feb-13	$2,316,922	$4,404,795
Mar-13	$3,412,069	$7,816,864
Apr-13	$2,532,266	$10,349,129
May-13	$3,245,624	$13,594,753
Jun-13	$5,081,069	$18,675,822
Jul-13	$4,896,354	$23,572,176
Aug-13	$3,333,964	$26,906,140
Sep-13	$4,532,909	$31,439,049
Oct-13	$4,795,813	$36,234,862
Nov-13	$3,312,130	$39,546,992
Dec-13	$4,075,487	$43,622,479
Total	**$43,622,479**	**$43,622,479**

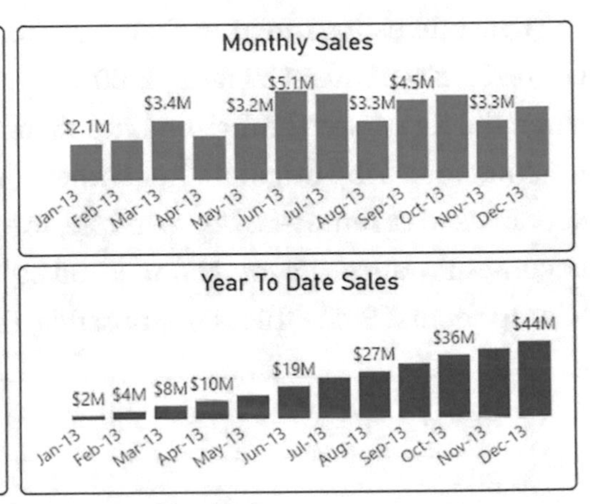

***Figure 2-61.** Year-to-date sales calculation with DATESYTD()*

In addition to DATESYTD(), there are also additional functions for MTD (month to date) and QTD (quarter to date).

In practice, you may need to deal with financial years, as shown in Figure 2-62. DATESYTD() also accepts a year-end date parameter. For example, to start accumulating from July, you can set the date to 06-30.

Month_Display1	Sales Total	Sales YTD
Jan-13	$2,087,872	$18,381,813
Feb-13	$2,316,922	$20,698,736
Mar-13	$3,412,069	$24,110,805
Apr-13	$2,532,266	$26,643,070
May-13	$3,245,624	$29,888,694
Jun-13	$5,081,069	$34,969,763
Jul-13	$4,896,354	$4,896,354
Aug-13	$3,333,964	$8,230,318
Sep-13	$4,532,909	$12,763,227
Oct-13	$4,795,813	$17,559,040
Nov-13	$3,312,130	$20,871,170
Dec-13	$4,075,487	$24,946,657
Total	**$43,622,479**	**$24,946,657**

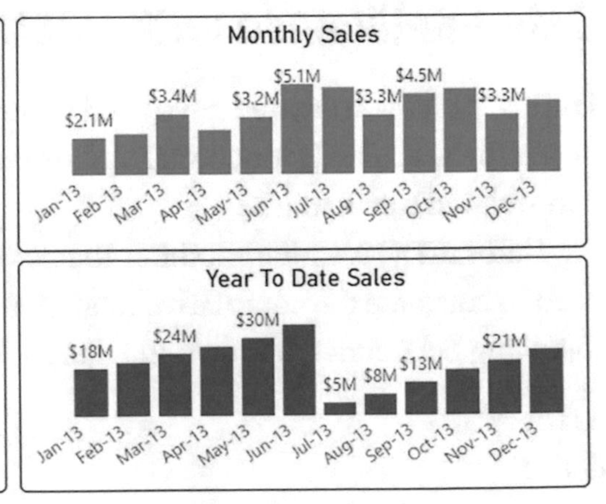

***Figure 2-62.** Year-to-date sales starting from July*

The **SamePeriodLastYear()** function is a filter modifier that calculates the value for the same period from the previous year. Figure 2-63 shows this can be useful in calculating change %. The following steps explain how to implement the measures.

First, create a measure for Sales Total SPLY (SamePeriodLastYear).

```
Sales Total SPLY =
CALCULATE(
    [Sales Total],
    SAMEPERIODLASTYEAR( 'DAX Calendar'[Date] )
)
```

1. Then, create a measure for Sales Change SPLY.

```
Sales Change SPLY = DIVIDE([Sales Total], [Sales Total SPLY])-1
```

Month_Display1

Jun-13

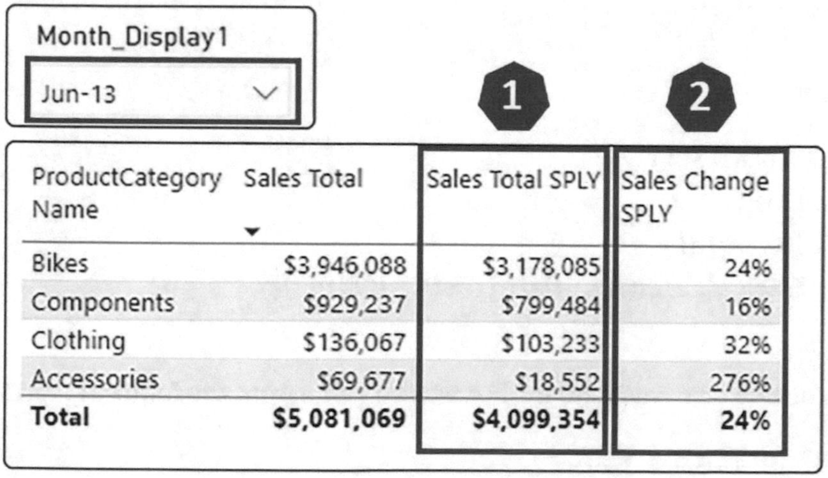

ProductCategory Name	Sales Total	Sales Total SPLY (1)	Sales Change SPLY (2)
Bikes	$3,946,088	$3,178,085	24%
Components	$929,237	$799,484	16%
Clothing	$136,067	$103,233	32%
Accessories	$69,677	$18,552	276%
Total	**$5,081,069**	**$4,099,354**	**24%**

Figure 2-63. *Same period as last year's calculations*

Warning Date intelligence functions require a proper date table. You can create a DAX date table using the formula from the previous session.

DATEADD() is a more generic function compared to SamePeriodLastYear(). For demo purposes, you have a few options to produce the SamePeriodLastYear() result, as follows.

```
Sales Total SPLY  =
CALCULATE(
  [Sales Total],
  DATEADD( 'Calendar'[Date], -1, YEAR )
)
```

Or

```
Sales Total SPLY  =
CALCULATE(
  [Sales Total],
  DATEADD( 'Calendar'[Date], -4, QUARTER )
)
```

Or

```
Sales Total SPLY  =
CALCULATE(
  [Sales Total],
  DATEADD( 'Calendar'[Date], -12, MONTH )
)
```

You can also calculate the last week's value from the following formula.

```
Sales Total Last Week =
CALCULATE(
  [Sales Total],
  DATEADD( 'Calendar'[Date], -7,DAY )
)
```

Variables in DAX

There are three main reasons you need to use variables in DAX.

- Simplify DAX
- Improve performance
- Resolve placeholder function errors

To simplify, start with a simple example to create a measure for the start of the last order month using the DATE() function. DATE() function accepts three parameters, year, month, and date, and returns a date value (e.g., `DATE( 2019, 12, 30 )` returns a date value).

You can calculate the last order month from the following formula.

```
Start of Last Order Month =
DATE(
  YEAR( MAX( 'monthly product sales data'[OrderDate] ) ),
  MONTH( MAX( 'monthly product sales data'[OrderDate] ) ),
  1
)
```

As you might have noticed, the MAX() function was repeated twice. This DAX formula can be rewritten using variables to make it more concise and easier to understand.

```
Last Order Month =
  VAR LastOrderDate = MAX('monthly product sales
  data'[OrderDate])
  RETURN DATE( YEAR( LastOrderDate ), MONTH(
  LastOrderDate ), 1 )
```

Note Whenever VAR is used, RETURN must be added.

Performance variables are helpful because performance is calculated only once, and the values are stored. Even though DAX has a lot of built-in efficiency smartness, it is still good to consider variables as a good optimization option.

A function PLACEHOLDER error. With the Last Order Month measure created above, you can create a new measure to calculate the sales total for the last month. Intuitively, you may want to use the Calculate() function with a filter modifier on the date, as shown in Figure 2-64.

```
1 Sales Total Last Month =
2 CALCULATE(
3     [Sales Total],
4     // add filter for last order month \\
5     'file02 monthly product sales data'[OrderDate] >= [Last Order Month]
6 )
```

A function 'PLACEHOLDER' has been used in a True/False expression that is used as a table filter expression. This is not allowed.

Figure 2-64. *Placeholder error*

You see the following error when you create the DAX measure shown in Figure 2-64.

A function 'PLACEHOLDER' has been used in a True/False expression that is used as a table filter expression. This is not allowed.

This error is very confusing. If someone translates it into plain English, it may look like the following.

Both sides of the comparison logic must be fixed values. At least one of the values is not fixed. This is not allowed.

The good news is that the fix is easy using a variable. When a variable is assigned a value in DAX, it holds the value. From the DAX perspective, the variable is considered a fixed value. The logic can be implemented using the following DAX measure.

```
Sales Total Last Month =
VAR LastMonth = [Last Order Month]
RETURN CALCULATE(
  [Sales Total],
  'file02 monthly product sales data'[OrderDate] >= LastMonth
)
```

Organize DAX Measures

More than 20 DAX measures were in the calendar table in the previous examples. It is starting to become difficult to keep track. Before you start to organize the measures, it is important to understand that the home table location for the measures is irrelevant. There are many ways to organize DAX measures.

When it comes to a report with multiple fact tables, you may want to organize them all under a single table, such as _Measures. The initial underscore ensures that the measure groups are displayed at the top.

You can use the following DAX formula to create a new table.

```
_Measure = DATATABLE( "_Test", STRING, {{""}} )
```

Once the table is created, migrate existing measures into this new table. Use the following steps, as illustrated in Figure 2-65.

1. Select **Model** view.
2. Choose the measures that need to be migrated (hold the Ctrl key to select multiple).
3. Go to the Home table drop-down and choose _Measures.
4. Enter a Display folder name (optional).

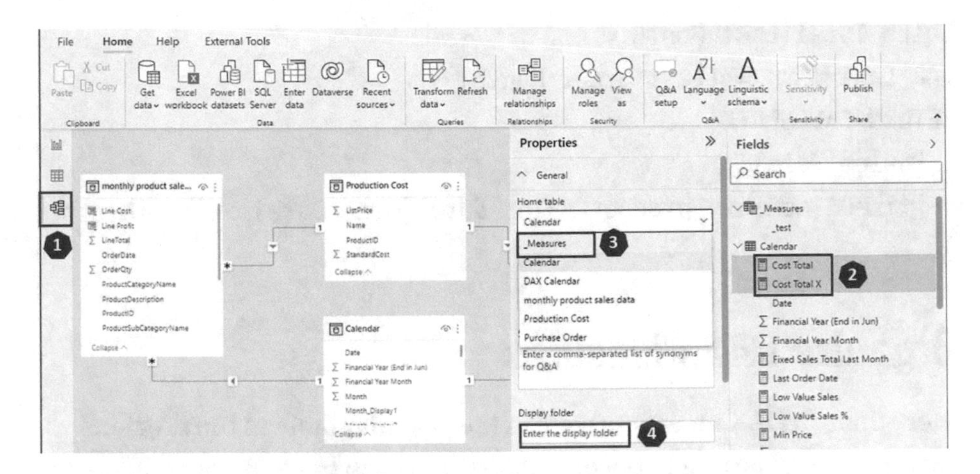

***Figure 2-65.** Change the home table for measures*

There could be over 100 measures for complicated reports. You can use the Display folder to further group measures into folders. One common practice is to group them under the fact table names. This makes it much easier to know the base table name for each measure, as shown in Figure 2-66.

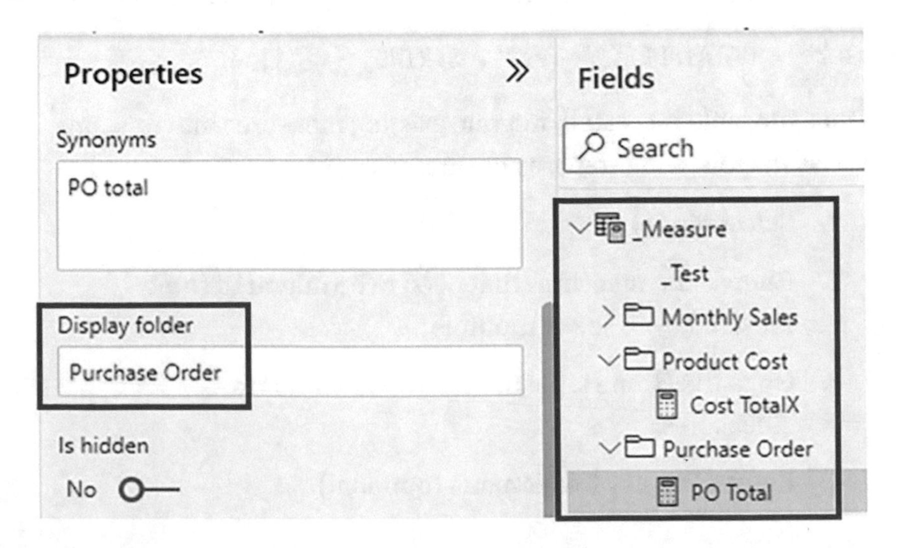

***Figure 2-66.** Organized measures in table and folder structure*

DAX Summary

You should now have a broad but basic understanding of the different aspects and practical uses of DAX. Some of the concepts mentioned in this session become clearer over time as you apply them in practice. Many other DAX functions are not covered here but have good practical uses, such as SUMMARIZE(), VALUE(), and CALCULATETABLE().

Conclusion

Congratulations on getting this far! This is a challenging chapter, but you have now gained a good practical understanding of the three key components of Power BI Desktop. With more practice, you can soon reach the intermediate level of Power BI development.

Developing report in Power BI Desktop is not the end goal. In the next chapter, you learn about the important Power BI services (cloud) to share reports and automate future data refresh.

CHAPTER 3

Power BI Service

Power BI Service is a cloud-based Power BI Server in the Microsoft Azure cloud. PBI Service contains everything you see on app.powerbi.com, including the workspaces and the published reports and datasets.

In this chapter, you gain important insights into reporting sharing through Power BI Server and learn ways to automate report refreshes.

Note Power BI Service is a fast-evolving beast. When writing this book in May 2022, Power BI had just announced the public preview of datamarts. This means all three core components of Power BI have been migrated into the cloud, including Power Query (dataflow), data model (datamart), and DAX (datamart).

Publishing and Sharing Reports

Publishing and sharing reports are not the most exciting part of report development, but it is full of traps for developers. It can also lead to a serious breach of company data policy. Sharing is not hard, so after reading this session of the book, you will have a good understanding of the preferred sharing option in Power BI.

D. Ding, *Transitioning to Microsoft Power Platform*,
https://doi.org/10.1007/978-1-4842-9239-6_3

Business Scenario

Back at the office, Kim thought the report was finally ready. She walked her manager Milo through the report, and Milo was also impressed. Milo now wanted her to share the report with a few more people within his team and a technology manager.

She understood that the data covers the full product range, including sales and cost data, which is very sensitive information. Only a few people in the technology, finance, and leadership teams can access this level of detail.

She is not sure how to share the report. She followed Power BI's documentation from the Microsoft website to publish and share reports.

Warning The following sharing method is not recommended.

Kim first published the report using the following steps, as illustrated in Figure 3-1.

1. Publish the report online.
2. Select My Workspace.
3. Open the report link.

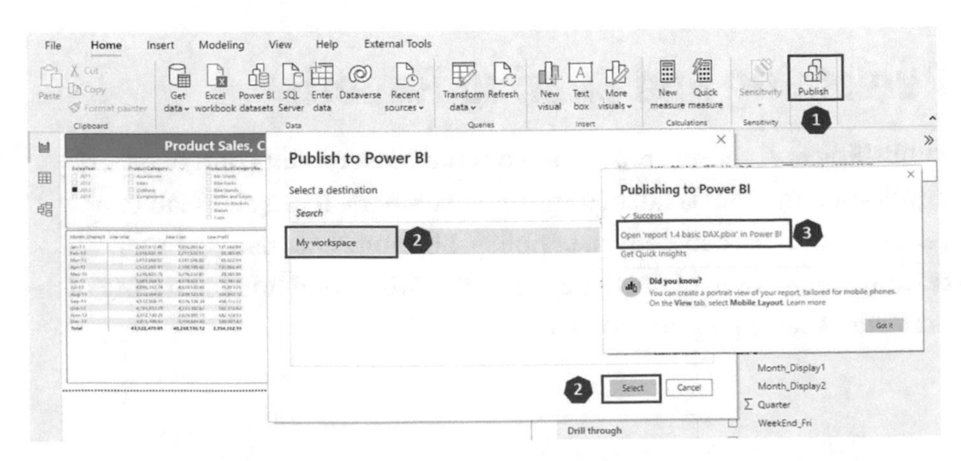

Figure 3-1. Publish a report in the workspace

Kim shared the report with a few users in the workspace. She did this using the following steps, as illustrated in Figure 3-2.

1. Click the Share button.
2. Enter the name or email of the receiver.
3. Click Send.

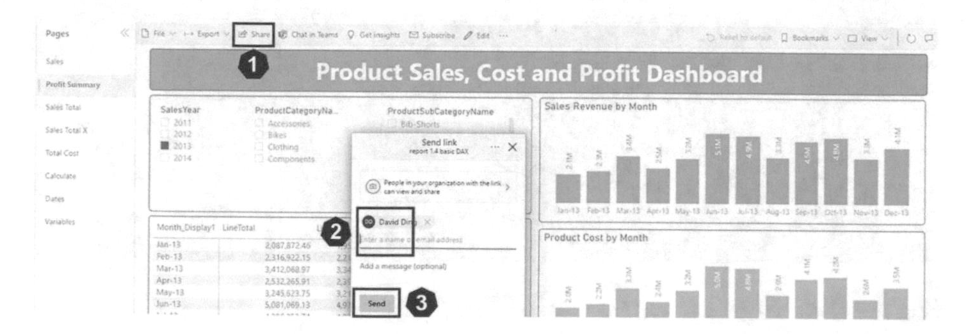

Figure 3-2. *Share report from online portal*

The user receives an email in the format shown in Figure 3-3.

David Ding has shared Power BI Report 'report 1.4 basic DAX' with you

MB **Microsoft Power BI <no-reply-powerbi@microsoft.com>**
9:49 PM

To: David Ding

Microsoft **Power BI**

David Ding shared this Power BI Report with you

report 1.4 basic DAX

Open this report >

Figure 3-3. *Email received by user*

Because the other colleagues do not have a Power BI Pro license, when signing up for the first time, they can sign up for a free trial license by selecting the **Try free** options from the website, as shown in Figure 3-4. Once the trial period starts, users can click the link and be taken directly to the report.

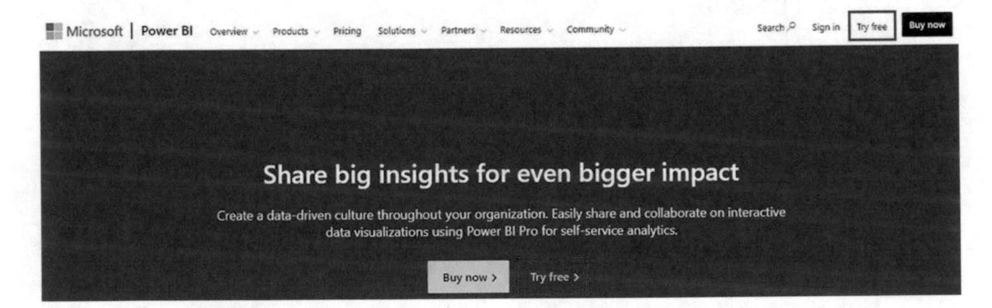

***Figure 3-4.** Sign up for Power BI Pro free trial*

Kim successfully shared the report with the specified users. Milo and the other users can also see the report Kim has created.

Sharing Risks

Microsoft's default method is not suitable for organizations with strict data policies. There are three issues with this sharing method.

- My Workspace
- Links
- *People in your organization* and *reshare*

My Workspace

A workspace is a folder in the cloud. My Workspace should only be used for development and testing, not for sharing.

My Workspace (personal workspace) is only accessible by the developer. It can be used to store reports and datasets. No one can access the reports in the personal workspace if the developer leaves the organization.

My workspace also does not have the collaboration and app compared to a normal workspace.

Sharing Links

Sharing links should be avoided. While it is the easiest way to share reports with other users, it increases the complexity of access control. You gain a better understanding by following these next steps, as illustrated in Figure 3-5.

1. Go to My Workspace and select **Manage permissions** against the published report.
2. You may see multiple sharing links containing different access groups and permissions.
3. Select **Manage Access** for each link.

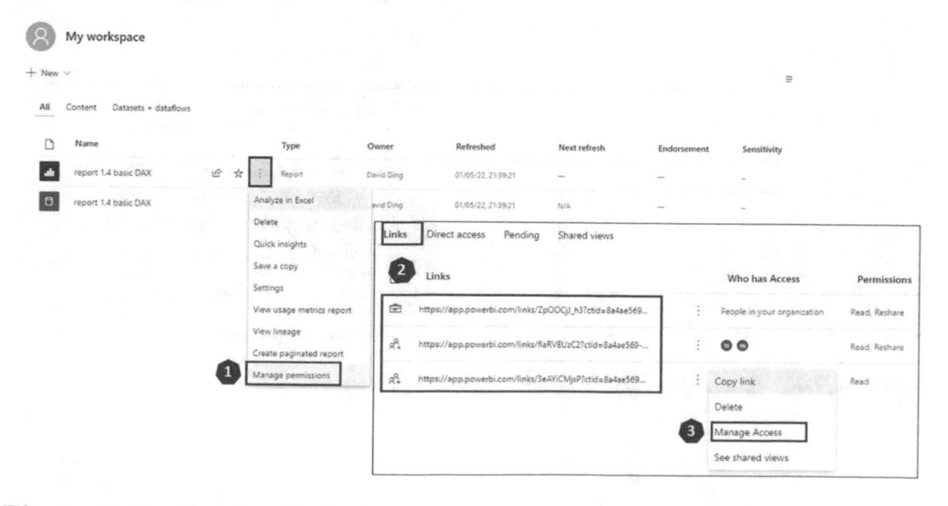

Figure 3-5. *Sharing links*

Each one of the links needs to be investigated separately to understand who has access to the report and the dataset, making the process more complicated than need be.

Share with Others in Your Organization

The default sharing option is to *share with the people in your organization.* This is often against the company data policy because anyone in the company can access the report via the link.

You can explore other link-sharing options by following these next steps, as illustrated in Figure 3-6.

1. Choose the sharing option.
2. Check the three sharing options.
3. Check the two setting options.

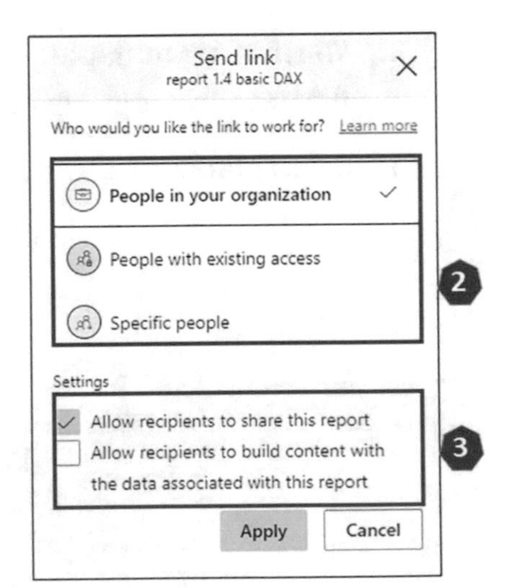

Figure 3-6. *Sharing link options and settings*

The following describes three link-sharing options.

- Option 1: Sharing with **People in your organization** means everyone can use the link to access the report.
- Option 2: Sharing with **People with existing access** means the link can only be used by people who already have access to the report via the workspace admin, direct access, or other options.
- Option 3: Sharing with **Specific people** means sharing the report with specifiedpeople.

In combination with the sharing option, there are also two sharing settings.

- **Allow recipients to share this report** combined with Option 3 allows the recipients to share the report and data without notifying the developer.
- **Allow recipients to build content ...** makes the report data visible to the recipients to have direct access to full data.

When sharing reports, it is good to understand the different options and adjust them to suit the environment and requirements.

A Better Way of Sharing

With a basic understanding of some issues with the sharing option, you may want to adopt the following process as the standard method shown in Figure 3-7. You may also customize this standard approach to better suit your environment.

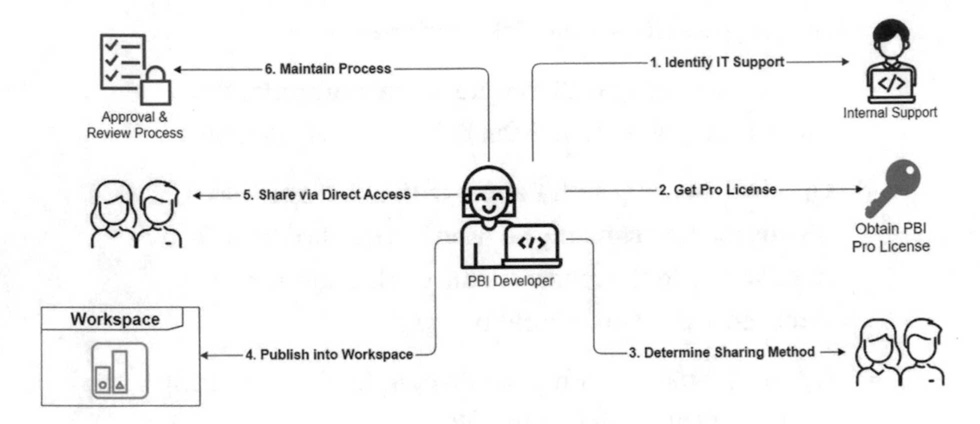

Figure 3-7. *Procedure for report sharing*

The following describes the standard approach to report sharing.

1. Determine IT support and person/teams responsible for the Power BI platform (initial sharing).

2. Obtain a Power BI Pro license or free trial (initial sharing).

3. Determine sharing method during requirement gathering.

4. Publish the report into a non-personal Power BI workspace.

5. Share via direct access.

6. Maintain separate approval and review processes.

Local Power BI Support

In many organizations, there is no dedicated local support for Power Platform. The team who looks after infrastructure or Office 365 would be able to grant the required access and control. They are also the team who may have access to Microsoft support. It is important to identify the support structure as early as possible.

The following questions may help during the initial meetings.

- What is the existing Power BI licensing arrangement?
- What teams (if any) are using Power BI?
- What workspaces should your reports be published into?
- Can you create new workspaces?
- Who do you contact for technical support?

Get Power BI Pro License

Both report developers and readers use the Power BI Pro license. You might be able to sign up for a 60-day free trial, but the local support team should guide you on what happens when the free trial expires.

Determine Sharing Method

Before starting development, you should know the answer to the two questions.

- How many people will be accessing the report?
- Should there be any access restrictions?

When asking the first question, note that the stakeholders tend to overestimate the number of users. You can assume the number of actual users would be much lower initially.

When asking about access restrictions, the stakeholder's initial response may be that it is not required. From experience, leaving room for this in the design is typically a good idea. This is discussed in Chapter 8.

Publish into Non-Personal Workspace

The web portal of Power BI is referred to as Power BI Services. It is a Power BI server running on a Microsoft Azure cloud server. It is being used mainly for sharing, collaboration, and monitoring purposes. Both the report and the data are published into this cloud server. A workspace is like a folder in the cloud server.

After discussing with the support team, there would hopefully be some guidance on workspaces. If you are one of the few who started using Power BI, you may need to create a new workspace. Chances are that IT has not blocked users from creating new workspaces yet. If that is the case, after notifying IT, the following steps explain how to create a workspace, as shown in Figure 3-8.

1. Go to app.powerbi.com and choose Workspaces.
2. Choose **Create a workspace**.
3. Name the workspace.
4. Choose the Pro license mode.

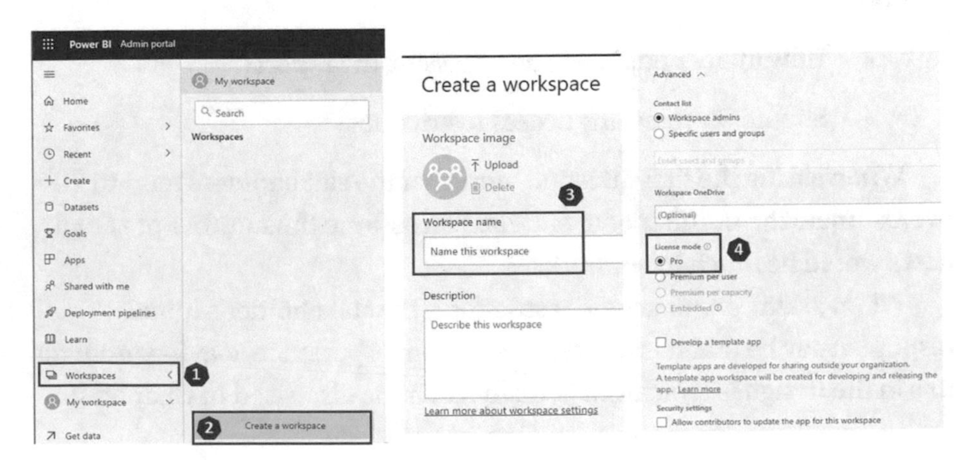

***Figure 3-8.** Creating a new Power BI Workspace*

When you create the workspace, you'll be the default admin for the workspace. This allows you to assign different roles to specific people or groups using the following steps, as illustrated in Figure 3-9.

1. Enter the person's email address.
2. Choose a role.
3. Click Add.
4. Find out more information about the different types of roles.

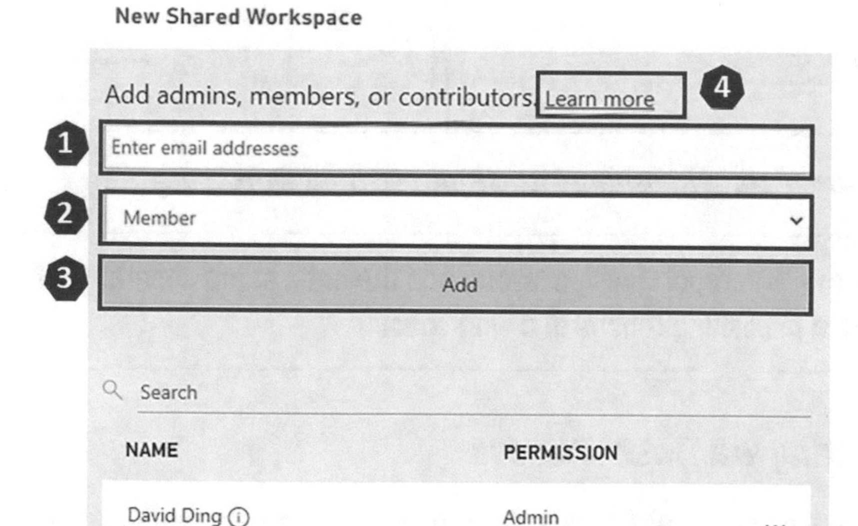

Figure 3-9. *Grant different roles to people or groups*

By giving access at the workspace level, the user can access all reports in the workspace. Assigning roles (admin/member/contributors) to other developers is a great way to improve collaboration. Sharing reports at the workspace level is generally not recommended.

With access to a non-personal workspace, you can republish the report into the workspace by following the steps shown in Figure 3-10.

1. When publishing from Power BI Desktop, choose the new workspace.

2. When publishing from Power BI Services (online), choose the workspace and check the published report.

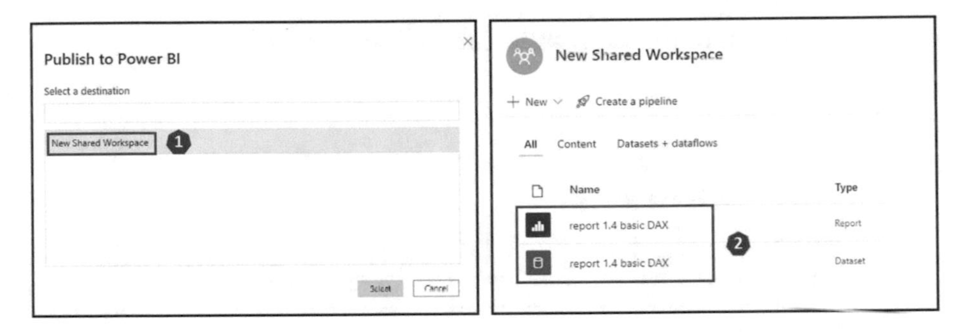

Figure 3-10. Publishing into a non-personal workspace

Note The report (design layout) and the dataset are separate items when publishing reports in a workspace.

Sharing via Direct Access

The preferred way of sharing reports in most scenarios is via *direct access*. The following steps explain how to share the report, as shown in Figure 3-11.

1. From the report drop-down, select **Manage permissions**.

2. In sharing view, choose **Direct access**.

3. Choose **Add User**.

4. Add username or email.

5. Untick the first two boxes and select **Grant access**.

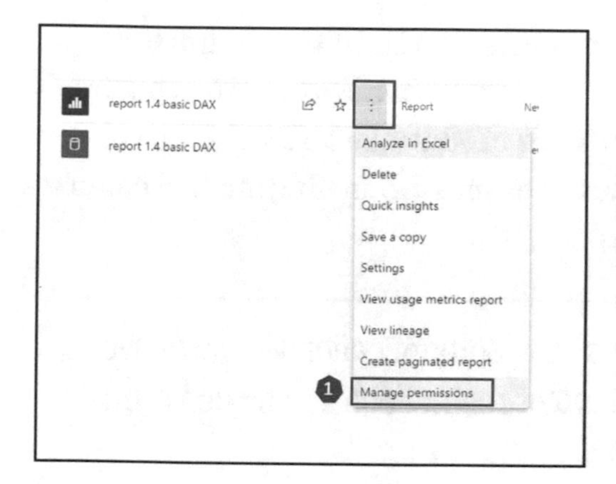

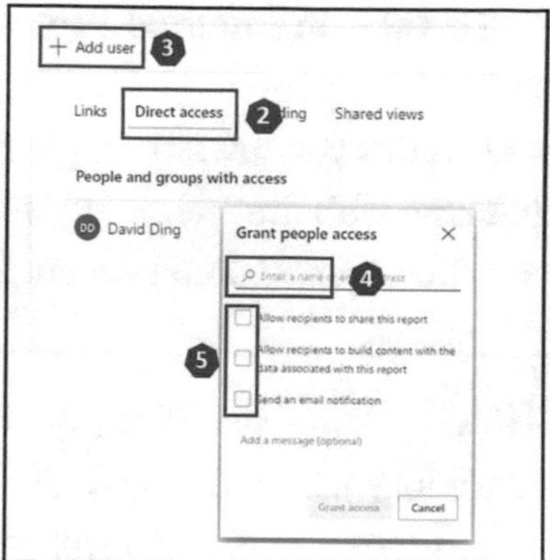

Figure 3-11. *Report sharing in Power BI Services*

You can also follow the steps shown in Figure 3-12 to obtain the link to manually send it to the user. This method only works for users with direct access to the report.

1. Select the report from the workspace.
2. Copy the web URL link and paste it into an email to send to the user.

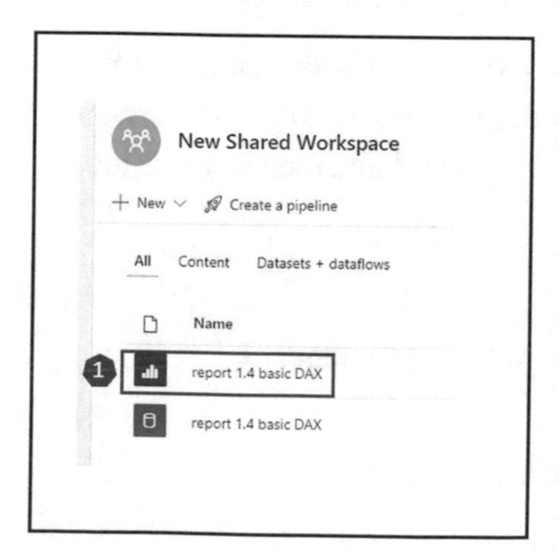

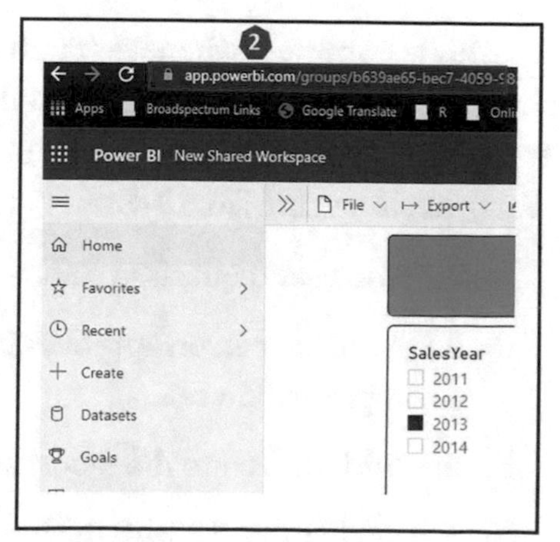

Figure 3-12. *Obtain direct access link*

This sharing method gives developers the maximum control over who can see which report. Power BI security is a huge topic; this book only guides you toward a higher degree of control. You are responsible for checking with the technology team to ensure the sharing method complies with the organizational data policy.

Note Power BI already covers user authentication via the Active Directory user login process. Further restrictions can be achieved using row and object-level security.

User Access Approval and Review Processes

The last step in sharing is keeping a record of the users who have access to the reports. This list is a critical step for control purposes. At any given time, you have a single table to

- Know what reports any user has access to
- Know what users have access to a given report

This is important whenever there is a need to demonstrate control for audit purposes. It is also a useful list during the report migration process. This process can be maintained in four steps and one data artifact (data table), as shown in Figure 3-13.

1. The user requests access.
2. A liaison person approves the access (manager or representatives).
3. Add the user to the report access control list.
4. Review this list with the liaison person regularly (normally six months).

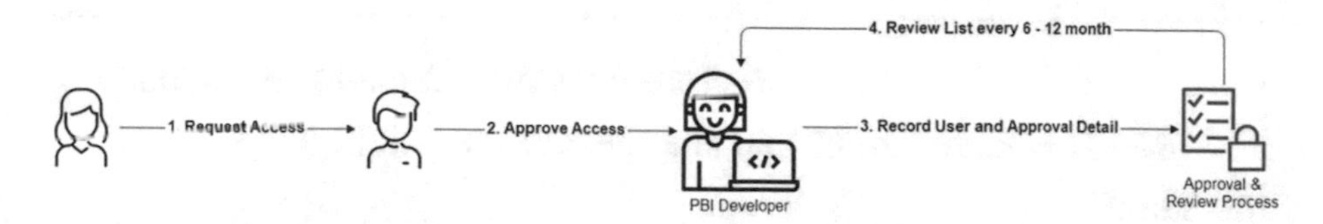

Figure 3-13. Report approval process

There are two ways to obtain the list of users.

- **Manual** uses an Excel spreadsheet to keep a list.
- **Automatic** uses Power BI API to obtain the list of users for reports and workspaces. (Power BI administrators must execute API requests with PowerShell commands.)

In actual practice, an Excel list in a shared folder (e.g., SharePoint folder) is the preferred method. The data artifact obtained through API can be used as a checkpoint. The key columns for this data artifact include the following.

- Workspace Link (where the report is contained)
- User ID/Name (the person requires access to the report)
- User role in the workspace (the person has been assigned a role in the workspace)
- User role in row-level security (see Chapter 8)
- Power BI Report Link
- Approval Date
- Approved By

Note Link IDs (web URL) are preferred over names for *workspace* and *report* because names can change, but the link ID stays the same.

Report-Sharing Summary

You now understand the two easy ways of sharing reports. In practice, different scenarios, data requirements, and resource restrictions require you to balance the approaches or add new ones. This is done by creating sharing links or by adopting the sharing process and using direct access (preferred).

Automation

In Excel, it is common to see the report owner going through the following steps regularly, as shown in Figure 3-14.

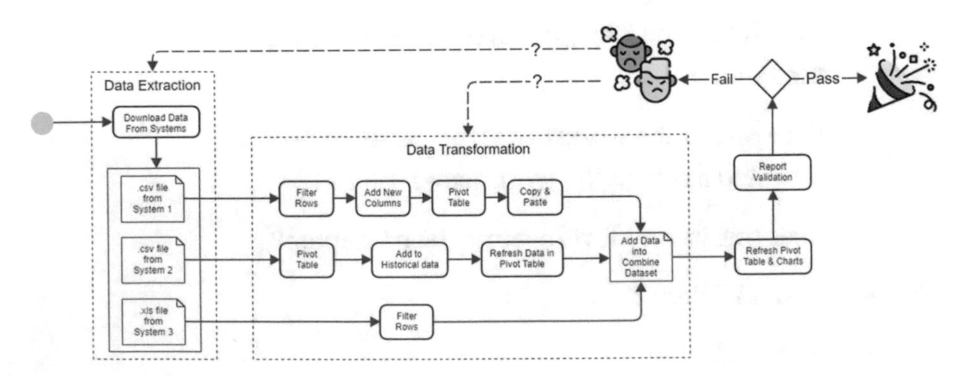

Figure 3-14. Excel report refresh process

Report automation has three main benefits.

- It saves time and resources.
- It reduces human error.
- It improves sustainability.

Save time and resources. This is the most obvious benefit. Once the report has been automated, it doesn't require additional resources to repeat the exercise. The freed-up resource can be allocated toward improvement and new development.

Reduce human error. Inside the data transformation box in Figure 3-14, there can be more than 100 clicks to achieve the desired outcome, including selecting, copying, and pasting in the right sheet/cell and changing the table range. This heavy manual process makes it easy for analysts to make mistakes. It can also take a lot of time to detect and correct the problems.

Improve sustainability. There is a resource limit on how many such reports can be manually maintained. The management team naturally aims to increase the limit by squeezing more reports with limited resources. This makes the analysts carry a lot of knowledge. This step-by-step process is also difficult for documentation and knowledge transfer. This results in many reporting continuities issues during team member change.

You can do an automation cost-benefit analysis, as shown in Figure 3-15. The goal is to obtain the maximum long-term benefit by making the most suitable automation decisions.

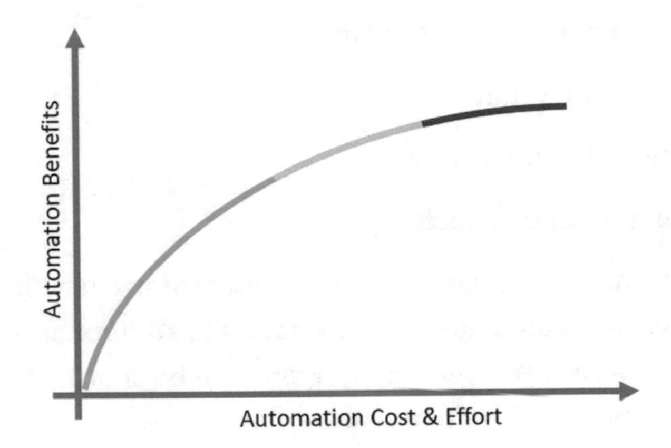

Figure 3-15. *Cost-benefit analysis on report automation*

It is also important to remember that some automation can be implemented in stages, but others need to be decided upfront.

The report automation activities can be grouped into three areas.

- data connection
- data transformation
- task scheduler

Data Connection

Power BI provides a connection to many data sources and systems. These sources can be grouped into the following areas.

- File: flat files, CSV, Excel
- Database (SQL Server, SAP, Oracle, MariaDB, IBM, AWS, Google)
- Power Platform sources (Dataverse, dataflow)
- Azure Cloud sources (blobs, data lakes, databases, databricks)

- Other services: SharePoint list, Salesforce, Twillo, Quickbooks, Zoho)
- Other (web (scraper), Bloomberg, OData)

The following steps take you to the data connection screen, as illustrated in Figure 3-16.

1. Select **More** in **Get data**.
2. Observe the different connectors in each connection group.

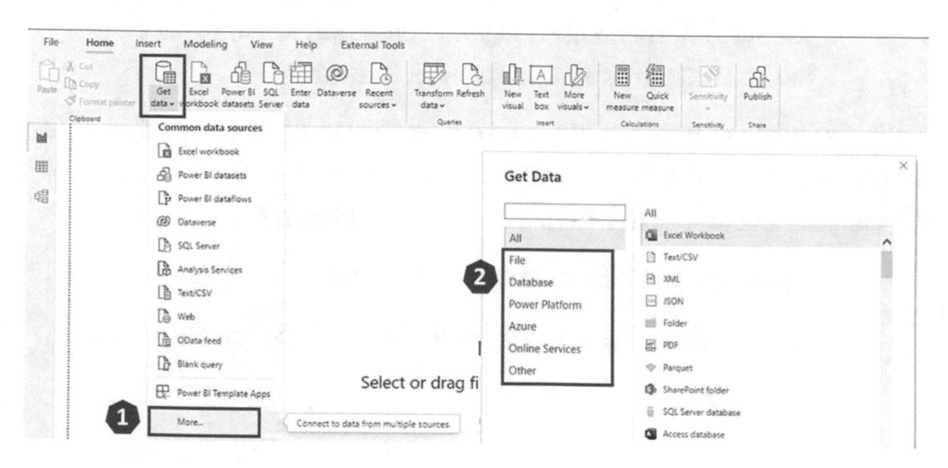

***Figure 3-16.** Power BI Connector Group*

In addition to the standard connectors, you also learn about two additional workarounds in connection.

- Flat files in a shared folder
- Personal Gateway
- ODATA feed

Export Flat Files in a Shared Folder

One of the common techniques in achieving automation is to export standard flat files into a SharePoint folder. The files can be created by manual data extraction or scheduling export jobs from source systems.

- **Manual data extraction**. The person/team with access to the system can follow a standard process to export the data in a consistent format every day/week/month.
- **Schedule export jobs**. The person/team with access to the system can create a regular execution job to export data consistently every day/week/month.

The result is to have a folder containing files in a consistent format (i.e., the same number of columns and column names). Power BI has two main ways to connect to files in a folder, as shown in Figure 3-17.

- Folder (local folder or network drive folder)
- SharePoint folder (document folder in SharePoint)

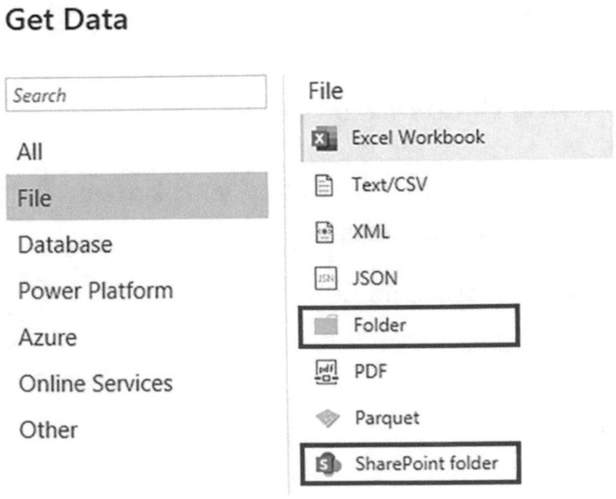

Figure 3-17. *Power BI Connector for folders*

The Monthly Data folder contains three separate monthly files. The following steps load all files from a folder, as shown in Figure 3-18.

1. Select **More** in Get Data on the Home ribbon.

2. Choose **Folder** from the File group.

3. Enter the folder drive link.

4. Choose **Combine & Transform Data.**

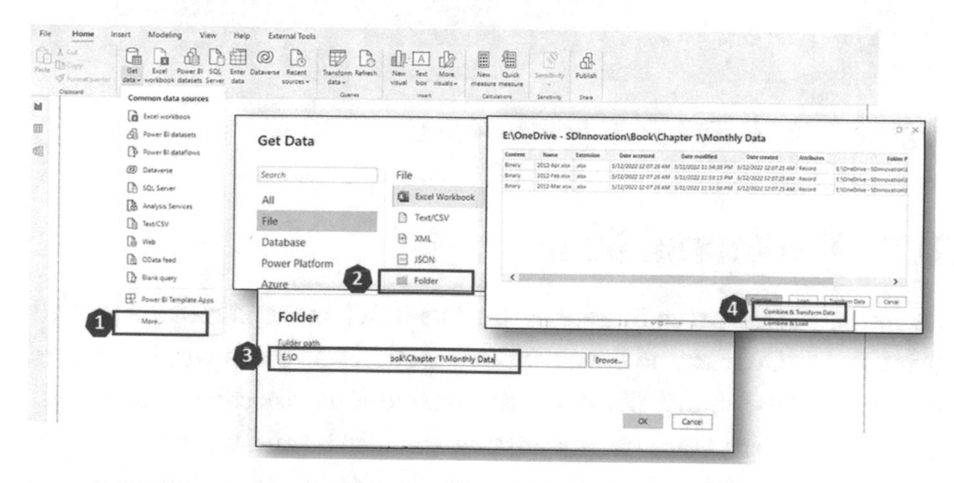

Figure 3-18. *Load data from folder*

The previous step creates a new "power query" that loads all files from the folder. You can now click the Close & Apply button to allow the data to be loaded into Power BI for front end report creation.

This is a good solution because the report can be refreshed when new data is added. This can be tested by adding a new 2012-May.xlsx file from the Files folder into the Monthly Data folder. Once the new file is added to the folder, you can click the Refresh button. Once the data has been refreshed the 2012-May data also show up as shown in Figure 3-19.

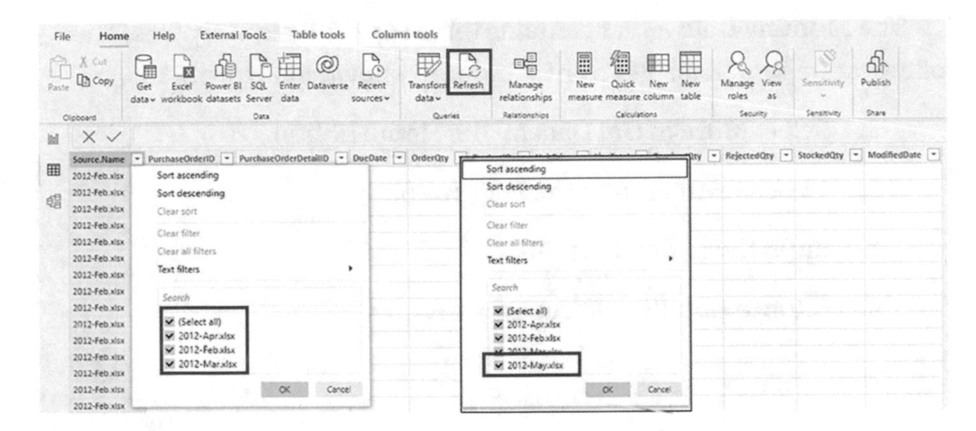

Figure 3-19. Data refresh with new file

Data Transformation

As you saw in the preceding steps, the data is first brought into Power BI using Power Queries and Power Query functions. As you have also seen in earlier sessions, almost all routine transformation tasks can be translated into Power Query. Once translated, they can also be automated.

As shown in Figure 3-20, there are two steps where the automatic transformation can occur when data is loaded from a folder: before and after combining data.

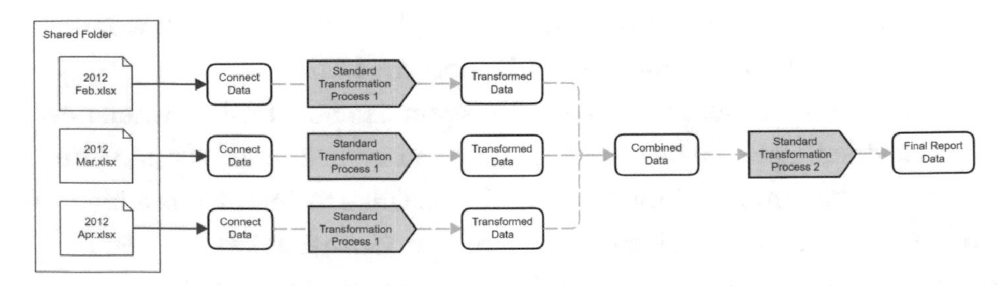

Figure 3-20. Data automation from folder

The two transformation steps inside the Power Query Editor are stored separately in two queries, as shown in Figure 3-21. You must change the Transform Sample File query to change process 1 before combining the datasets. To make a change in process 2 after datasets have been combined, you need to change the Monthly Data query.

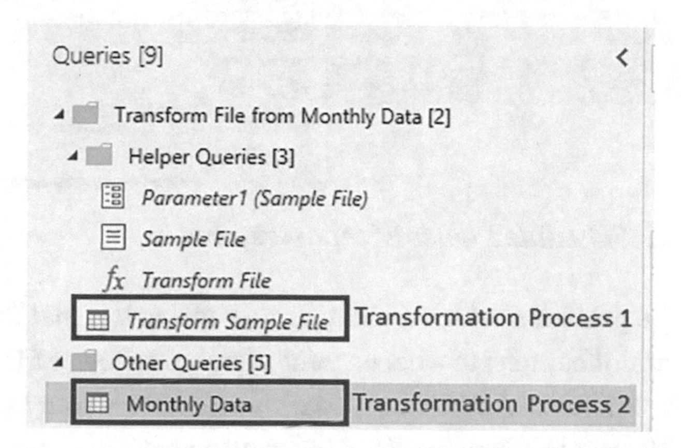

Figure 3-21. *Data transformation steps when loading from folder*

Scheduling with Power BI Gateways

Once the data connector and the transformation steps are properly configured, it is time to trigger the process automatically. The most common trigger is the scheduler in Power BI Services. The following steps describe how to schedule a report, as shown in Figure 3-22.

1. Log in to app.powerbi.com, choose the workspace, and click the data refresh button for the report.

2. Expand **Scheduled refresh**.

3. Turn on auto refresh.

4. Add a time and configure the frequency and time.

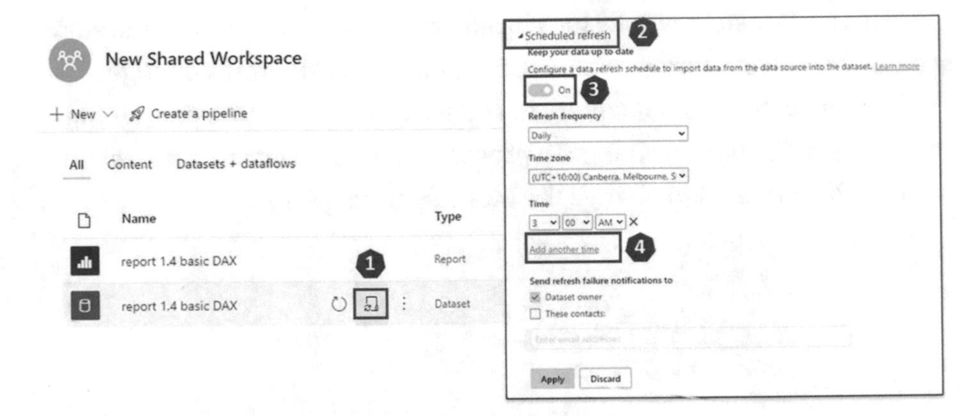

Figure 3-22. *Schedule Power BI report refresh*

One of the problems with the share drive approach is that Power BI Service cannot connect to a file on your computer. Power BI Service resides in the Microsoft Azure cloud, so it cannot access your local machine. This problem can be solved by using a gateway to create the connection and allow Power BI Service to extract data from your local machine.

Power BI Gateway

When you publish a Power BI report, you upload the report and dataset into the Azure cloud. Once the report is published, you can configure the scheduler to use the Power BI gateway to automatically refresh the reports.

There are two types of gateways, personal and standard. A personal gateway is a simplified version of the standard gateway. Figure 3-23 shows the following steps to install the personal gateway.

1. Go to `https://powerbi.microsoft.com/en-us/gateway/` to download and install personal mode.

2. Enter the organization account.

3. Log in to establish the connection.

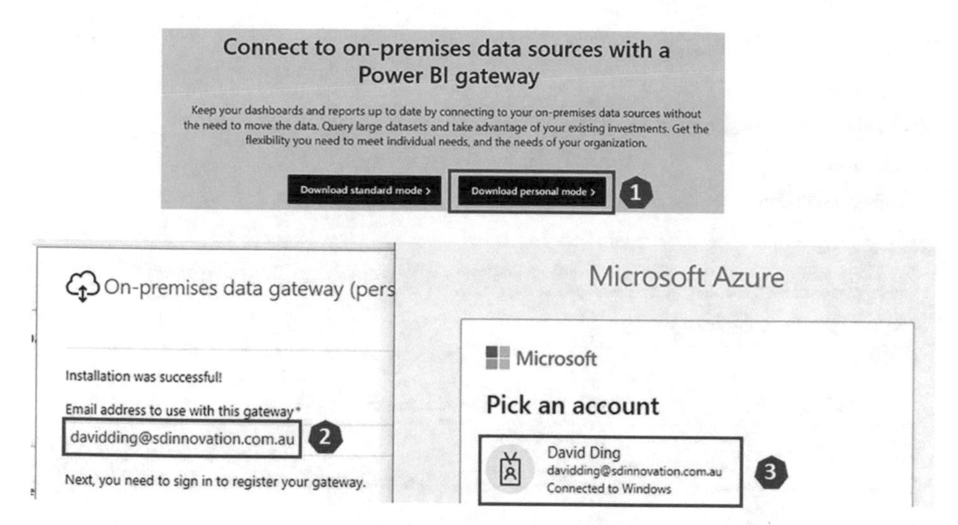

Figure 3-23. *Connect to Power BI gateway personal mode*

Once connected, you can allow the gateway to connect to your data sources through your local machine, as shown in Figure 3-24. Some sources (such as SharePoint) may also require additional login credentials. You can check the data source credentials area to make sure there is no error. Once completed, you may continue to configure the scheduler.

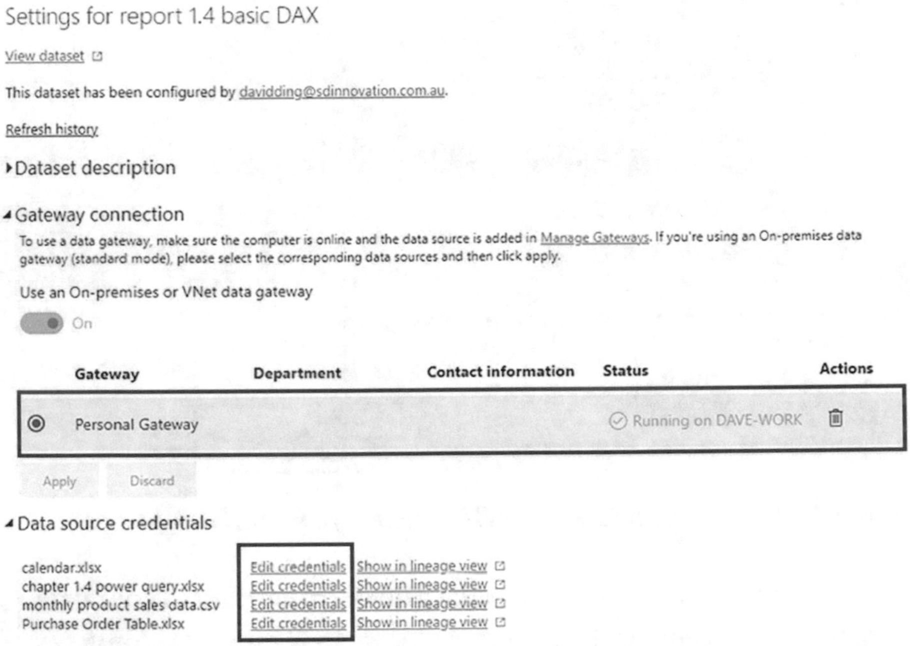

Settings for report 1.4 basic DAX

View dataset

This dataset has been configured by davidding@sdinnovation.com.au.

Refresh history

▸Dataset description

◂Gateway connection

To use a data gateway, make sure the computer is online and the data source is added in Manage Gateways. If you're using an On-premises data gateway (standard mode), please select the corresponding data sources and then click apply.

Use an On-premises or VNet data gateway

On

	Gateway	Department	Contact information	Status	Actions
◉	Personal Gateway			Running on DAVE-WORK	🗑

Apply Discard

◂Data source credentials

calendar.xlsx	Edit credentials	Show in lineage view
chapter 1.4 power query.xlsx	Edit credentials	Show in lineage view
monthly product sales data.csv	Edit credentials	Show in lineage view
Purchase Order Table.xlsx	Edit credentials	Show in lineage view

Figure 3-24. *Configure Power BI gateway and credentials*

Conclusion

This chapter completes the final piece of the puzzle for Power BI basics. You have gained a practical understanding of the two main reasons for using Power BI Service, sharing and automation. The following chapters dive into more practical dashboard design and report governance aspects.

Mini-Hackathon

There is a mini-hackathon challenge for the milestone chapters. These mini hackathons are designed to help you solidify your learning about Power BI. This first mini-hackathon covers the first three chapters of the book.

Note Feel free to google; treat it as an urgent work task.

Hackathon Data

The data source is public M3IT/COVID-19_Data in Australia at `https://github.com/M3IT/COVID-19_Data/tree/master/Data`.

Objectives

If you can access the Power BI service, follow objective B; otherwise, follow objective A.

- **Objective A**: Create a Power BI Desktop reporting dashboard.
- **Objective B**: Create, share, and automate a reporting dashboard via Power BI Service.

The following describes additional requirements for objectives A and B.

- The dashboard must contain *at least* one reporting page and four interactive Power BI visuals.
- The dashboard must have a theme (e.g., Covid-related death).
- Connect directly to the web source (i.e., do not download any file).
- Use DAX measures where possible.

- There must be at least one Calculate() measure.
- There must be a separate date table.

Time Limit

Complete the objective within two hours.

CHAPTER 4

Data Visualization

When thinking about data visualization, it is easy to associate it with modern computing and data processing. It might be surprising to most people that the first chart was created in 1765 by Joseph Priestley, and the bar, line, and pie charts were all invented around 1800. One of the best data visualizations was produced in 1869 by Charles Minard, a map of Napoleon's disastrous Russian campaign of 1812, as shown in Figure 4-1.

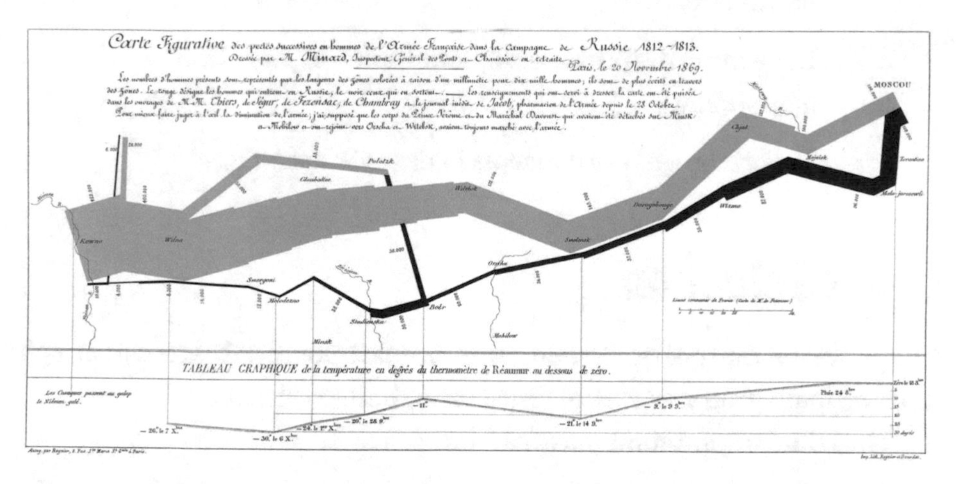

***Figure 4-1.** Map of Napoleon's disastrous Russian campaign of 1812*

In Figure 4-1, you can sense a strong concept of design and storytelling. The chart showed that the French army entered Russia with 422,000 troops in 1812. The army was down to only 100,000 troops when they reached Moscow and began to retreat. By the time they left Russia in 1813, only 10,000 troops remained.

D. Ding, *Transitioning to Microsoft Power Platform*,
https://doi.org/10.1007/978-1-4842-9239-6_4

There are three main elements in the visualization.

- A title with a description
- A main chart consisting of a pink line (troops move in) and a black line (troops retreat)
- A line chart with temperature and weather conditions placed at the bottom to provide additional information about the return trip

Design is the art or process of deciding how something looks or works.

This chapter focuses on the thinking behind effective reporting dashboards with the following topics.

- Chart selection, the basic element that makes up the dashboard
- Determine the purpose, exploration vs. explanation (storytelling)
- Report design, key concepts in data visualizations

Chart Selection

Charts are the most basic element of report dashboards. Charts can be split into two main groups: classical visuals and modern visuals.

The following are some classical charts.

- Line chart
- Pie chart
- Scatter chart
- Box and Whisker diagram
- Column chart and bar chart

- Map
- Table and matrix visuals

The following are examples of modern charts.

- Sunburst diagram
- Network diagram
- Chord diagram
- Sankey diagram (parallel coordination diagram)

Figure 4-2 places the diagrams in a scatter diagram, with the X-axis representing *visual attractiveness* and the Y-axis representing the *understandability* of the chart.

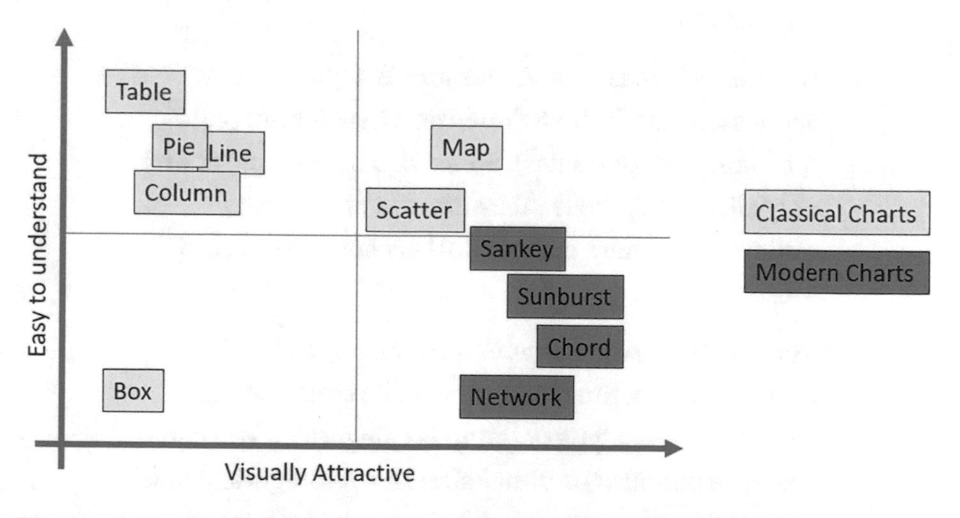

***Figure 4-2.** Understandability and attractiveness of the charts*

Note Part of the reason the classical charts are easier to understand is that they are more widely used. Most users are good at understanding the underlying message of these charts.

Even though attractive and easy-to-understand charts are preferred, these charts may not be suitable for the dataset or the user requirements. As a developer, it is good to understand the following common attributes of the charts before you explore the different visualization options.

- **Number of dimensions**: A dimension is an attribute of a dataset that refers to a specific aspect of the data (e.g., sales date, product category, and country of origin). Some visuals (e.g., pie chart) take only one dimension and others (e.g., stacked column chart or multiple line chart) takes more. More dimensions allow more information to be displayed. More dimensions may also complicate the visual and make it harder to understand.
- **Number of measures**: A measure is a quantitative value associated with a dimension in a dataset (e.g., total cost and total sales). Some visuals (e.g., pie chart) only allow for a single measure, and others (e.g., multiple line chart, scatter plot) accept two or more measures.
- **Type of dimensions** (*continuous* vs. *categorical*). A numeric or datetime measure can be continuous, where a product type or group is categorical. You can change a continuous dimension into a categorical one, but not the other way around. Geolocation, latitude, and longitude are other common types of dimensions.

You examine the attributes when exploring the following five visuals. Some practical improvement tips have also been included.

Pie and Donut Charts

Pie and donut charts use one dimension and one measure.

As shown in Figure 4-3, pie and donut charts can be used to represent different categories.

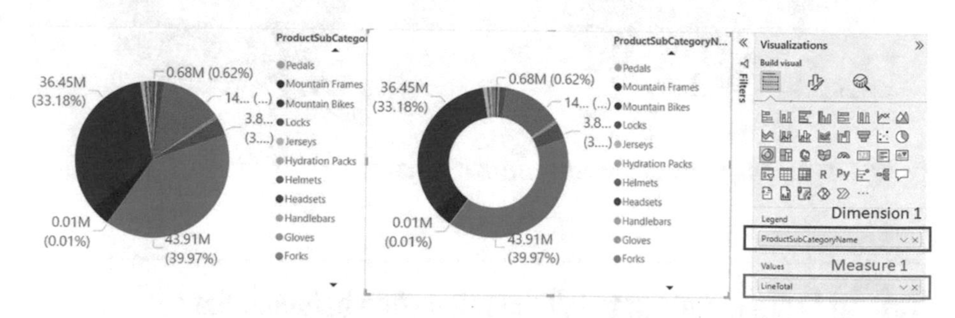

***Figure 4-3.** Pie and donut charts*

These are not good examples because they can be very confusing. There are a few ways to improve the effectiveness of pie and donut charts.

- Sort by value. This puts the most significant category first and follows a clockwise sequence.
- Remove the legend and add a category to the data label. This makes it easier to match categories and values.
- Limit the number of categories to five and group the rest in *Other*. This helps to simplify the chart as the report user cannot see the values anyway.
- Add a total value () in the middle of the donut chart. This extra piece of information provides additional context for the chart.

The improved charts are shown in Figure 4-4. This improved version is much easier to understand compared to the original charts.

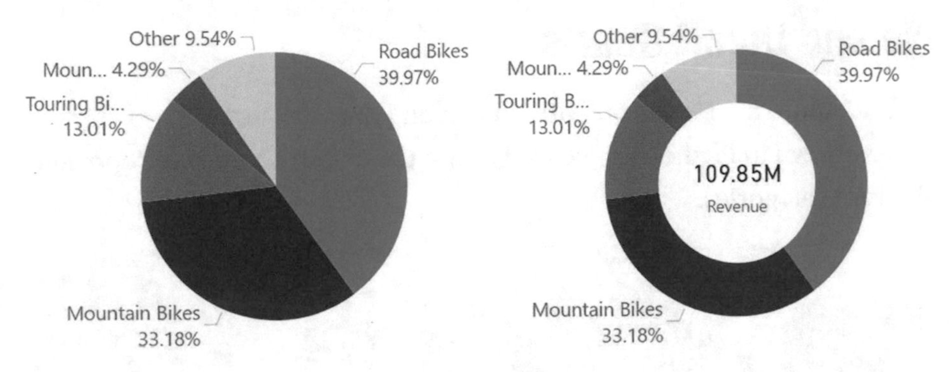

***Figure 4-4.** Improved pie and donut charts*

Tip It is often difficult to tell the difference between the category values. In the example in Figure 4-4, the top two categories—Road Bikes and Mountain Bikes—look similar in size, even though there is a 6% gap. Data labels can help to clarify the difference.

Column and Bar Charts

Column charts and bar charts take two *dimensions* and one or more *measures*.

There are two display options for the column and bar charts.

- **Stacked** means the split is on top of each other.
- **Clustered** means the split is next to each other.

On the surface, there is no difference between a column and a bar chart. They are just rectangles in different directions, as shown in

Figure 4-5. This view is not true in practice. It is better to follow the following rules.

- **Column charts** are used for continuous variables, such as date/time.
- **Bar charts** are used for categorical variables, such as category names.

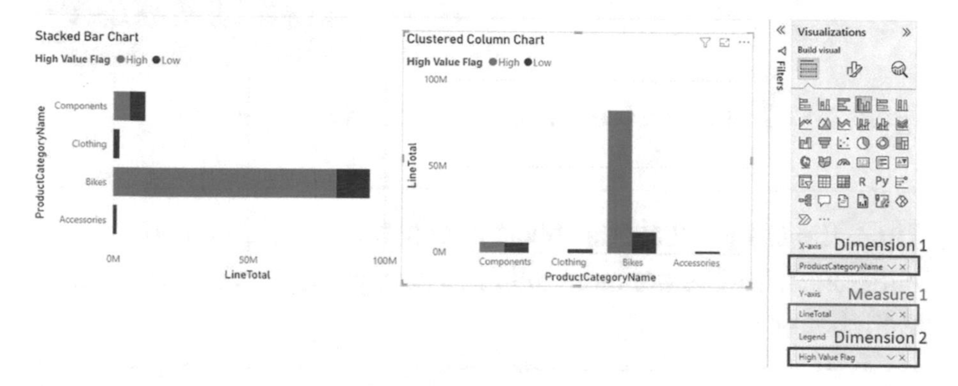

Figure 4-5. *Bar and column charts*

There are a few ways to improve the column and bar charts.

- Sort by value to place the most significant category up the top or to the left.
- Remove the title and value axes to clean up unnecessary information in the chart.
- Add a data label to make information easier to read.
- Change the column chart X-axis to a year-month value by adding the calendar table.

The improved charts are shown in Figure 4-6. These improvement actions resulted in a better presentation of the information.

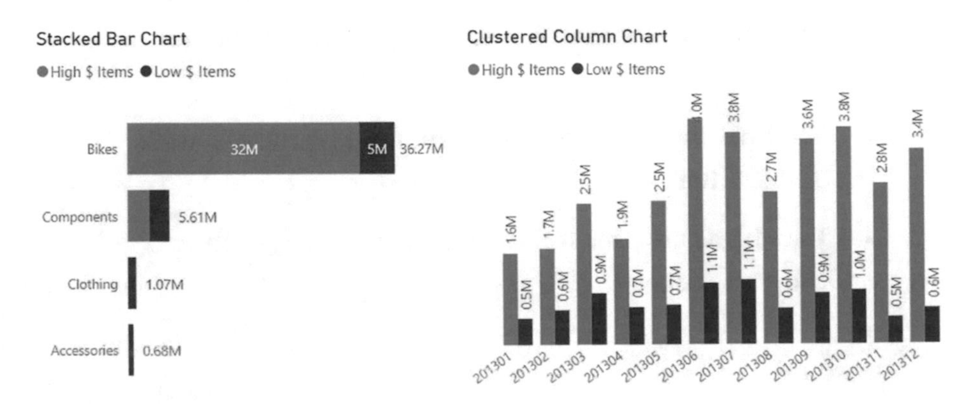

Figure 4-6. Improved bar and column charts

Tip The clustered charts should not contain more than two categories. If there are more than two categories, line charts normally work better.

Line Charts

Like the bar and column charts, a line chart also takes a maximum of two dimensions and one or more measures. The main benefit of a line chart compared to a column chart is that it can display more categories, as shown in Figure 4-7.

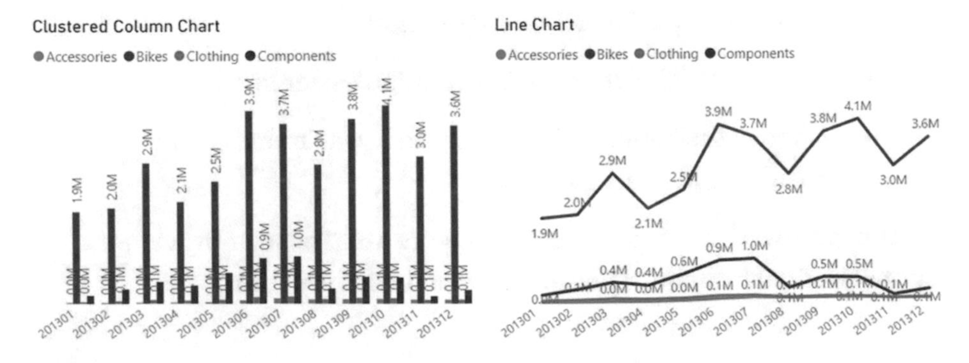

Figure 4-7. Column chart vs. line chart

Note When a chart has many data points, having an axis instead of a data label may simplify it, as shown in Figure 4-8.

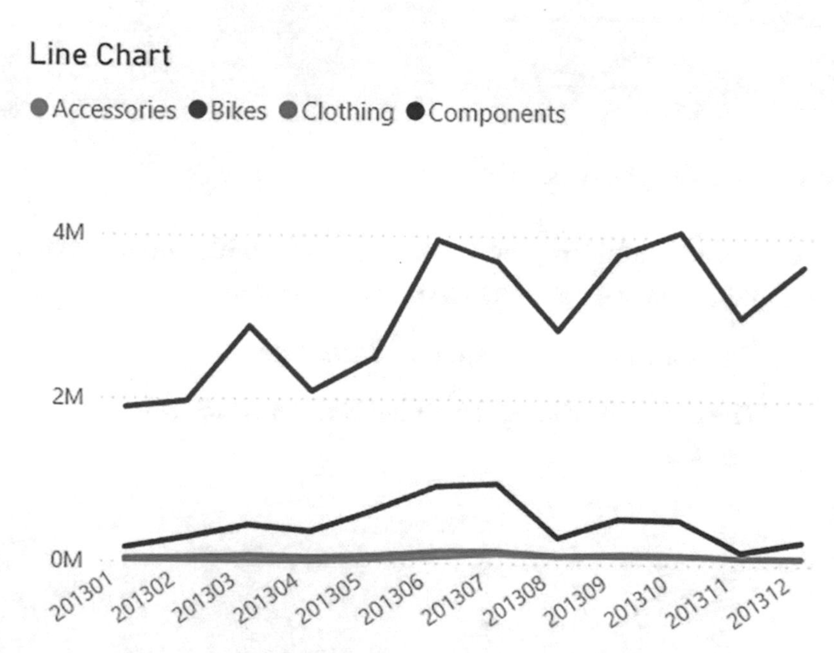

Figure 4-8. *Improved line chart*

Table vs. Matrix Visuals

Power BI offers table and matrix visuals for presenting tabular data. Each takes multiple *dimensions* and multiple *measures*.

Small value categories hidden from view are a major issue for the bar, column, and line charts. If you want the users to see these small value categories, consider using a table/matrix visual, as shown in Figure 4-9.

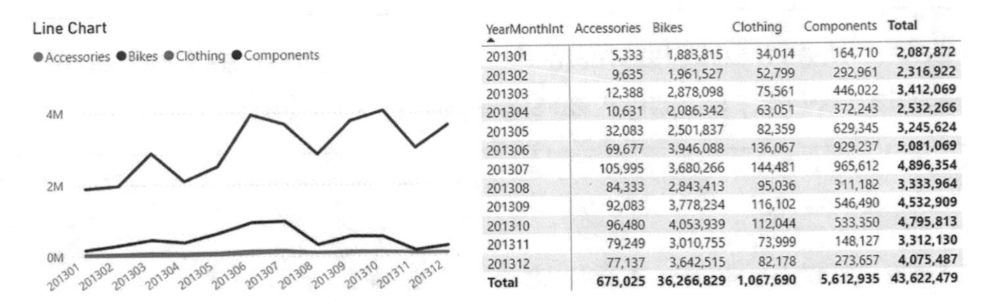

YearMonthInt	Accessories	Bikes	Clothing	Components	Total
201301	5,333	1,883,815	34,014	164,710	**2,087,872**
201302	9,635	1,961,527	52,799	292,961	**2,316,922**
201303	12,388	2,878,098	75,561	446,022	**3,412,069**
201304	10,631	2,086,342	63,051	372,243	**2,532,266**
201305	32,083	2,501,837	82,359	629,345	**3,245,624**
201306	69,677	3,946,088	136,067	929,237	**5,081,069**
201307	105,995	3,680,266	144,481	965,612	**4,896,354**
201308	84,333	2,843,413	95,036	311,182	**3,333,964**
201309	92,083	3,778,234	116,102	546,490	**4,532,909**
201310	96,480	4,053,939	112,044	533,350	**4,795,813**
201311	79,249	3,010,755	73,999	148,127	**3,312,130**
201312	77,137	3,642,515	82,178	273,657	**4,075,487**
Total	**675,025**	**36,266,829**	**1,067,690**	**5,612,935**	**43,622,479**

***Figure 4-9.** Line chart vs. matrix*

There are additional improvement settings for table and matrix visuals. For example, Figure 4-10 shows two such improvements.

- The matrix on the left enabled ***Data bars***.
- The matrix on the right enabled ***Background color*** and ***Icons***.

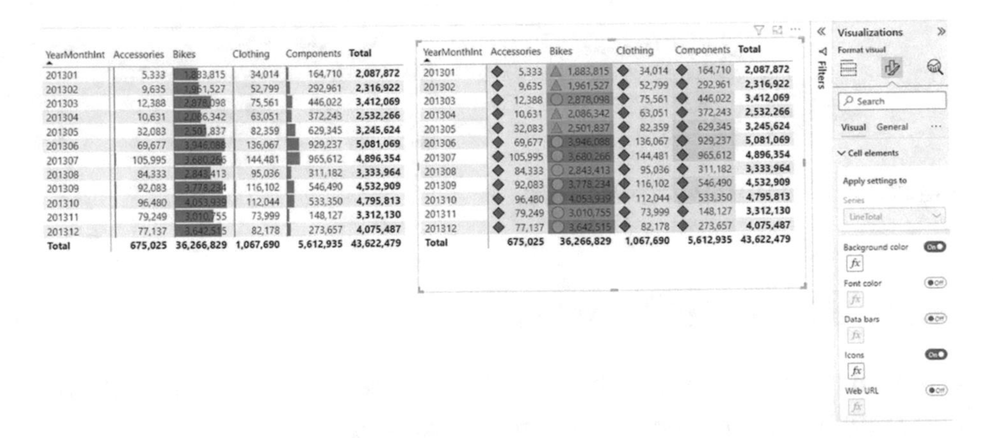

YearMonthInt	Accessories	Bikes	Clothing	Components	Total
201301	5,333	1,883,815	34,014	164,710	**2,087,872**
201302	9,635	1,961,527	52,799	292,961	**2,316,922**
201303	12,388	2,878,098	75,561	446,022	**3,412,069**
201304	10,631	2,086,342	63,051	372,243	**2,532,266**
201305	32,083	2,501,837	82,359	629,345	**3,245,624**
201306	69,677	3,946,088	136,067	929,237	**5,081,069**
201307	105,995	3,680,266	144,481	965,612	**4,896,354**
201308	84,333	2,843,413	95,036	311,182	**3,333,964**
201309	92,083	3,778,234	116,102	546,490	**4,532,909**
201310	96,480	4,053,939	112,044	533,350	**4,795,813**
201311	79,249	3,010,755	73,999	148,127	**3,312,130**
201312	77,137	3,642,515	82,178	273,657	**4,075,487**
Total	**675,025**	**36,266,829**	**1,067,690**	**5,612,935**	**43,622,479**

YearMonthInt	Accessories	Bikes	Clothing	Components	Total
201301	5,333	1,883,815	34,014	164,710	**2,087,872**
201302	9,635	1,961,527	52,799	292,961	**2,316,922**
201303	12,388	2,878,098	75,561	446,022	**3,412,069**
201304	10,631	2,086,342	63,051	372,243	**2,532,266**
201305	32,083	2,501,837	82,359	629,345	**3,245,624**
201306	69,677	3,946,088	136,067	929,237	**5,081,069**
201307	105,995	3,680,266	144,481	965,612	**4,896,354**
201308	84,333	2,843,413	95,036	311,182	**3,333,964**
201309	92,083	3,778,234	116,102	546,490	**4,532,909**
201310	96,480	4,053,939	112,044	533,350	**4,795,813**
201311	79,249	3,010,755	73,999	148,127	**3,312,130**
201312	77,137	3,642,515	82,178	273,657	**4,075,487**
Total	**675,025**	**36,266,829**	**1,067,690**	**5,612,935**	**43,622,479**

***Figure 4-10.** Improved matrix*

Table and matrix visuals are also the most mobile-friendly charts, as shown in Figure 4-11. This is because it can display a lot of information in a limited space. This is also known as *information density*.

Note Once the Mobile layout is configured, it becomes the default layout whenever the user opens the report on a smartphone.

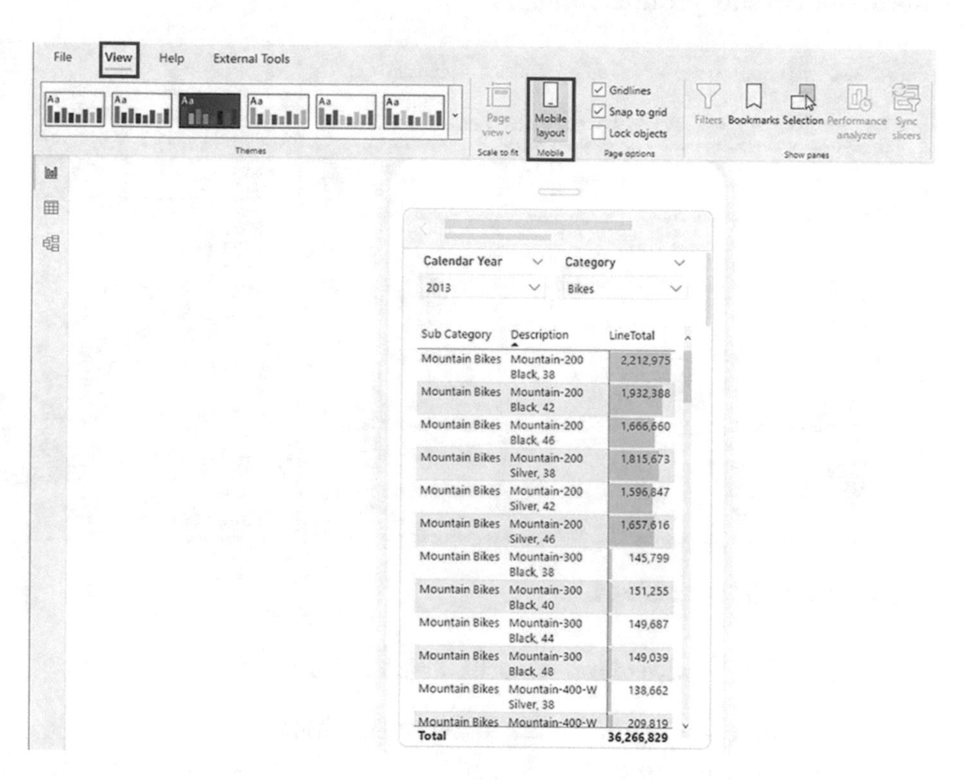

Figure 4-11. *A table in a Mobile layout*

Scatter Charts

A scatter chart takes a maximum of two *dimensions* and three *measures.*

The scatter chart takes more measures and dimensions than most other charts except for tables and matrices. More consideration is often required to determine the right combination of dimensions and measures in a scatter chart.

Figure 4-12 shows one example of using the scatter chart with the monthly product sales data. The purpose is to identify product groups with *high sales* and *high volume* and use the *number of products* as additional information to gain product insights.

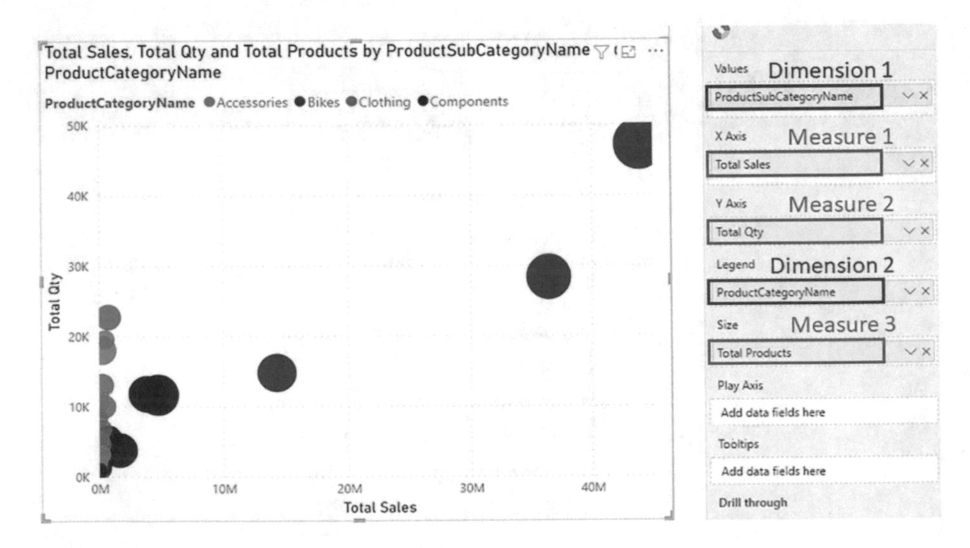

Figure 4-12. *Scatter chart*

There are a few ways to improve this scatter chart.

- Change the X- and Y-axes to log scale. This makes the circles closer together.
- Add a middle line in the X- and Y-axes to divide the chart into four quadrants.
- Change the legend color to lighter colors.
- Add a color border under Markers and adjust the marker size.
- Tidy up the chart by simplifying titles.

The improved chart is shown in Figure 4-13. This improved diagram can be powerful for decision-making. You may observe that mountain bikes are high in sales and quantity while having fewer products than road bikes and road frames. This may indicate good efficiencies in this product category.

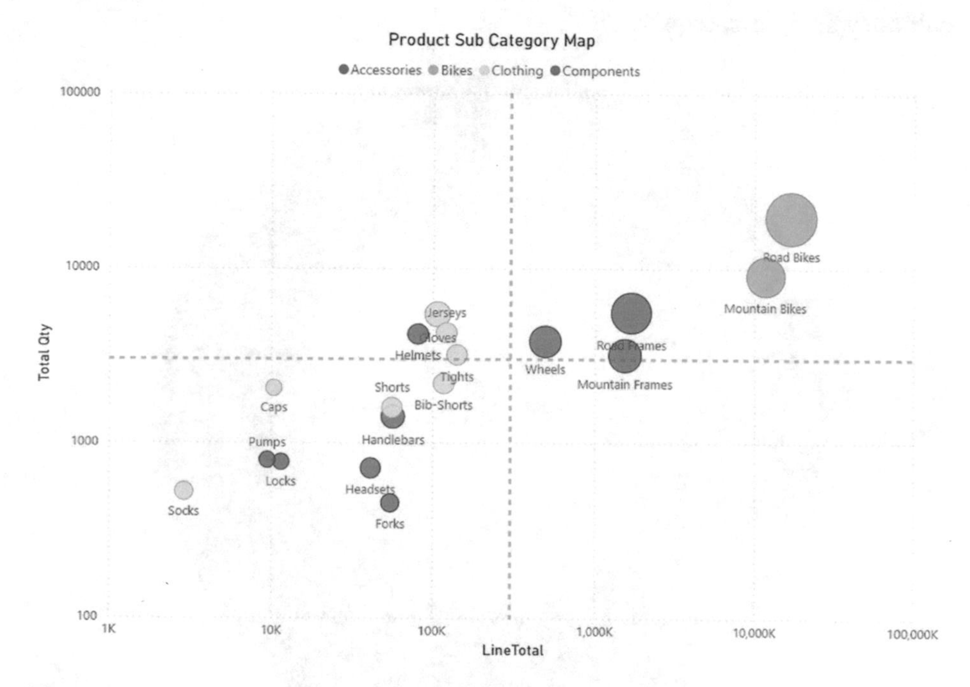

Figure 4-13. *Improved scatter chart with quadrant analysis*

Tip Many consulting firms utilize this quadrant approach to derive easy-to-understand business strategies. The challenge is to identify the right combination of dimensions and measures.

Sunburst Diagrams

A sunburst diagram is considered a modern chart. It takes one *hierarchy* (multiple layers) and one *measure*.

Figure 4-14 shows a sunburst diagram with a hierarchy layer for category and subcategory.

Figure 4-14. *Sunburst diagram with two layers*

A sunburst diagram is not a standard chart in Power BI, which means you must download it from the marketplace. The following steps, as shown in Figure 4-15, explain how to do it.

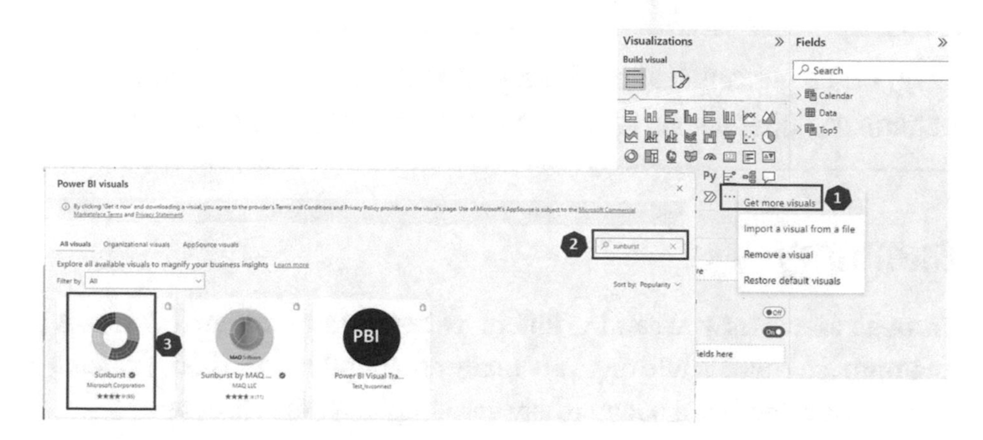

Figure 4-15. *Import custom visuals*

The sunburst diagram hierarchy can have many layers. It can be customized to improve its visual attractiveness by making some categories white, as shown in Figure 4-16. The diagram shows a three-layer sunburst diagram.

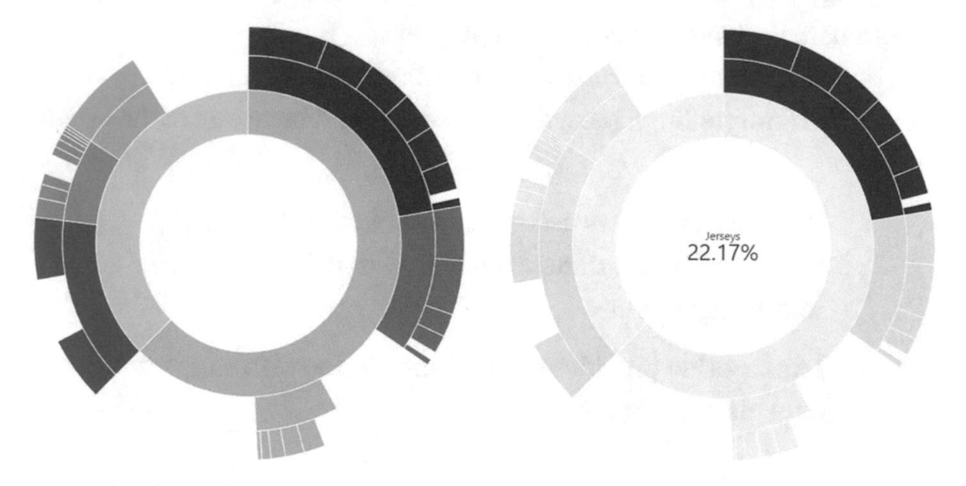

Figure 4-16. *Sunburst diagram with three-layer hierarchy and hidden categories*

Tip Data presentation can bore your audience. Some modern charts can offer a good way to "spice up" the presentation.

Summary

Through examples, you saw the difference between a standard chart and an improved chart. While there are many other options, the core principles remain the same. It is a balance between simplicity, storyline, and information density.

Data Explanation vs. Exploration

A reporting page is created when you combine multiple visualizations into a single page in Power BI. Before you start to design the layout, it is important to understand the report's purpose.

There are two main purposes for reporting: exploration and explanation (storytelling). Most dashboards should have one of these two purposes.

- **Exploration** allows users to self-service and explore topics with multi-dimension data with drill-down and roll-up capabilities.
- **Explanation** provides reasons and evidence to support or explain a concept or problem.

Data Exploration

Self-service BI solutions such as Power BI have greatly enhanced the user experience in self-service data exploration.

When designing the exploration reports, starting with a main theme is good. The dashboard shown in Figure 4-17 has a sales performance theme. The following dimensions and facts are used.

- Dimensions (year, month, category, subcategory)
- Facts (sales, quantity)

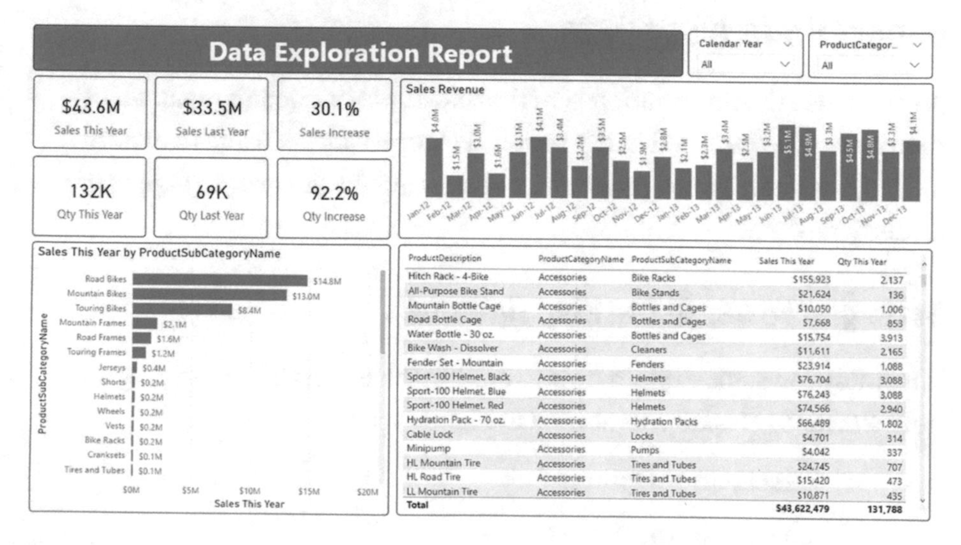

ProductDescription	ProductCategoryName	ProductSubCategoryName	Sales This Year	Qty This Year
Hitch Rack - 4-Bike	Accessories	Bike Racks	$155,923	2,137
All-Purpose Bike Stand	Accessories	Bike Stands	$21,624	136
Mountain Bottle Cage	Accessories	Bottles and Cages	$10,050	1,006
Road Bottle Cage	Accessories	Bottles and Cages	$7,668	853
Water Bottle - 30 oz.	Accessories	Bottles and Cages	$15,754	3,913
Bike Wash - Dissolver	Accessories	Cleaners	$11,611	2,165
Fender Set - Mountain	Accessories	Fenders	$23,914	1,088
Sport-100 Helmet, Black	Accessories	Helmets	$76,704	3,088
Sport-100 Helmet, Blue	Accessories	Helmets	$76,243	3,088
Sport-100 Helmet, Red	Accessories	Helmets	$74,566	2,940
Hydration Pack - 70 oz.	Accessories	Hydration Packs	$66,489	1,802
Cable Lock	Accessories	Locks	$4,701	314
Minipump	Accessories	Pumps	$4,042	337
HL Mountain Tire	Accessories	Tires and Tubes	$24,745	707
HL Road Tire	Accessories	Tires and Tubes	$15,420	473
LL Mountain Tire	Accessories	Tires and Tubes	$10,871	435
Total			**$43,622,479**	**131,788**

Figure 4-17. *Data exploration report*

Dimensions are used to slice and dice the information. You can add multiple facts to further explore the sales performance theme (e.g., quantity and unit price).

This exploration report should contain the key components to help the viewers answer the following questions.

- What caused the spike in some months?
- What is the high and low season for each product?
- How are the top product subcategories performing?

The granular data for the dashboard in Figure 4-17 is at a sales item level. The charts and tables are rolling up from the detail level. Having granular data with multiple dimensions in the underlying table is the key to exploration analysis, *bringing in as much data at the most granular level as possible.*

Export Underlying Data

Just because the information is in the data table, it doesn't mean all of them must be displayed in the report. Power BI allows users to export *underlying data.* Two settings must be aligned for the user to export the underlying data.

Setting 1: Power BI Desktop Options

The following steps explain how to enable the export of underlying data from Power BI Desktop, as shown in Figure 4-18.

1. Select File ➤ Options.
2. Select **Options**.
3. Select **Report settings**.
4. Choose underlying data under export data.

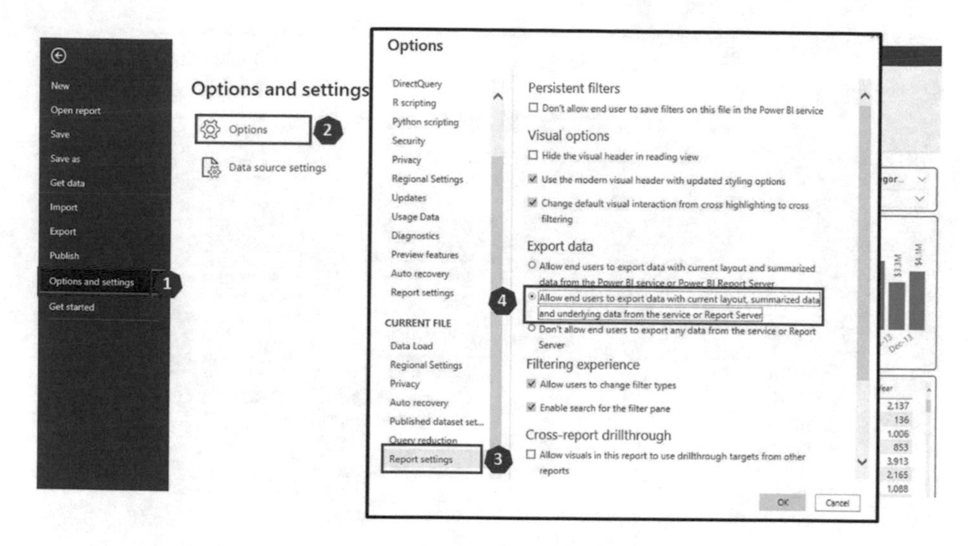

Figure 4-18. *Setting to enable export underlying data*

Setting 2: Power BI Service Options

The following steps explain how to change the settings in Power BI Service, as shown in Figure 4-19.

1. Choose the report drop-down and select settings.
2. Choose the current layout and underlying data.

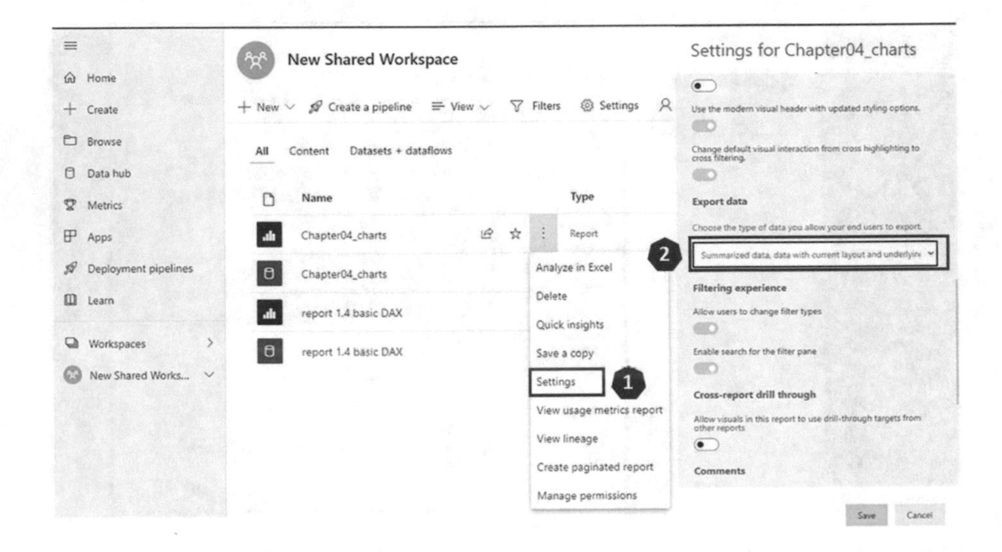

***Figure 4-19.** Settings to enable export of underlying data in Power BI Services*

Once the settings are selected correctly, report viewers can export the underlying data. The following steps explain how, as illustrated in Figure 4-20.

1. Choose the drop-down *after* applying the filters.
2. Choose **Export data**.
3. Select **Underlying data**.
4. Note the maximum number of rows.

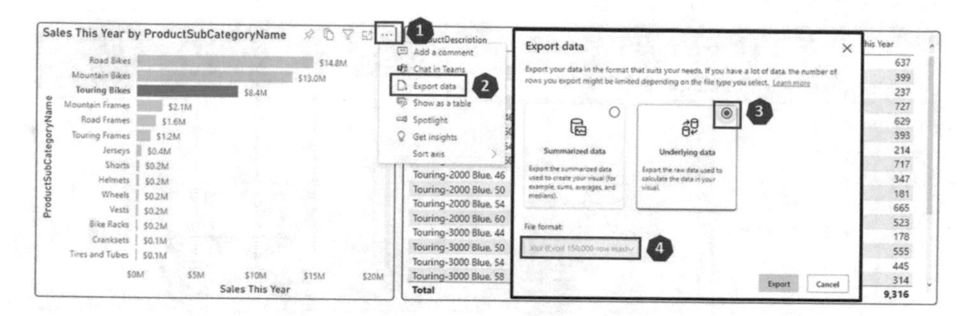

Figure 4-20. *Export data from an online report*

The user can use the underlying to conduct further analysis. It is also common for people to use Power BI as a data export tool to connect to the underlying data.

Data Explanation

Data explanation is different from data exploration in the following ways.

- Data explanation starts from a conclusion.
- Data explanation limits or prevents report interactivity.
- Data explanation provides less information.

An example of the data explanation report is in Figure 4-21. After some exploration of the data, you have reached the following recommendation. Once you reach a conclusion, you can design the report to explain the following data story to the users.

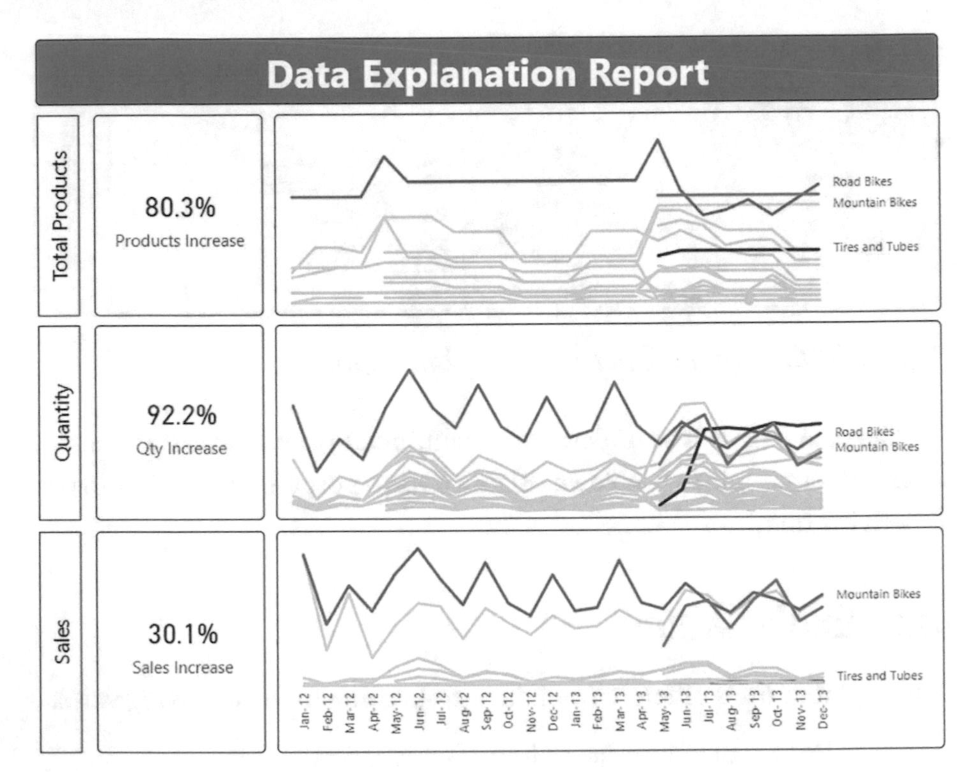

Figure 4-21. *Data explanation for product reduction*

AWM has the strategy to expand its product range in 2013. The total number products increased by 80%. This has resulted in a 90% increase in sales volume. Unfortunately, the sales only increased by 30%. "Mountain bikes" is the only new subcategory contributed to sales performance. While further analysis is required on the profitability of the new products, the obvious recommendation is to reduce the number of products.

The recommendation's audience members may question the findings and want to interrogate the data. In contrast to the exploration report, the aim is to **display as little information as possible** to support the recommendation while keeping the audience focused.

Tip A gray color is an effective way to display background (context) information in reports. Intuitively, the viewer understands to focus on the colors and ignore the grays, but the information is still there to provide context and comparison.

Summary

Because of the underlying differences between the explanation and exploration reports, it is strongly recommended that the developer understand the report's purpose early in the process.

Report Design

At the beginning of this chapter, you have explored some useful data visualizations, which are the fundamental building blocks of the reports. In the previous session, you learned the two common purposes of reports. This session focuses on report design.

One key objective with report design is to improve the user experience by making the relevant information easy to understand while allowing the user to get to the information with minimal click and training. The following areas are explored.

- The structure of multiple reports
- A design template for data exploration reports
- Color selcction

Business Scenario

AWS production team requires several reports to guide production decisions. The production planning team has heard about the good work Kim has done in the finance area. In the requirements meeting, the production manager John Greentree asked Kim to start with three reports: product cost, sales volume, and stock level.

Structure the Reports

The business often requires multiple reports. Before designing the individual reports, it is critical to think about how to make it easier for the users to arrive at them. One common method is a hierarchical structure, as shown in Figure 4-22.

Figure 4-22. *Report structure example*

Caution The key principle in a hierarchical report structure is that all reports are built from the ground up. That means starting with the operational reports that contain the most detailed data. This ensures the leadership, management, and operations team sees the same information and avoids misalignment.

The reports in different levels should follow the following guidelines, as shown in Figure 4-23. The strategic report seen by the leaders contains little detail but covers many key areas. The operational report contains a lot of detail but focuses on a single or very few areas.

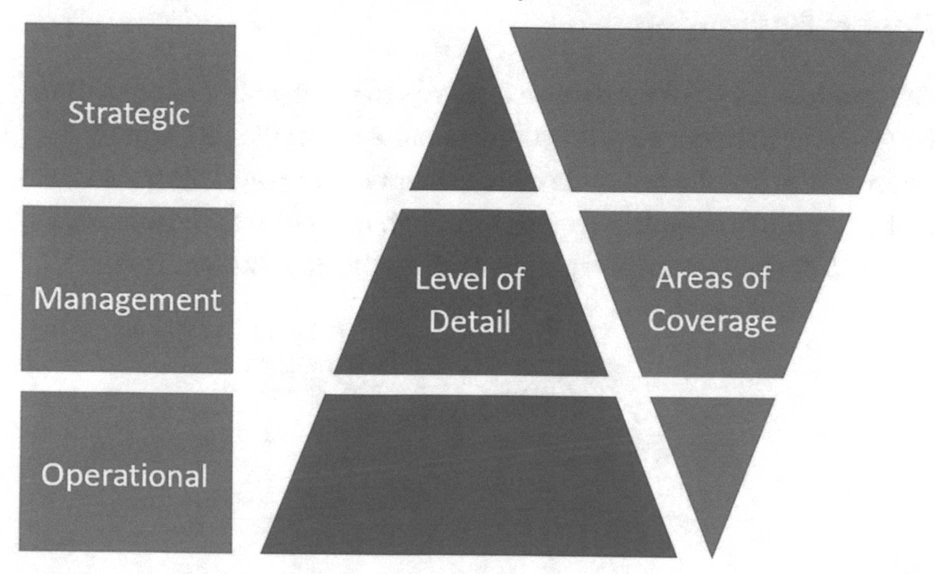

***Figure 4-23.** Report structure, level of detail, and areas of coverage*

The production and inventory reports are combined into the same Power BI report file in the example shown in Figure 4-24. The report file contains the management report and three operational reports.

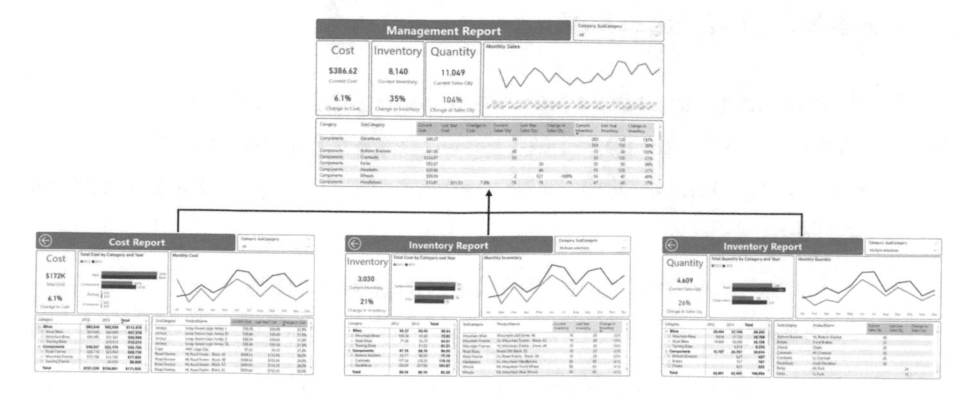

***Figure 4-24.** Production and inventory reports*

The three operational reports are designed in the same format to ensure consistency. This consistency is very important in driving a better user experience.

Report Navigation

One intuitive design for report navigation is to use the buttons to navigate between the different reports within the same report file. In the prior example, the users arrive at the management report page and can navigate to the individual operational reports through the buttons. The users can return to the management report by clicking the Back button. This is shown in Figure 4-25.

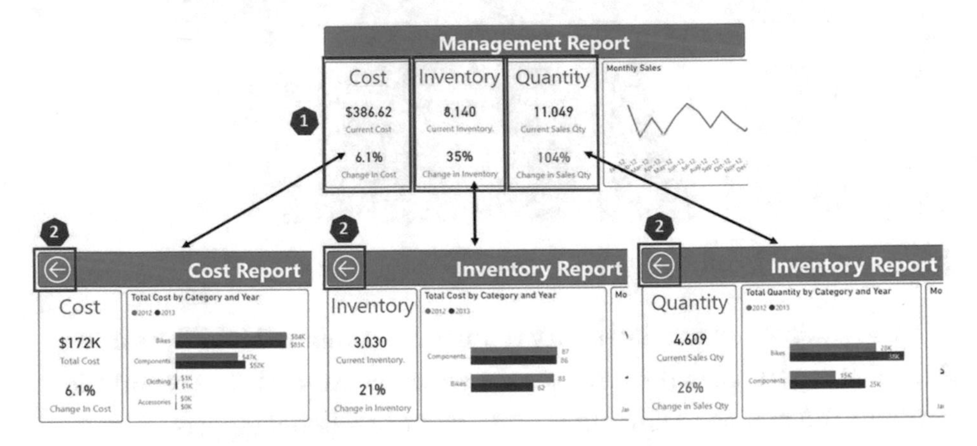

***Figure 4-25.** Navigation buttons*

Buttons to navigate to the operational reports are necessary. A Back button to return to the management report page is also important.

The following steps explain how to select and configure the buttons, as shown in Figure 4-26.

1. Go to Insert ➤ Buttons ➤ Blank, and add a title (e.g., cost).

2. Configure Actions type to **Page navigation**.

3. Choose the destination on the Cost page.
4. Add a Back button to the Cost, Inventory, and Quantity pages.
5. Hide the Cost, Inventory, and Quantity report pages (optional).

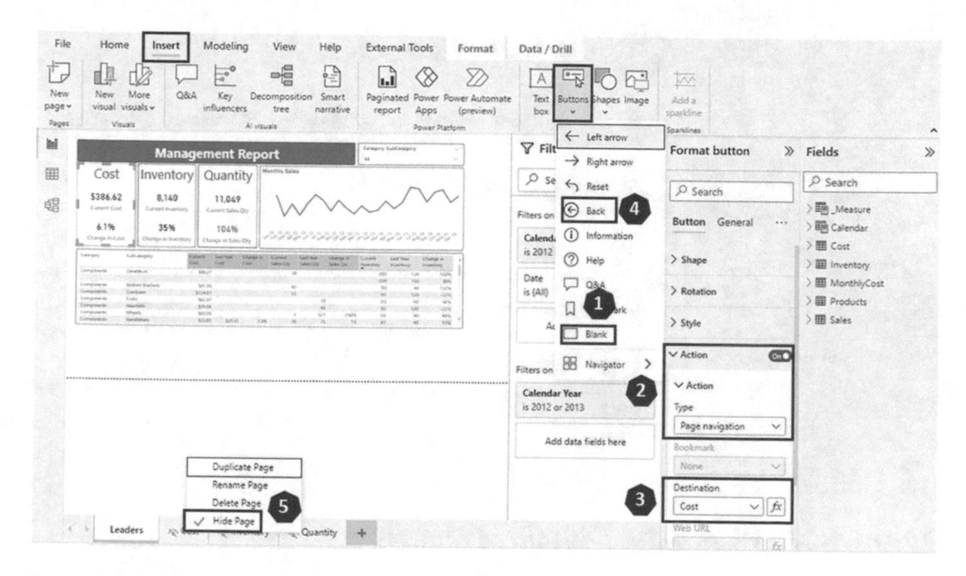

Figure 4-26. Configure navigation buttons

Power BI Dashboard

In theory, it is possible to combine all reports and data into a single Power BI report file; this is often not practical for two main reasons.

- Too much data slows down the developer and user experience.
- There are collaboration challenges in Power BI and between teams.

As a result, the report files can only cover some operational and management reports.

Figure 4-27 shows a workspace that contains three reports. You want to create a Power BI dashboard containing information from each report. A Power BI dashboard is one way to achieve this.

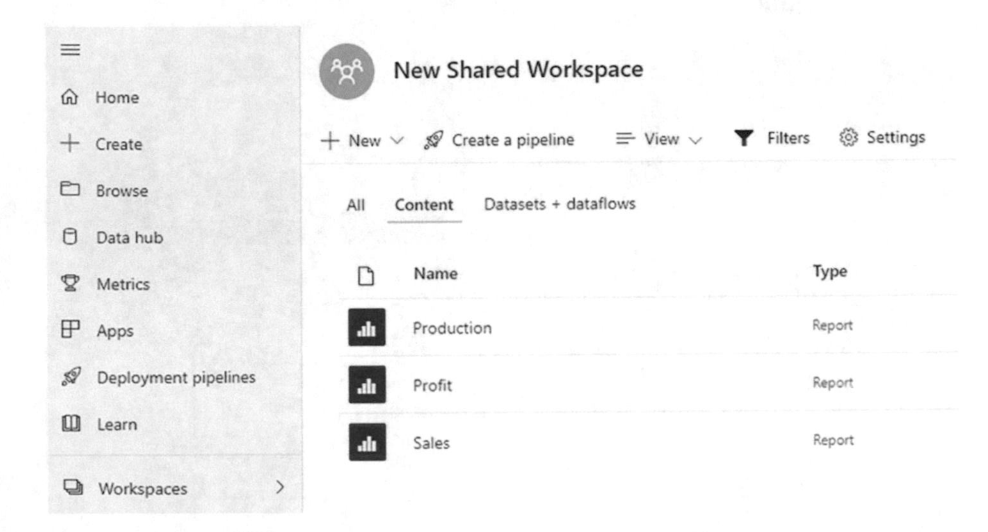

***Figure 4-27.** Workspace with three reports*

The following steps explain how to create a dashboard, as shown in Figure 4-28.

1. Hover a visual and click the ***Pin*** button.
2. Choose **New dashboard**.
3. Enter an appropriate name.

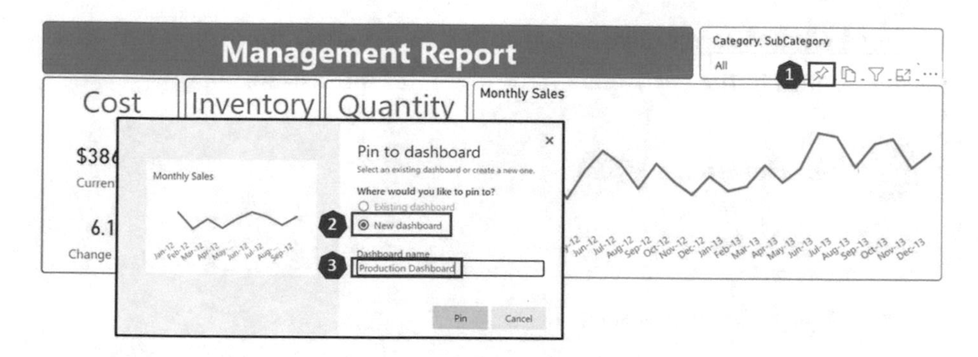

Figure 4-28. Pin visual to a new dashboard

This process does two things: create a new dashboard and pin the selected visual to the dashboard.

The following steps add other visuals from a different report to the same dashboard, as shown in Figure 4-29.

1. Hover over a visual and click the Pin button.
2. Choose **Existing dashboard**.
3. Select the name of the existing dashboard.

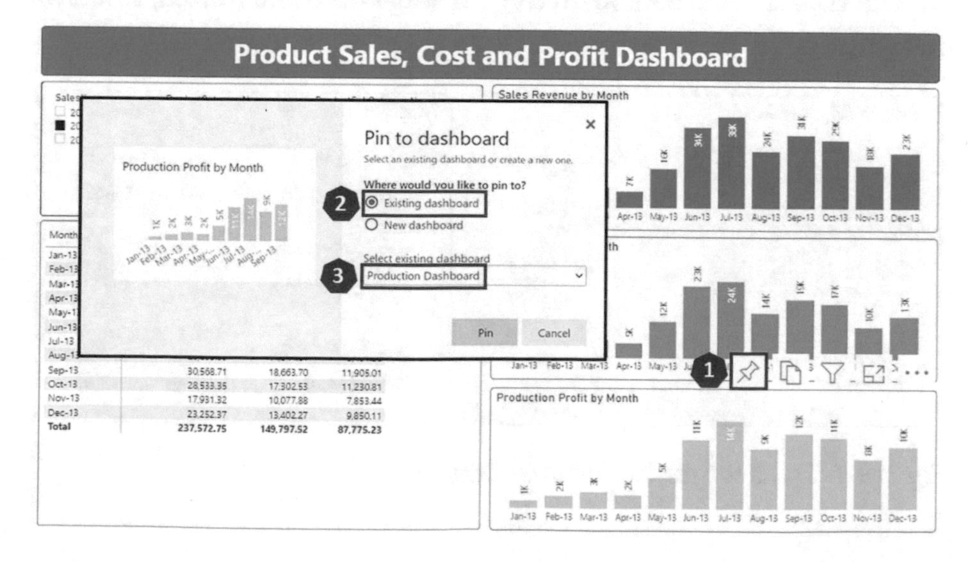

Figure 4-29. Pin additional visuals into the existing dashboard

Once all the visuals have been pinned to the dashboard, you can view the dashboard in the workspace, as shown in Figure 4-30.

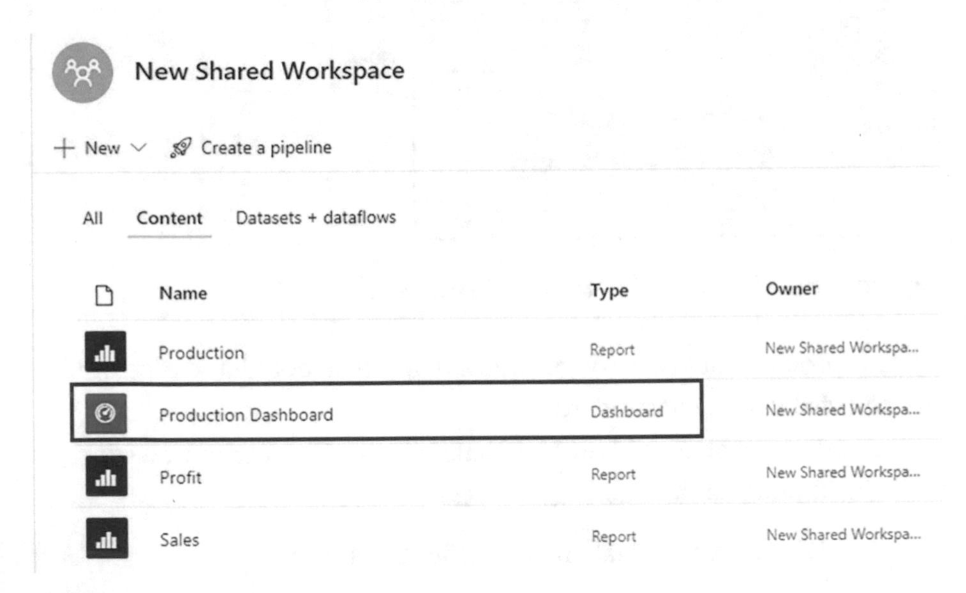

***Figure 4-30.** Production dashboard in the workspace*

The visuals in the dashboard can be reorganized and resized, as shown in Figure 4-31.

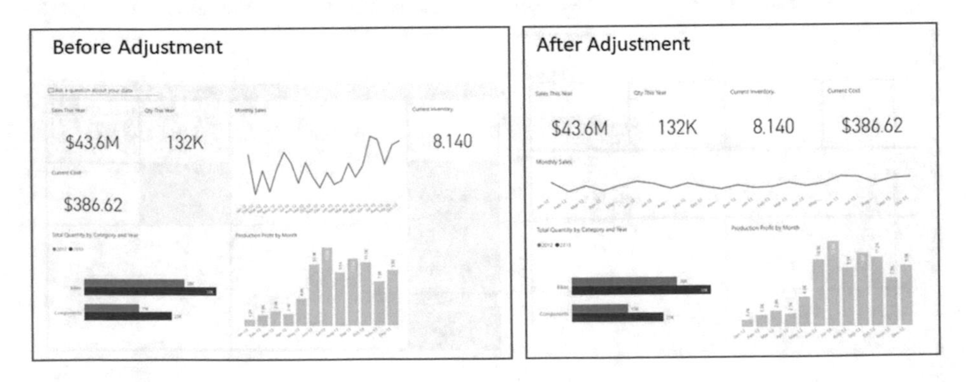

***Figure 4-31.** Reorganize the dashboard*

When the user clicks any charts in the dashboard, it navigates them to the relevant report. The Power BI dashboard allows report developers to bring visuals from different reports into a single place.

Compared to bringing data into a single report file, the dashboard has the benefit of limiting the file size and complexity for each report. The shortfall of this approach is that it cannot interact or filter data across different reports.

Design Template for Exploration Reports

A good design always starts from the user's perspective. Two well-known eye-tracking studies help to form the foundation of design principles.

The first study, presented by the Nielsen Norman Group, was done in 2006. It found that "People don't usually read every word on a webpage, in an app, or even in an article or text passage. Instead, they often scan."

Figure 4-32 shows the F-shaped pattern when scanning information on the screen. The user pays more attention to the information at the top left corner, and less attention to the bottom-right corner.

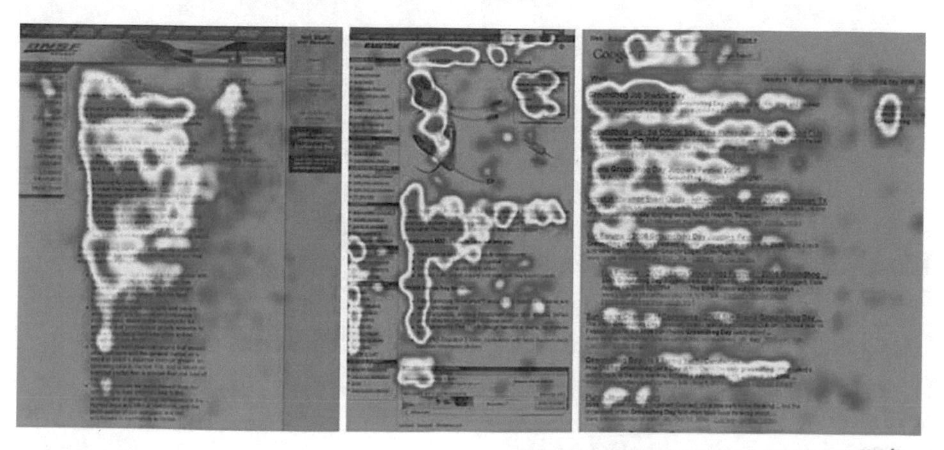

Figure 4-32. *F-shaped pattern*

An eye-tracking second study introduced the *layered cake pattern*, indicating that proper use of titles and subtitles can also guide the scanning process, as shown in Figure 4-33.

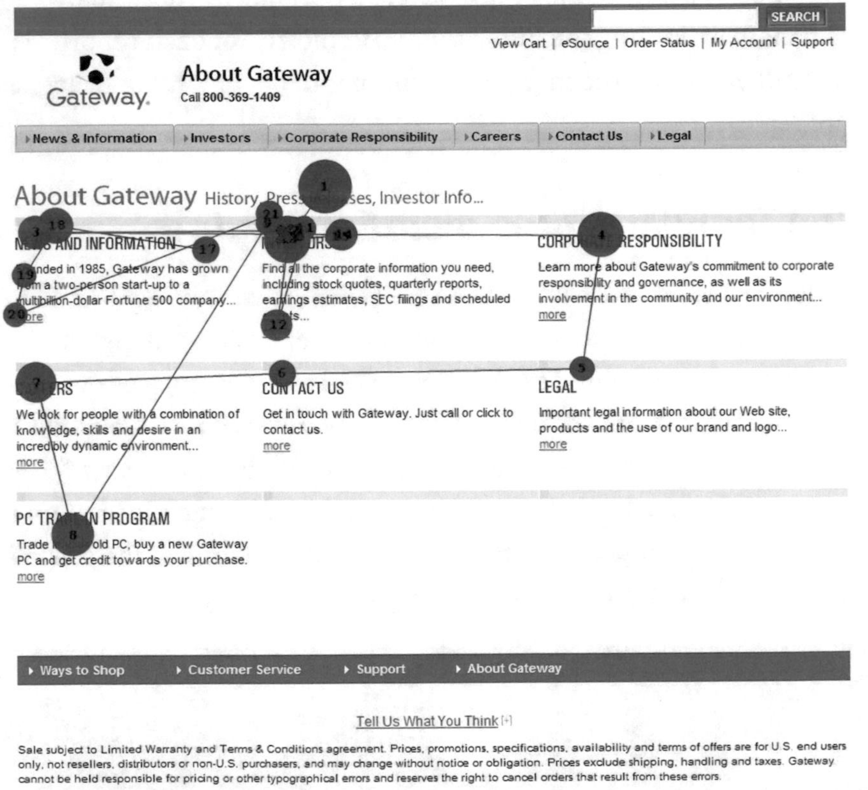

Figure 4-33. *Layered cake pattern*

Combining the two eye-tracking studies, consider the template in Figure 4-34 for most data exploration reports.

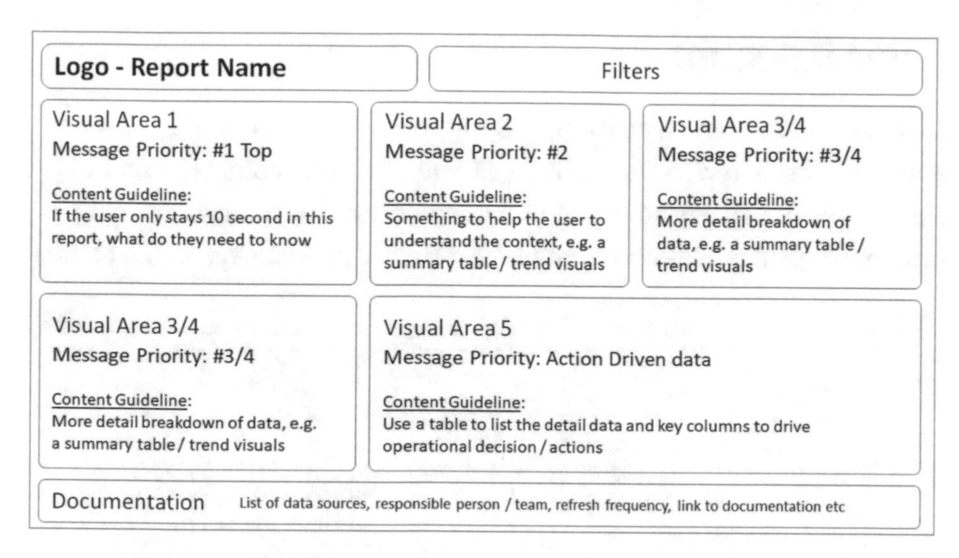

Figure 4-34. *Report template*

Note Look at the details shown in Figure 4-34. I have run this experiment with many users. Most people only scan through the titles. You almost have to force yourself to read line by line. What do you think this experiment is telling you?

While you don't need to follow the same design, adopting a standard design for the team would be beneficial. A simple and consistent design help users to quickly adapt to a new report.

The other reason for a team to adopt a standard design is because of efficiencies. A standard report design removes a lot of design effort from the equation, making the final output more consistent and reducing development time.

Color Selection

Color selection is one interesting but often overlooked topic. If you search *Power BI dashboard* in Google Images, you see many colors shown in Figure 4-35. If you pay attention to the search results, you can normally separate a good dashboard from the rest simply by focusing on the colors.

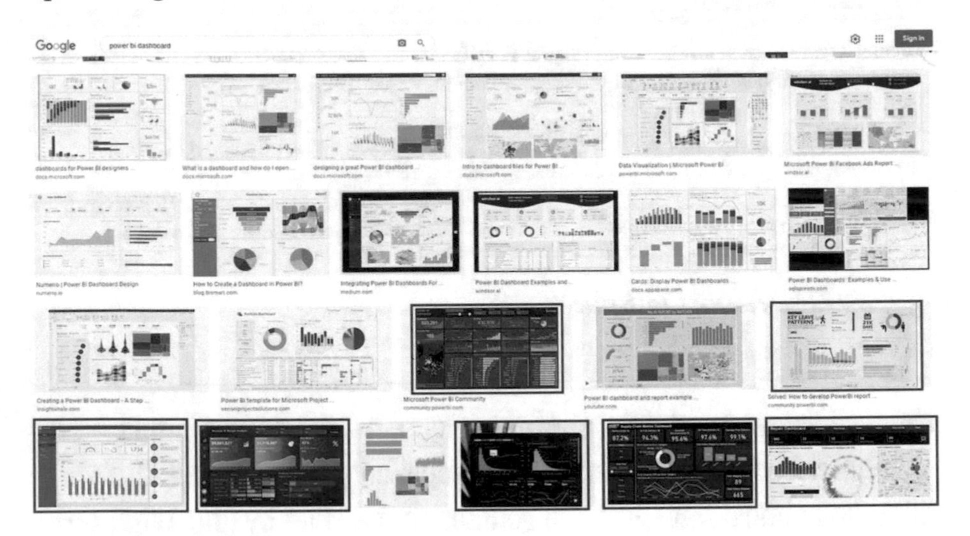

Figure 4-35. *Identify the good reports*

The following are four simple rules for color decision-making.

- The 60-30-10 rule
- Color association
- RAG (red, amber, green) colors
- Color blindness

The 60-30-10 Color Rule

The 60-30-10 rule is mostly used in interior design, but it can be applied to any color mix to achieve a pleasing visual balance. There is the base color, second color, and vibrant color.

- **60%** is the base color (e.g., wall and ceiling paint)
- **30%** is the second color (e.g., sofa and carpet)
- **10%** is the vibrant color (e.g., artwork and cushion)

You can use the cost report in Figure 4-36 as an example.

- Base color: white
- Second color: blue
- Vibrant color: amber and red

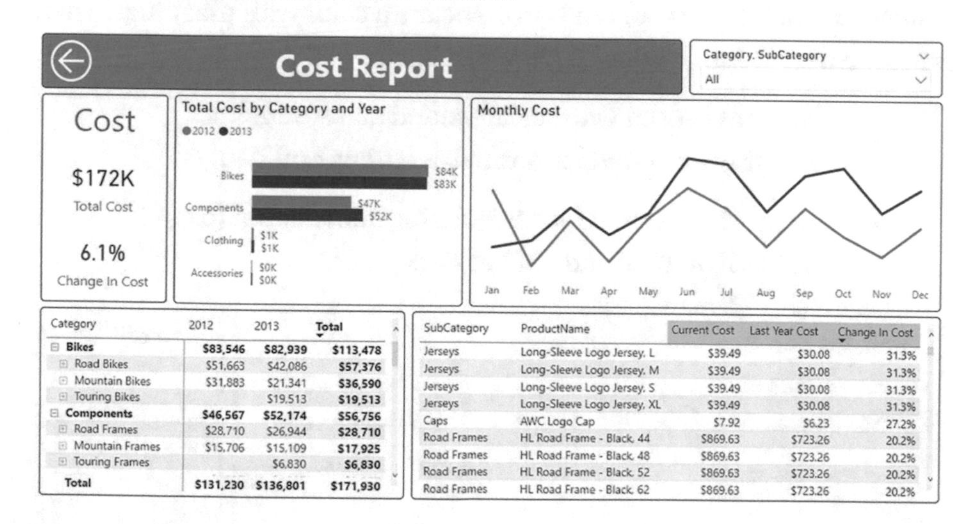

Figure 4-36. *Cost report color mix*

This color combination makes the information clear to the audience. You can also easily spot the red and green colors and the message they try to highlight.

There are two key benefits of adopting this rule.

- Keeps the report clean and tidy
- Makes the important information more visible to the audience

---

Caution For the base color, you may want to consider white color where feasible. When printing, a white background saves 85% of ink compared to a colored background.

Color Association

Color association is another concept that is natural to the audience. Color association means the user tends to associate a color with meanings. There are two types of associations, as shown in Figure 4-37.

- **Gradient** is used with scalar-valued measurements (e.g., gray–pink–red for normal, warning, and danger)
- **Categorical** is used for categorizing information (e.g., 2012 as light blue and 2013 as dark blue).

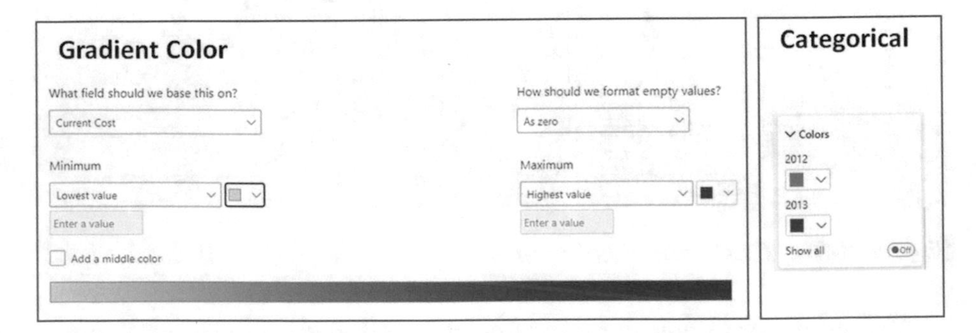

***Figure 4-37.** Color associations*

Once the association is formed in the user's mind, the user could be confused if the color association changes. It is important to keep it consistent throughout the report. For example, whenever the user sees light blue, it always represents the year 2012.

RAG Colors

Red, amber, and green are known as the RAG colors. Each color has a clear meaning. It is strongly recommended that you reserve these colors for specific use as vibrant colors to reinforce some information in the report.

- **Red** means *bad* or *problem.*
- **Amber** means *warning* or *attention.*
- **Green** means *good.*

Note There are exceptions to this. In China, red means *good,* and green means *bad.* Be aware of the audience.

Some organizations use red, amber, or green as their corporate colors. You can perhaps find other corporate colors to use and reserve the RAG colors to their universal meaning.

Color Blindness

Color blindness is a red/green color vision deficiency. It's a common problem that affects around 1 in 12 men and 1 in 200 women. Most people with color deficiency have difficulty distinguishing between red, yellow, and green shades. The other side of the problem is that no one knows which audience has vision deficiency. When choosing colors, you may want to factor this in.

ColorBrewer (colorbrewer2.org) and similar sites can assist when choosing colors. The following steps explain how to pick the colors for the dashboard, as shown in Figure 4-38.

1. Select **colorblind safe**.
2. Choose the number of colors required.
3. Select **data nature**.
4. Pick a color.
5. Choose a HEX color and use it in the Power BI color selection.

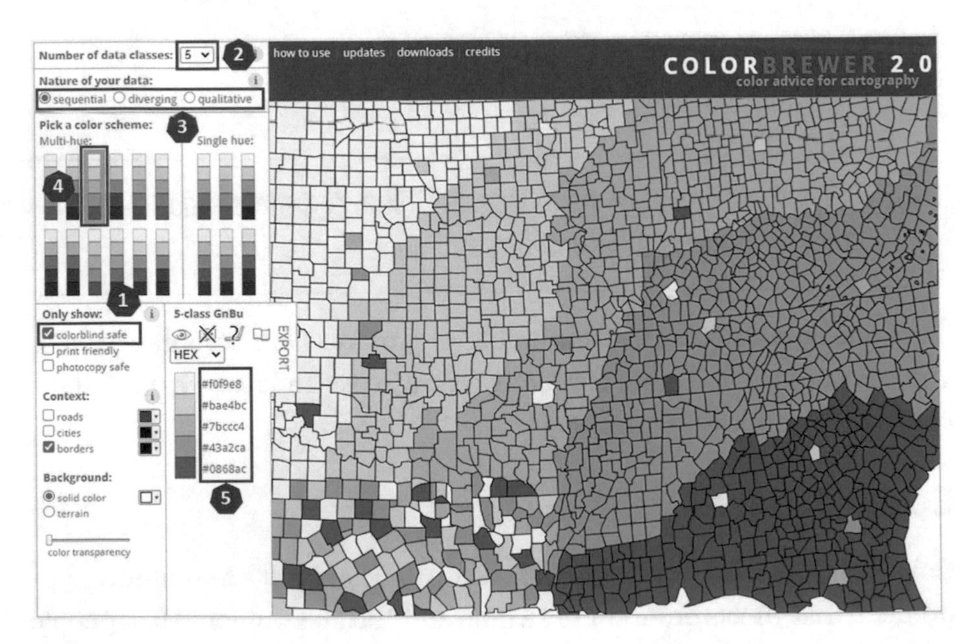

***Figure 4-38.** Color picker from ColorBrewer2.org*

Conclusion

Even though this chapter did not cover the user experience, every session focuses on a different area of user experience, from chart options to report purpose to the final design. As you start to work on Power BI reports, the key to improvement is to build as many reports as possible while seeking user feedback constantly.

Tip Report developers are often emotionally attached to their reports. The audience may not provide honest feedback out of politeness. To constantly improve the design, the developers must focus on non-verbal cues like body language and facial expressions.

You should follow a few rules to guide you into good dashboard design habits.

- Determine if the dashboard is meant to explain or is for exploration.
- Use classic charts where possible because everyone understands them.
- Choose color carefully; 60-30-10 and RAG are your friends.
- Keep it clean and keep it simple.

Mini-Hackathon

There is a mini-hackathon challenge for the milestone chapters. These mini hackathons are designed to help you solidify learning about Power BI. This second mini-hackathon focuses on practicing your data visualization skills.

Hackathon Data

The data source is Covid Data in Australia; see `https://github.com/M3IT/COVID-19_Data/tree/master/Data`.

Objectives

Continuing from the first mini-hackathon, you need to produce two reports with the additional information you have gained in this chapter.

- Report 1 is an exploration report. Find a theme from the data and create a self-service exploration report to allow users to analyze data from multiple dimensions.
- Report 2 tells a data story. Based on some correlations identified in report 1, create a report to tell a data story.

Time Limit

The objectives should be completed within three hours.

CHAPTER 5

Power BI Governance

In the previous chapters, you acquired the practical skillsets for building and sharing Power BI reports. In this chapter, you gain insights into Power BI reporting governance.

This chapter explores the following topics on governance.

- Why governance?
- Activity data
- Standardization
- Development operations (DevOps)
- Development support
- Data security

Why Governance?

Introducing a reporting governance program may result in a lot of discussion between the leadership and the development teams. Organizations often invest in governance programs for many reasons, such as more consistent reporting quality and reduced risk on a single point of failure. It also gives the management team a sense of control. While the development team also appreciates the structure these programs bring, they are also aware that governance programs can be demanding and introduce restrictions that slow down delivery.

D. Ding, *Transitioning to Microsoft Power Platform*,
https://doi.org/10.1007/978-1-4842-9239-6_5

Governance is all about striking the right balance for your unique organization. Like report design, governance is more of a form of art than science. It is difficult to directly adopt a governance framework that works in another organization for the following reasons.

- **Reporting maturity**: Some organizations are more mature than others in data culture, Power BI skillset, and security controls.
- **Resource allocation**: Some organizations have fewer report developers and resources than others that can be allocated toward governance and control. A smaller team also means less effort in governance.
- **Leadership support**: Some leaders understand the power of governance better than others. Sponsorship from business leaders is a key determinator in the success of governance programs.

A Business Scenario

Kim has developed five reports for the finance and production teams. The leadership team sees the potential in Power BI and has decided to invest more. Kim had convinced two other team analysts to work on Power BI. Kim was excited about the endorsement from the leadership team because she believed this would at least triple their delivery capacity and dramatically speed up the transformation from Excel into modern BI reports for the entire organization.

The average daily effort graph for the three analysts is summarized in Figure 5-1. Note that there is a total of eight hours per person per day. Two new analysts joined in week 5, which takes the total effort to 24 hours. These ten weeks can be broken into the following four stages.

- **Stage 1 (Solo Player)**: During weeks 1 to 4, Kim was the only person working on the reports. As she built up more reports, the maintenance effort increased as expected.
- **Stage 2 (New Team)**: During weeks 5 to 7, two new analysts joined to develop more Power BI reports. After an initial burst in the development effort, the ongoing maintenance work increases.
- **Stage 3 (Storming)**: During weeks 8 to 10, the new reports start to clash, the reconciliation workload kicks in, and businesses complain about the confusion in numbers and where to get information.
- **Stage 4 (Coordination)**: In week 10 and beyond, analysts from different teams collaborate and reach a new operating rhythm.

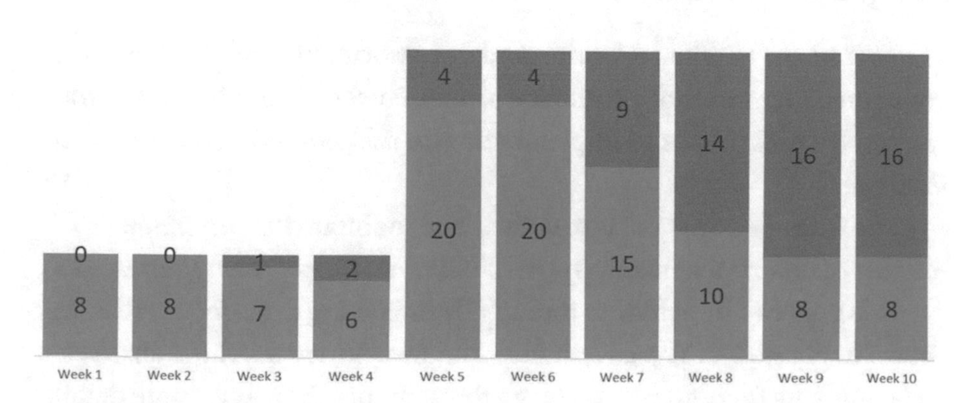

Figure 5-1. *Effort graph with development vs. maintenance*

During stages 3 and 4, the following issues start to become apparent.

- **Lost single source of truth**: The reports on the same metric show different numbers. This resulted in a lot of confusion in the business and additional time and effort in reconciliation for the analysts.
- **Lack of transparency**: There is no documentation, and many reports contain confusing DAX formulas. This makes it stressful to maintain and troubleshoot.
- **Duplication of work**: The three analysts occasionally compete against each other to produce the same report. This not only resulted in a waste of resources but also in additional time for reconciliation.
- **Lack of support**: The three analysts often face development challenges but cannot obtain etiquette support. They end up wasting a lot of time on small problems.

Adoption-Driven Governance Framework

The four stages of initial adoption and the associated challenges are very common in organizations without a proper Power BI adoption plan. You can resolve most issues by implementing an *adoption-driven governance framework*.

The framework's goal is to deliver and maintain the *minimum number of reports* that meets the *maximum number of user requirements*.

This goal can be achieved through the following three focus areas, as shown in Figure 5-2. The governance framework may be a little abstract and confusing for now, but as you go through this chapter in more detail, you can always come back and compare it against the diagram.

- **Drive user adoption** establishes a standard approach to measure and drive adoption (blue boxes).
- **Track benefits** quantifies benefits based on report usage (green boxes).
- **Improve development efficiency**. Improve the user experience, adoption, and development processes (yellow boxes).

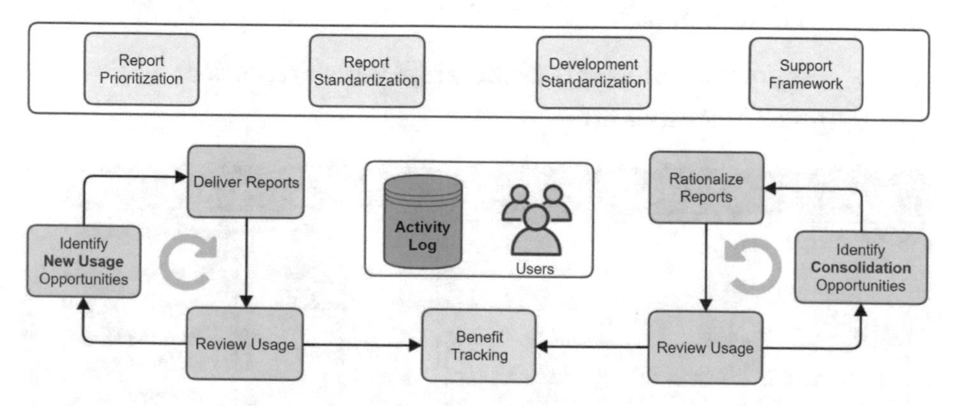

***Figure 5-2.** Adoption-driven governance framework*

The adoption-driven governance framework is an evidence-based framework. The core of this framework is the Workspace Activity Log dataset and user engagement.

Note Everything in the framework is designed to be measurable because it is hard to improve what can't be measured.

Usage Data

You can work through the governance process and dive into each process in more detail soon. First, you can focus on an example usage report in Power BI (see Figure 5-3). This report conveys the following information.

- **Report usage**: Who is using the report, and how often?
- **User group reports**: What are the areas of interest for various groups?
- **Reports of interest to leadership**: Which reports do the leaders look at?

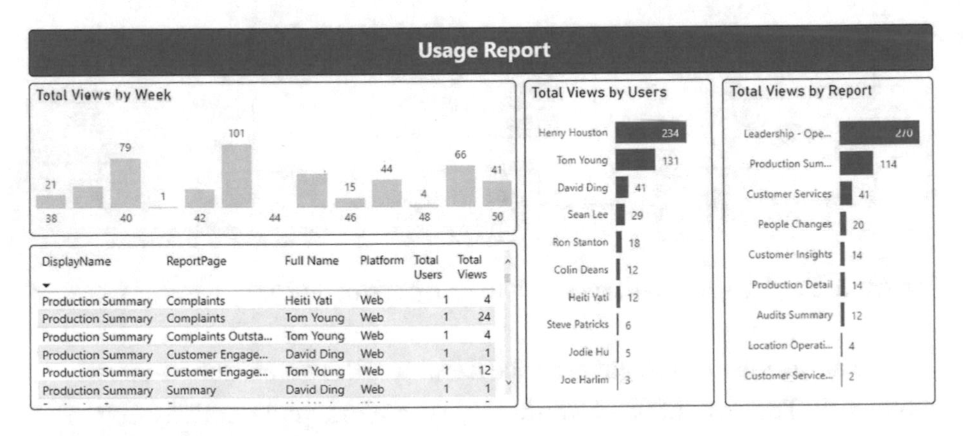

Figure 5-3. *Workspace usage data*

There are two ways of obtaining the data to develop such a report: via a workspace usage report or the PowerShell Power BI API.

Workspace Usage Report

A workspace usage report can be created by following a few steps. It is strongly recommended that you collect the usage report for every production workspace.

First, generate a usage metrics report for any report in the workspace. The following steps explain how to do this, as shown in Figure 5-4.

1. Select **View usage metrics report**.
2. Choose **View usage metrics**.

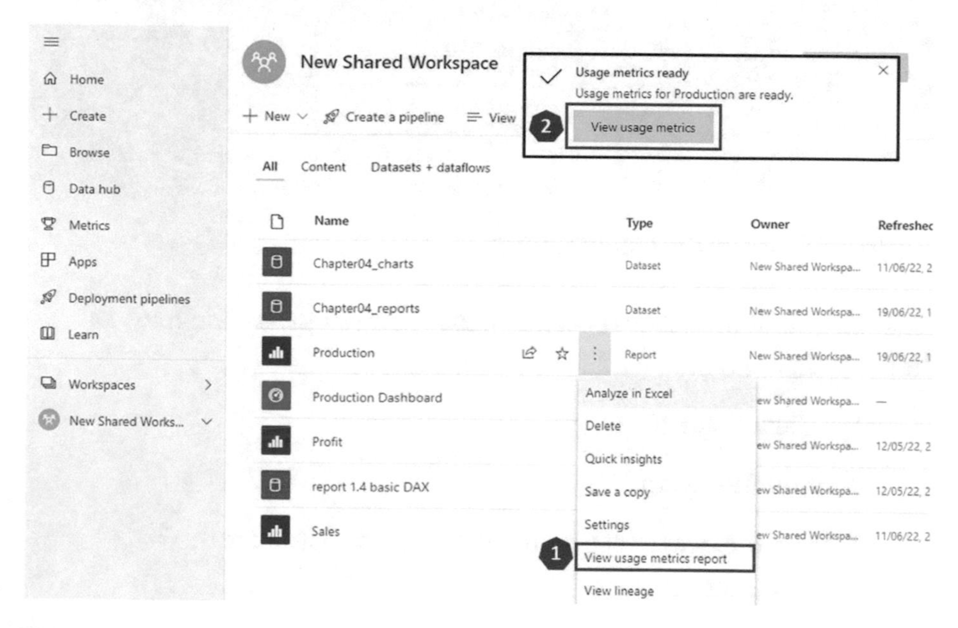

Figure 5-4. *View usage metrics report*

Next, save the usage metrics report by following these steps, as shown in Figure 5-5.

1. Select **Save a copy**.
2. Name it **report _Consumption**.
3. Go to report.

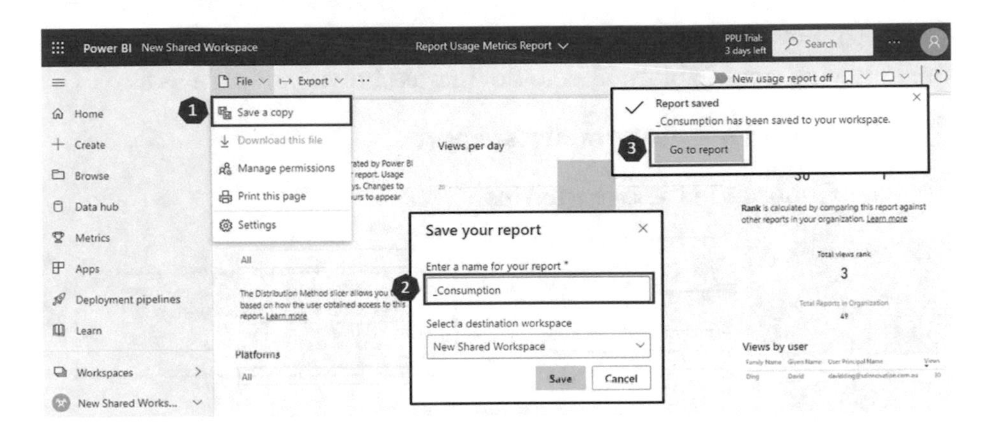

__Figure 5-5.__ Save the usage report

Finally, edit the usage report. The following steps explain how, as shown in Figure 5-6.

1. Clear ReportGuid.
2. Edit the report.
3. Use the data tables in the edit view to customize the usage report.

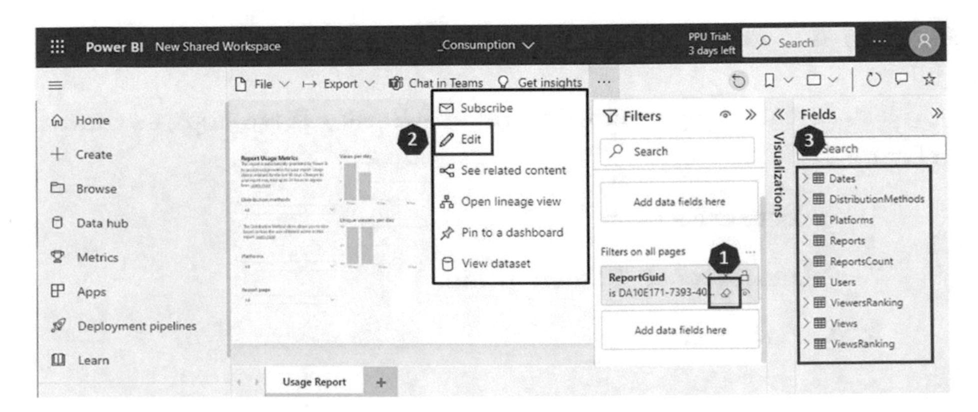

__Figure 5-6.__ Data for a new report

You can also download the underlying usage data from the detail table.

There are three problems with using workspace usage report to gain user insights.

- **Not scalable**: One consumption report only covers a single workspace. This makes it time-consuming for a user to manage many workspaces.
- **Limited report functions**: The report is online and lacks data modeling and Power Query capabilities.
- **No development activities**: The report only covers viewer activities, not admin or development activities.

PowerShell cmdlet and API

The PowerShell cmdlet and API are the preferred approaches because it solves the problems mentioned earlier. The PowerShell script that contains the cmdlet and API can be executed regularly to capture the data required.

This method is not widely adopted because the development teams require the following privileges.

- **Power BI administrator**: This privilege gives the user a lot of power over the company environment. It should not be granted to developers who lack knowledge of the Azure tenant.
- **Local machine administrator**: This privilege allows users to download, install, and run PowerShell scripts on the computer. This access can be granted according to company policies.

The technical details are not explained here. There are three steps to developing the usage report using the PowerShell script.

1. Identify the team with Power BI Administrator privilege and determine how to access Power BI API.
2. Develop and agree on the process for executing the PowerShell script regularly.
3. Develop the activity report in Power BI.

The following is an example PowerShell script.

```
## Login
Login-PowerBI

## Set initial parameters
$outFilePath = "C:\...\Usage\activities"
$nbrDaysToGoBack = 7
$consolidatedDataset = @()

for ($i = 1; $i -le $nbrDaysToGoBack; $i++)
{
    ## set boundaries
    $startDateToExtract = (((Get-Date).AddDays(-$i)).
    ToUniversalTime()).ToString('yyyy-MM-ddT00:00:00.000')
    $endDateToExtract = (((Get-Date).AddDays(-$i)).
    ToUniversalTime()).ToString('yyyy-MM-ddT23:59:59.999')
    $activities = ''

    ## Run GetActivityEvent API
    $activities = Get-PowerBIActivityEvent -StartDateTime
    $startDateToExtract -EndDateTime $endDateToExtract
```

```
    ## prepare output file name in folder
    $displaydate = (((Get-Date).AddDays(-$i)).
    ToUniversalTime()).ToString('yyyy-MM-dd')
    $outfile_name = "$outFilePath" + "_$i.json"
    $activities | Out-File -FilePath $outfile_name
}
```

This script generates a daily activity file that covers all workspace activities for the entire organization (if executed by the administrator) as shown in Figure 5-7.

Name	Status	Date modified	Type
activities_2022-06-26.json	⊘	2022-07-03 2:37 PM	JSON Source File
activities_2022-06-27.json	⊘	2022-07-03 2:37 PM	JSON Source File
activities_2022-06-28.json	⊘	2022-07-03 2:37 PM	JSON Source File
activities_2022-06-29.json	⊘	2022-07-03 2:37 PM	JSON Source File
activities_2022-06-30.json	⊘	2022-07-03 2:37 PM	JSON Source File
activities_2022-07-01.json	⊘	2022-07-03 2:37 PM	JSON Source File
activities_2022-07-02.json	⊘	2022-07-03 2:37 PM	JSON Source File

Figure 5-7. *Daily workspace activity files*

The daily JSON files can be loaded into Power BI using the folder option. The following steps explain how to do it (see Figure 5-8).

1. Go to **Get data** and select **More**.
2. Choose **Folder**.
3. Enter the folder location.
4. Select **Transform Data**.

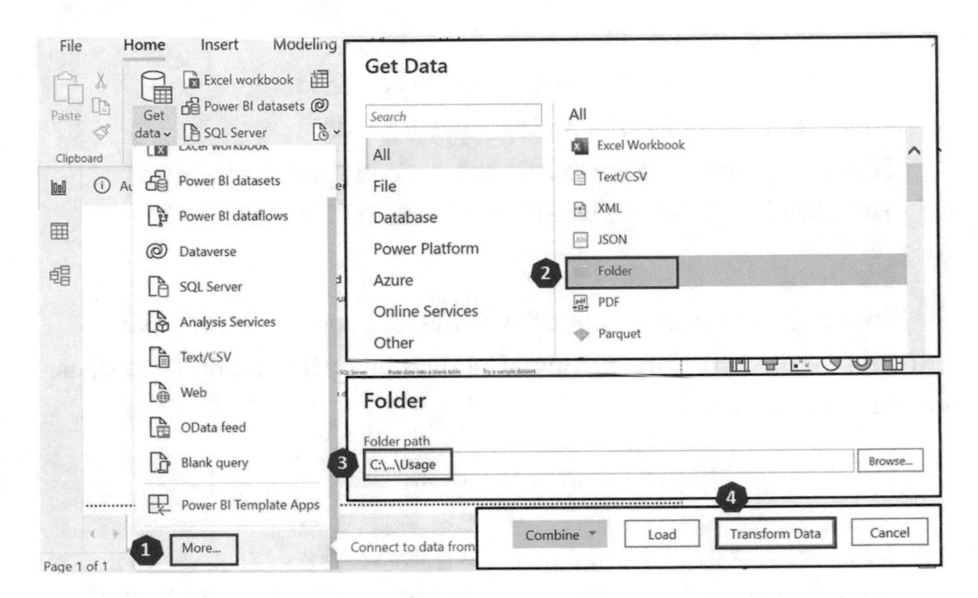

Figure 5-8. Loading activity data from Folder

When you click the Expand button shown in Figure 5-9, Power Query automatically detects the JSON file, load, and combine the files accordingly.

	Content	Name	Extension	Date acce
1	Binary	activities_2022-06-26.json	.json	2022-
2	Binary	activities_2022-06-27.json	.json	2022-
3	Binary	activities_2022-06-28.json	.json	2022-

Activity	ItemName	WorkSpaceName	DatasetI
GetGroupsAsAdmin	null	null	
GetGroupsAsAdmin	null	null	
GetGroupsAsAdmin	null	null	
ExportActivityEvents	null	null	
ExportActivityEvents	null	null	
ExportActivityEvents	null	null	
ExportActivityEvents	null	null	

Figure 5-9. Expand and consolidate individual files

Caution The file may contain personal information, which can be removed from the script or during the Power Query step.

Drive User Adoption

There is a difference between usage and requirements. The business always has new reporting ideas or requirements, but the ideas do not always translate into reporting usage. Occasionally, a report is developed but has very little usage.

Note It doesn't matter how wonderful this report looks and how useful it may be; it is a waste of time if no one is looking at it. This framework forces some responsibilities of report adoption onto the reporting team.

A reporting team normally has two task categories: *development pipeline* and *maintenance tasks*. The user adoption processes create and prioritize the items in the development pipeline.

There are two processes for creating the items in the development pipeline, as shown in Figure 5-10.

- A **new usage process** identifies usage opportunities to develop new reports.
- A **rationalization process** consolidates and simplifies usage patterns.

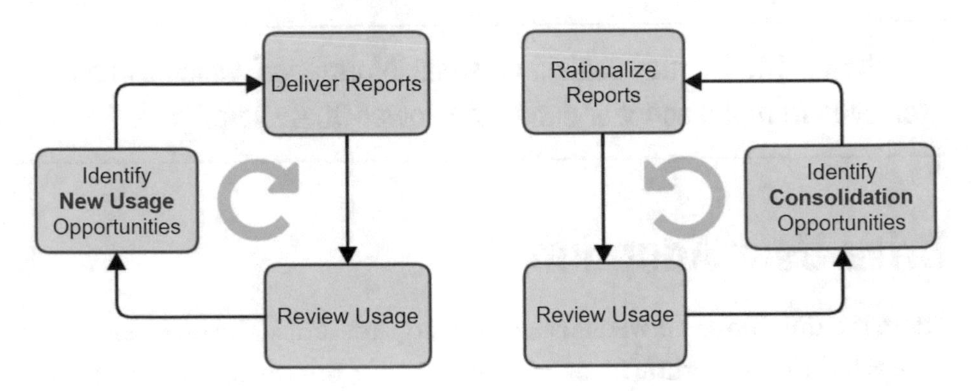

Figure 5-10. *Two processes to drive user adoption*

Identify a New Usage Process

The first process is to *identify new usage*. This starts from the initial engagement process. In addition to collecting the requirements, the developer must also estimate the usage. As shown in Figure 5-11, three key measurements are required to estimate the usage: *number of users*, *frequency*, and *seniority of user*.

The final priority score (145 in the example) is calculated by combining the priority scores for each seniority group. The priority score for each seniority group is calculated as follows.

Priority Score = Number of Users × Usage Frequency × Seniority Multiplier

Seniority Group	Seniority Multiplier	Number of Users	Weekly Usage Frequency	**Priority Score**
Operational Team	1	20	5	**100**
Management Team	3	5	2	**30**
Leadership Team	5	3	1	**15**
Chief Executives	10	0		**0**
Total		**28**		**145**

Figure 5-11. *Estimating report usage score for each report*

You can adjust the seniority multiplier value, but it should be consistent across the reports.

- **Deliver reports**: Once you have a few new requirements with priority scores, you can combine them with the report complexity to determine which report to work on next. The delivery process is closely linked to the development standards explained later.
- **Review usage**: Once the report is delivered into production, its usage is tracked regularly (typically monthly). This is compared against the original estimation. If the usage falls short, the reporting team must initiate conversations with the business to understand the reasons for low adoption and drive improvements.

The review process also encourages the development team to understand the business operating rhythm (e.g., how often the business reviews the figures, what they discuss in meetings, etc.). This often leads to additional new usage opportunities but may also lead to consolidation opportunities.

Identify Consolidation Process

In addition to identifying new usage opportunities, there are often good reasons to consolidate reports and rationalize the usage pattern. For example, instead of having the users visit three separate reports to collect information, the three reports can be combined to improve the user experience and efficiency.

The consolidation can start from the activity data to understand what reports a particular user group is looking at. This drives better engagement with the business.

The delivery and review processes following the identification process are similar to the new usage process.

Track Benefits

It is often difficult for reporting teams to initiate benefits tracking. This is partly because it can result in the reporting team having to constantly justify their existence. Although this session only aims to present a tracking method, it is almost unavoidable if the adoption-driven governance framework is adopted. Two main benefits can be tracked: the effort saved and the value added.

Effort Saved

The effort saved can be tracked from the usage data. It can be calculated using the following simple formula.

Effort Saved = Hours Saved × Hourly Rate

If this is tracked regularly, it encourages the reporting team and individuals to care about user adoption of their reports. Over time, this enhances the integration of reports into every business process.

Value Adds

Often, a report is created to enable business improvements (e.g., compliance, performance, or usage). Every improvement that is measured needs to be tracked over time. This tracking also makes it possible to understand the improvement since the report's inception.

If you can put a dollar value on the improvements, this can be tracked as a value add over time; for example, a 1% improvement from baseline equates to $10,000.

Improve Development Efficiency

There are two main objectives of the reporting team that embrace the adoption-driven governance framework: *increase usage* and *improve efficiency*. Figure 5-12 shows the development effort before and after optimization. In an optimized scenario, the per-report effort decreases as more reports are being developed.

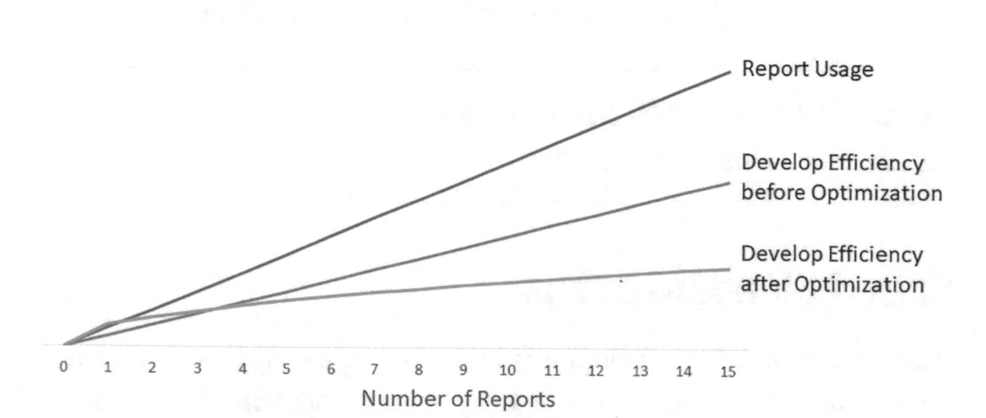

Figure 5-12. *Report development efficiency*

Under the traditional approach in Excel-based teams, the reports contain many manual steps, making it harder to scale up. The automation options in Power BI allow the team to scale up much faster than traditional methods. The adoption-driven governance framework also covers four focus areas to further improve the development efficiencies, as shown in Figure 5-13.

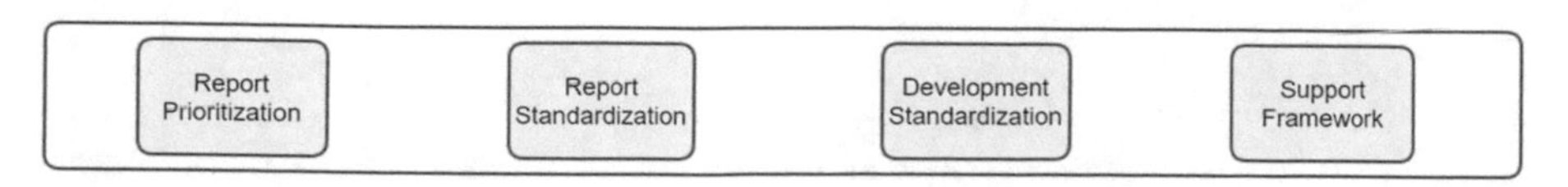

Figure 5-13. *Four areas of focus for development efficiency*

Report Prioritization

This chapter has already covered this topic in the drive user adoption session. The usage report should drive the prioritization conversation instead of purely relationship driven.

Both development and report maintenance should be categorized into different categories with different source allocation and service level agreements (SLA). For example, the report with senior leaders may be granted top priority in development and receive the best SLAs.

Note The prioritization categories should be regularly reviewed against actual usage data, not estimations.

Report Standardization

Design effort in report development refers to the time and resources spent on the look and feel and the interactivity of the report. Sometimes the report developer can go out of the way to meet user requirements. While it is good to have report developers that care about the report designs, it may not be the best use of the developer's time. Figure 5-14 shows that the effort follows the diminishing return graph.

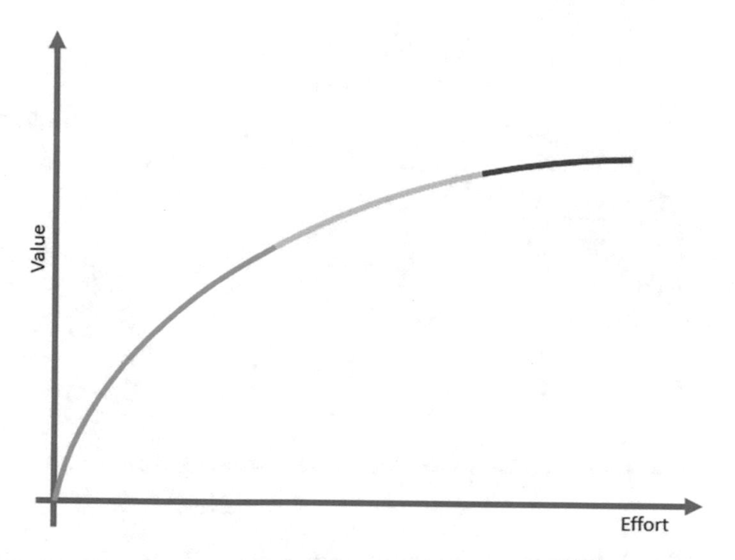

Figure 5-14. *Diminishing return graph for report design effort*

The diminishing return graph shows that a small amount of development effort creates a lot of value for the user at the beginning of the development process. The amount of value for the same amount of effort shrinks over time. When it gets into the red zone, delivering a small amount of value takes a lot of effort.

A standardized report design is one effective way of preventing the developer from entering the red zone. The standardized design helps in two primary ways.

- It prevents the developer from spending too much time trying different design options.
- It prevents the users from inserting customized design requirements.

In addition to preventing the developer from entering the red zone, it also improves the development efficiency, so the developer can deliver more value with the same amount of effort, as shown in Figure 5-15.

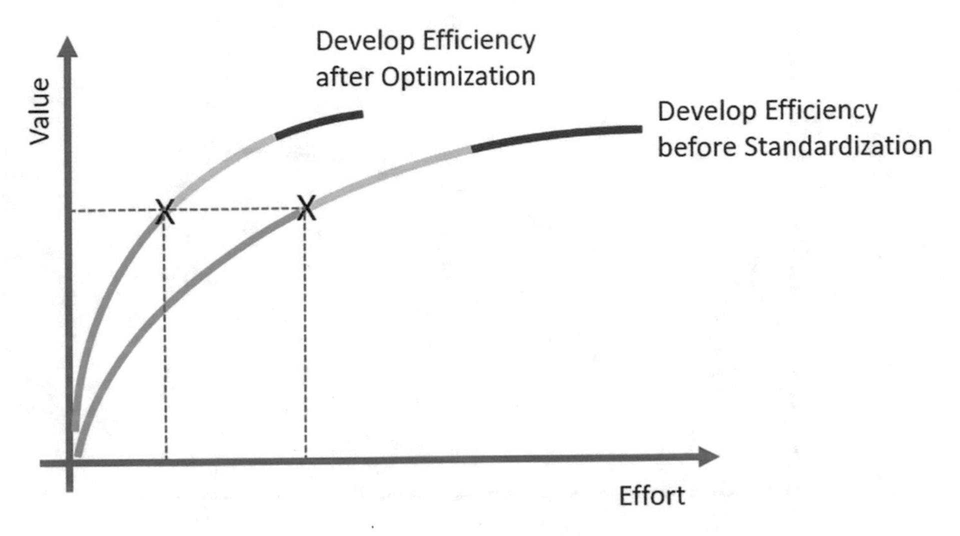

Figure 5-15. *Deliver the same value with less effort*

Once the information is presented, most people nod in agreement. The real challenge begins with the discussions on what to standardize. There are the following areas for consideration.

- Navigation method
- Data connection
- Standardize DAX measures
- Colors
- Report layout and background

Navigation Method

This is covered in Chapter 4. A clear structure makes it much easier for users to navigate the required information. Figure 5-16 shows a simple three-layer reporting structure.

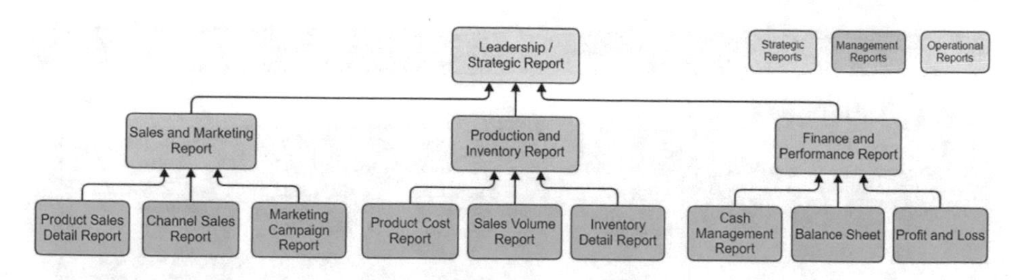

***Figure 5-16.** Reporting structure*

Before the developer starts working on a new report, he/she needs to determine where the report fits. If it is an operational-level report, there needs to be a navigation method from the management report layer to help users to find the new report. This can be a text link or a button. It should be displayed clearly to the user where to get more detailed information.

Data Connection

Most reporting teams have less than ten data sources and three to four types of SQL servers, data marts/lakes, shared drives, SharePoint, web sources, and APIs. A standardized approach makes it much easier for developers to connect to the different data sources.

Power BI service has a helpful solution called a *dataflow* that can standardize data sources. A dataflow runs a cloud-based Power Query and stores the query output in a Power BI workspace. This is helpful when for example, if some data tables are frequently used in different reports, they can be loaded and scheduled in a Power BI dataflow. As shown in Figure 5-17, the following steps explain how to do it.

1. Go to a Power BI workspace and choose **Dataflow**.
2. Select **Define new tables**.
3. Select **Azure SQL database** (or any other listed data sources).

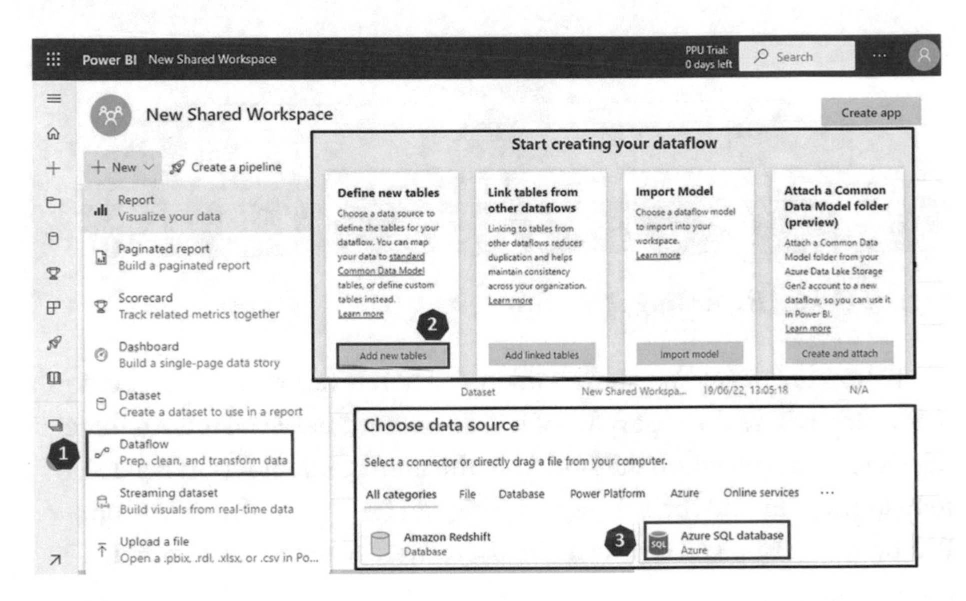

***Figure 5-17.** Add new dataflow*

The following screens should look familiar to you as dataflow is the same as Power Query in the cloud. Figure 5-18 shows the steps required to configure the SQL query.

1. Enter the server and database names.
2. Enter the query that returns the required table.
3. Enter the SQL Server credentials.

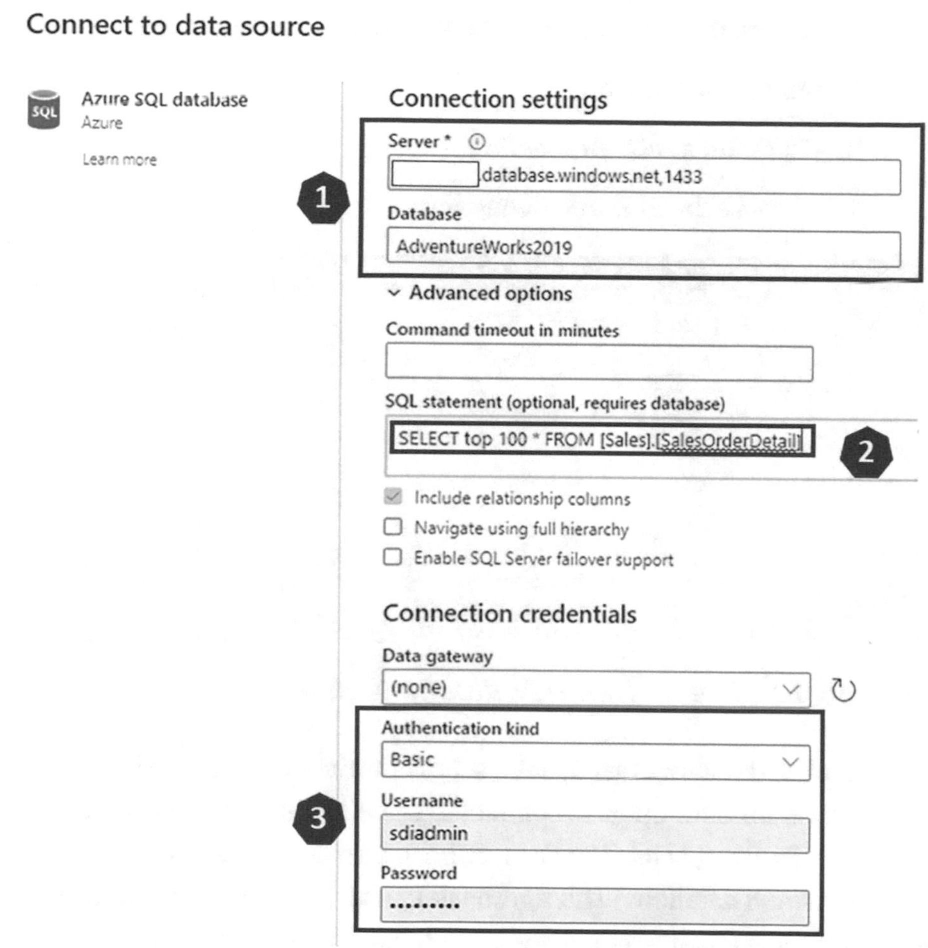

Connect to data source

Azure SQL database
Azure
Learn more

Connection settings
Server *
.database.windows.net,1433
Database
AdventureWorks2019
Advanced options
Command timeout in minutes
SQL statement (optional, requires database)
SELECT top 100 * FROM [Sales].[SalesOrderDetail]
Include relationship columns
Navigate using full hierarchy
Enable SQL Server failover support

Connection credentials
Data gateway
(none)
Authentication kind
Basic
Username
sdiadmin
Password
•••••••••

***Figure 5-18.** Configure Azure SQL Server connection*

Note A data gateway may be required for on-prem and non-Azure Cloud servers.

Once you confirm, the table is displayed as required. The following steps finalize the dataflow and schedule refresh time and frequency, as shown in Figure 5-19.

1. Check the table and click **Save & close**.
2. Name the dataflow.
3. Click **Set a refresh schedule**.
4. Choose the time and frequency.

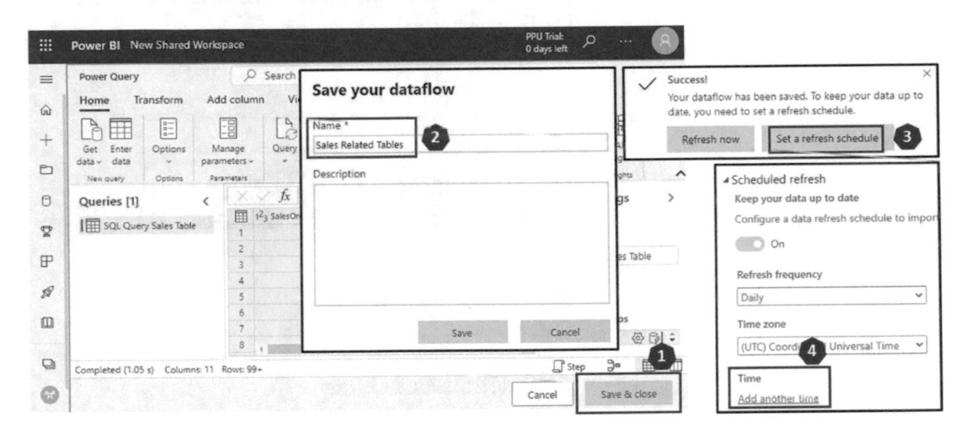

Figure 5-19. *Save and schedule dataflow refresh*

Once the dataflow is set up for a set of common tables, it makes it much easier for developers to connect to these standard transformed tables. (In Desktop, select Get Data, choose Dataflows, and navigate to the workspace and dataflow.) This approach can also be useful in common reference tables such as employees, products, and hierarchy tables.

Standardized Data Model

The standardization in the data model mainly refers to the reference tables and the connections. For different reports to display consistent information on the same metrics, you must consider standardizing the backend tables, data models, and table relationships. Only then can you standardize DAX measures.

Standardize DAX Measures

When you build reports for different teams, you might notice that 50% of the measures are reusable. Once the reference tables and data models are standardized, you can store DAX measures in the documentation and copy and paste them into the different reports.

Colors and Background Image

Most organizations have corporate colors and PowerPoint background images. These can be used to standardize the look and feel of the reports.

A good way to create a customized theme is shown in Figure 5-20.

1. Select **Customize current theme** from View.
2. Customize the theme.
3. Select theme colors.

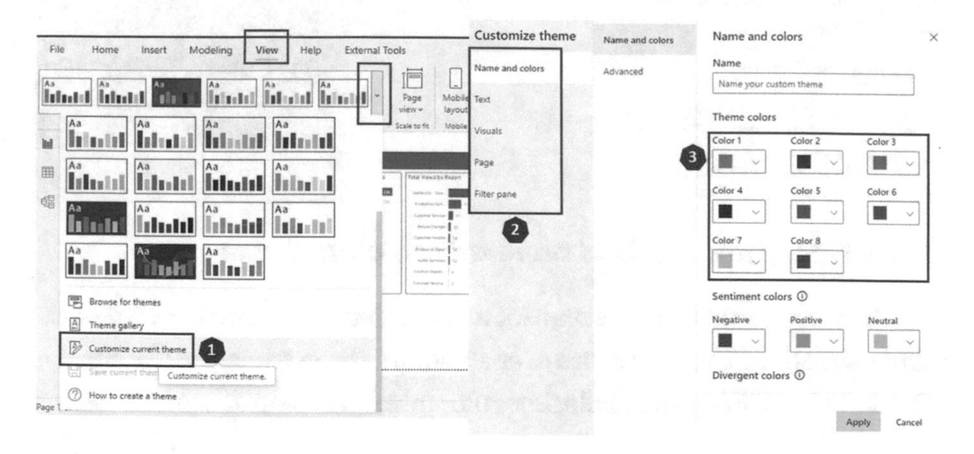

Figure 5-20. Customize a theme

The following steps save the theme as a standard corporate theme for other reports, as illustrated in Figure 5-21.

1. Select **Save current theme** to save the theme.
2. Select **Browse for themes** to load the theme.

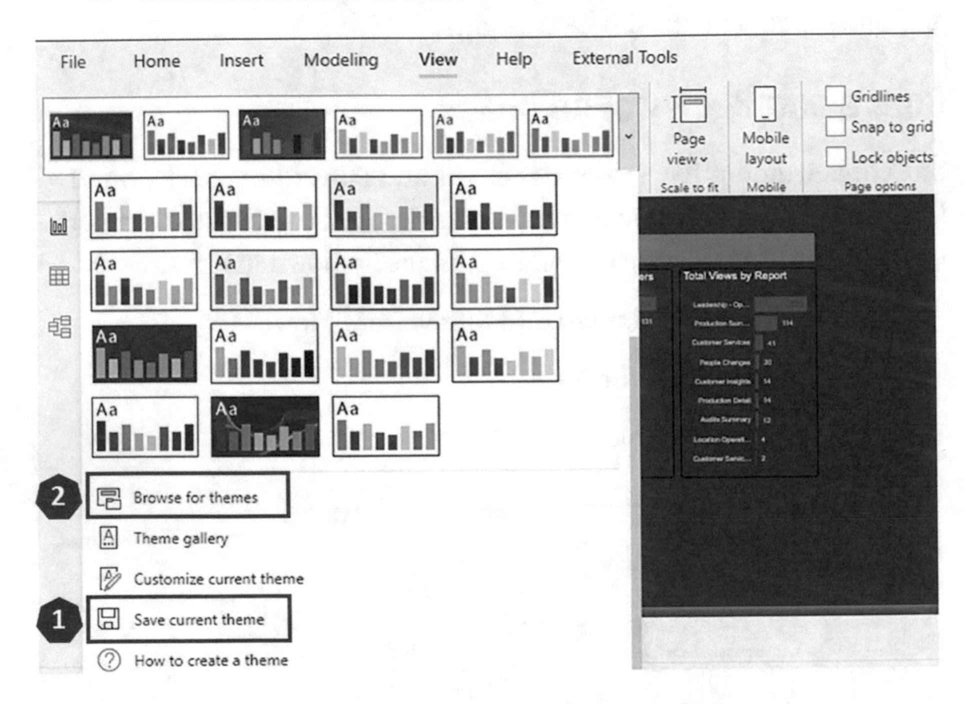

Figure 5-21. *Save and load standard corporate theme*

While standardization makes a lot of sense, it is important to **avoid over-standardization.** This is because over-standardization can harm the creative spirit and may also result in inflexible rules that hurt the user experience.

Development Standardization

In addition to standardizing designs, there is also a need to control the development process. One key difference between Power BI and normal BI development is that development, testing, and production can all happen in the same environment. This is where simplicity can lead to problems.

This session discusses the following topics to take reports to the next level.

- Agile MVP approach
- The whole-of-life cost and optimization step
- Workspace separation
- Jira for report development and maintenance
- Confluence for report documentation

Agile MVP Approach

Most software development and advanced analytics teams adopt the *Agile methodology*. The key characteristic of the Agile approach is the iteration cycle and task prioritization method.

The *minimum viable product (MVP)* is a simple report that meets some of the most basic requirements. The concept is originally suggested by Eric Ries (2011). The MVP becomes the starting point iterations.

Combining MVP and the Agile method results in a process shown in Figure 5-22.

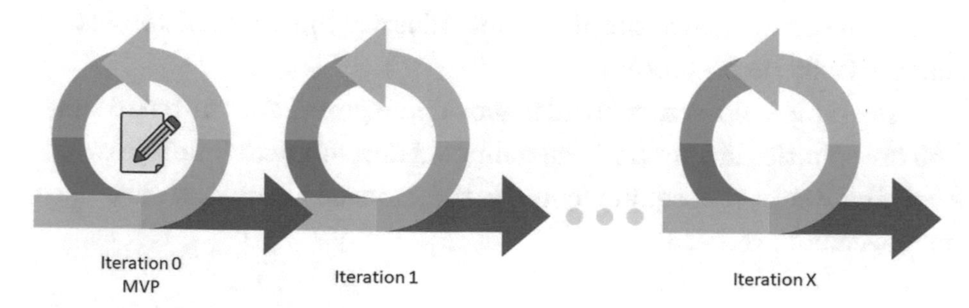

Figure 5-22. *MVP Agile iteration processes*

Before kicking off the Agile MVP process, a collaboration team needs to be created to include report developers, business representatives, and data specialists. This group meets regularly to discuss and prioritize requirements and to track progress.

The MVP report typically starts with one fact table (e.g., sales) and two- or three-dimension tables (e.g., dates, products, employees). The goal of the MVP is to set up an initial product for user engagement discussions.

Once MVP is ready, the team can discuss the requirements for the next iteration. The high-value requirements are implemented first. This process ensures the business is always in the loop and the report delivers value even from an early stage.

The report can be launched after some iterations, but the launch does not always mark the end of the development cycle. The report can continue to be refined based on user feedback after launch.

Jira for Report Development and Maintenance

The Agile MVP report development process can use either a Kanban or a Scrum board to plan the development tasks for each cycle or sprint. A sprint is a fixed-length period (e.g., weekly) where the team can prioritize tasks to be completed during the sprint. Atlassian Jira is one of the best tools to facilitate the process.

An example Jira scrum board is shown in Figure 5-23. The board shows the tasks/stories in a sprint. If the estimated development time is less than four weeks, the requirements can be managed as sub-tasks to reduce management overhead.

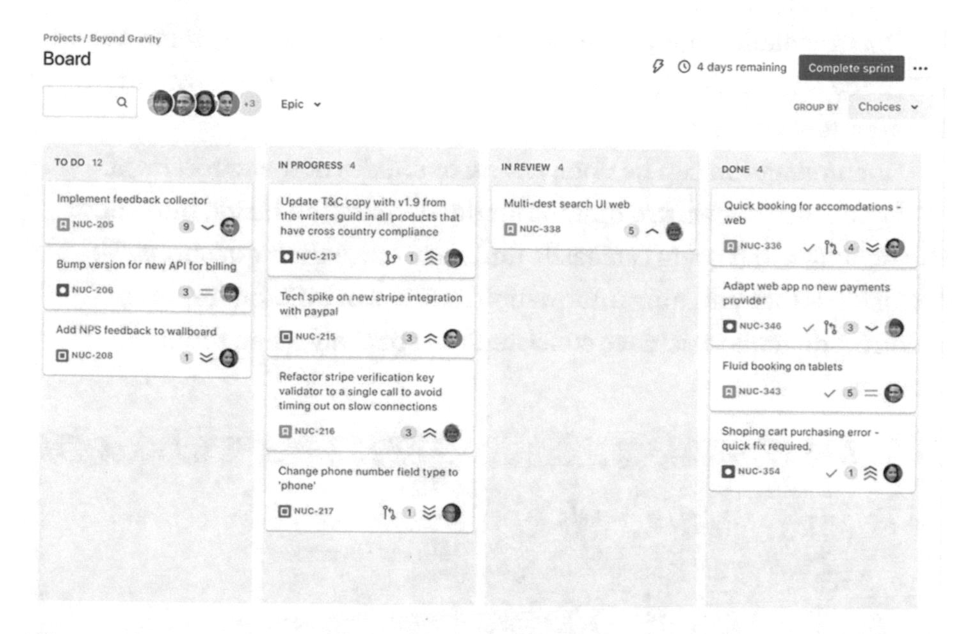

Figure 5-23. *Tasks in a Jira scrum board*

Tracking report maintenance tasks are critical in an adoption-driven governance framework. Sometimes users can abuse the development resources to work on low-priority reports. Other times, some reports cause a lot of maintenance effort. The tasks become important facts to drive improvement conversations.

Report Documentation

Report documentation is one area that most teams agree on, but very few teams do it well. Using *total sales* as an example, the logic is often not as simple as adding the Sales column. It may require some product categories to be excluded. It may also have a few different date options: date of sales, payment date, and delivery date. The choices are often not obvious to other developers or users.

Documentation allows the user to better understand the information presented and for developers to replicate the same logic consistently in different reports.

Documentation can be done *within* or *outside* of the report itself.

Figure 5-24 shows two options for documentation inside the report. The benefit is that users can easily find answers to their questions. The downside is that the same information appears in different reports, making it difficult to achieve consistency, especially when required changes.

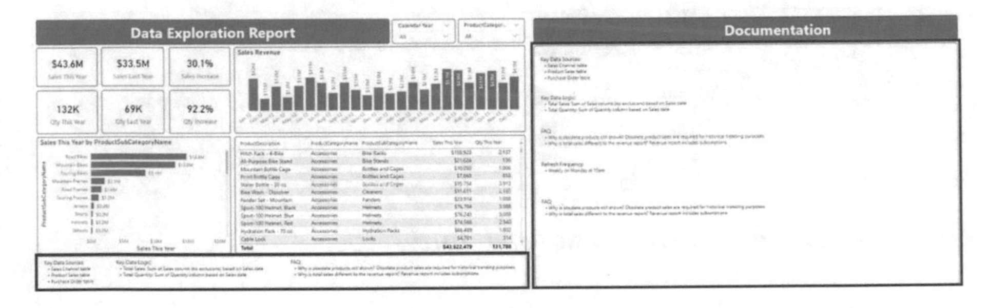

***Figure 5-24.** Report documentation options*

Figure 5-25 shows documentation outside the report using Confluence. You can use Microsoft Word for the same purpose. There are some advantages to using Jira and Confluence (both by Atlassian) in combination, as referencing between the tasks in Jira and documentation in Confluence is seamless. The benefit of external documentation is that it is more comprehensive and easier to share. The downside is that the users need to go through extra clicks to find the documentation. The user must then search through the documentation to find the required information.

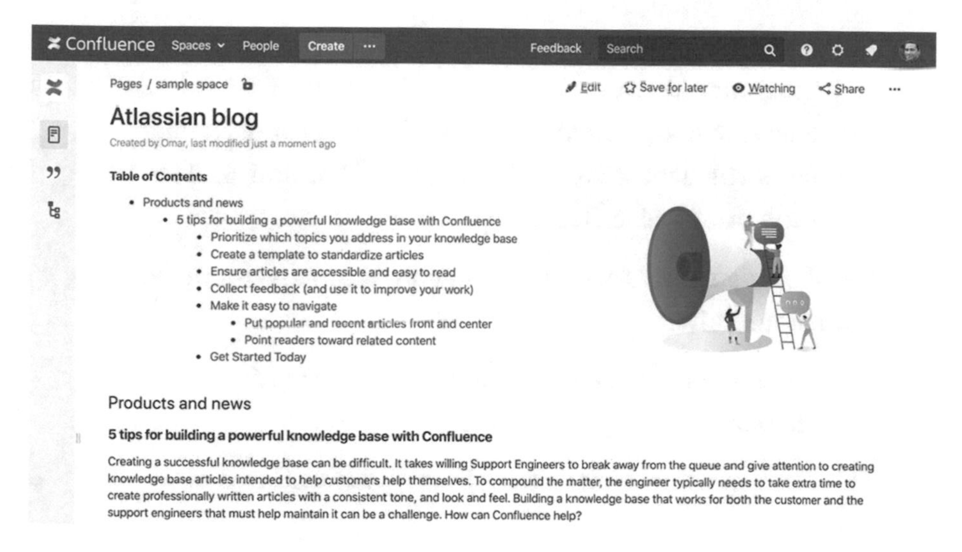

Figure 5-25. *Confluence documentation*

You may consider adopting a *mixed approach* by displaying some basic information inside the report and detailed documentation in an external document.

Business-owned documentation is an approach some teams adopt to keep the documentation up to date over time. The documentation may contain the following information.

- **Purpose**: Why is the report created? The purpose will almost surely change when the user sees the first draft. Documenting the purpose is not to prevent deviation from the original purpose but to capture the changes over time.

- **Data sources**: What data sources should be used? This could be used to capture existing datasets and future datasets. Again, it is important to capture the change over time.

- **Definitions**: It is important to clearly define what different term means in a report. For example, Total Sales means different things to different groups. It may cover the data sources, filter logic, date period, and so forth. It may also change over time.

The following summarizes the engagement process for business-owned documentation.

1. A business representative highlights the required changes in the documentation.
2. The reporting team raises a Jira ticket linked to the request.
3. The reporting team reviews the request.
4. The reporting team makes the changes and updates the documentation.

The goal is to have the business responsible for documentation, and the change process starts with a change in documentation. The developer is normally the person who creates and updates the documentation without ownership. Figure 5-26 shows that the business-owned documentation process ensures that the documentation, the change tracker (Jira), and the reports are always in sync.

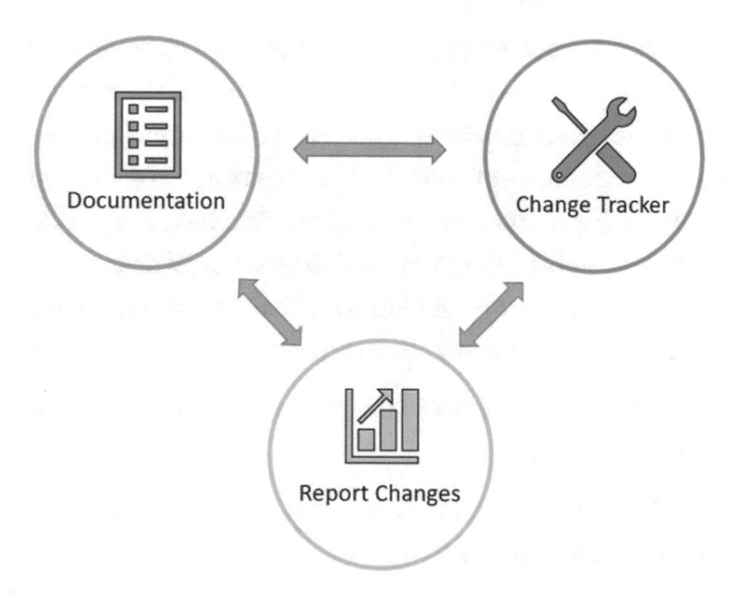

Figure 5-26. *Synchronized documentation, change tracker, and report*

Whole-Life Cost and Optimization Steps

One issue with the Agile approach is the constant changes. With every change, there is a small amount of leftover code or logic. It could be a table or a few outdated measures that are no longer required. Optimization is required to keep the report simple and reduce overall maintenance effort.

It is difficult to determine the frequency and timing for this optimization step, but the initial optimization should happen before deployment into production.

Workspace Structure

Power BI reports are contained inside workspaces. Each workspace can have many reports. There can be many workspaces in the same organization. Each workspace has a different level of access and control. For example, the finance analytics team has a workspace separate from the sales analytics team because the two teams operate independently. There could also be two workspaces for the compliance and the manufacturing team because the two teams have different access requirements.

If the workspaces are not controlled, it can result in every report developer having his or her own workspace. This can result in chaos with conflicts between reports and increased data security risks.

Three levels of structure can help control the workspaces: global structure, regional structure, and local structure, as shown in Figure 5-27.

- **Local structure** refers to the workspaces controlled by a specific team.
- **Global structure** refers to all workspaces across the entire organization.
- **Regional structure** refers to the workspaces across many teams inside a large business unit.

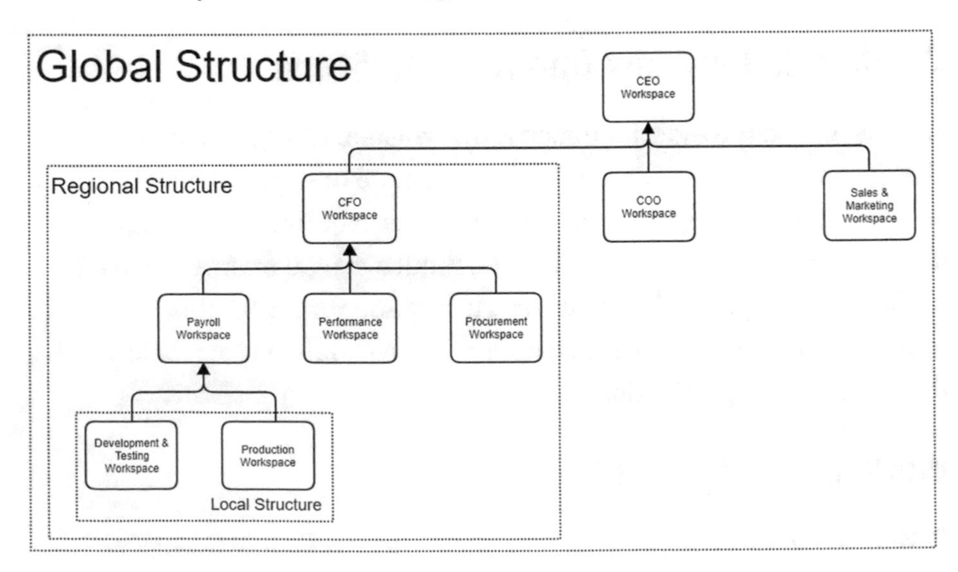

Figure 5-27. *Workspace structure*

A local structure should contain three workspaces, development, test, and production. The development and testing workspaces are often combined into a single workspace. This means the local structure should contain at least two workspaces. The following describes the benefits of separation.

- It keeps the production workspace to a minimal number of reports.
- Usage tracking can be restricted to production workspaces.
- It allows stronger access control for the production workspaces.

A regional structure requires the collaboration of multiple teams. The key characteristic of the regional and global structures is that it follows a top-down approach. This approach ensures that leaders receive consistent numbers in a controlled environment.

Tip Dataflows contained in local structures can be used in other workspaces. This can greatly reduce data duplication across multiple workspaces and contribute to establishing a single source of truth.

The workspace structure often overlaps with the navigation method discussed earlier, but note that the navigation structure can all reside in a single workspace. While two structures may look similar in a diagram, but is also independent.

Support Framework

Hopefully, you have understood the standardization opportunities when adopting Power BI into your organization. Now, let's move on to another important and often forgotten area about supporting the developers.

Many organizations do not have a dedicated in-house support team for self-service report developers. Technology teams often treat self-service report tools as glorified Excel. Leaders assume they can hire skilled developers who bridge knowledge gaps through Google, YouTube, and training.

Support is critical to ensure data security, user experience, and efficiency.

Support Levels

The developers require three channels of support to achieve their development objectives.

- Level 1: Online support such as documentation, Power BI community, blogs, and training
- Level 2: In-house support by Power BI admins and center of excellent teams
- Level 3: External support by Microsoft and partners

All organizations understand the importance of Level 1 support for the daily operation of reporting teams. Organizations that only deploy Level 1 support typically face the following challenges.

- Slow improvements for developers: the developers may be stuck at a level for a long time because they don't know what they don't know.

- Inefficient solutions: the developers only use their limited knowledge to solve business problems. This can lead to a lot of efficiency issues.
- Inconsistent approach: different developer adopts different approaches to solve the same business problems. This can lead to a chaotic environment with strong internal competition and confusing users.

Level 2 support requires an internal Power BI center of excellence (CoE) team to support and mentor the rest of the business. While supporting the business, the CoE team is tasked to drive compliance with the governance framework.

Level 2 support is typically consistent with three approaches.

- **Internal forum**: Internal developers can post questions and issues to the CoE team and the rest of the developer community for assistance. The internal forum is better than external forums because it prevents information leakage and allows developers to share problems and solve them in a secure environment.
- **Internal training**: CoE can run regular training to provide practical solutions to common problems. Because of the proximity to the business challenges, the CoE team can design training to suit different levels of internal users.
- **Internal consulting**: CoE can also be asked to provide consulting work to internal business units to help design and implement more complicated solutions.

- **Power BI Admin and Gateway**: CoE can use its administrator privilege to assist the internal developer community, such as running usage data reports, automating routine tasks, and controlling data gateways.

Level 3 support is required when CoE from Level 2 may need help for various reasons. There is a direct support channel for large organizations with Microsoft enterprise agreements. These support resources should be utilized to the maximum benefit of the organization. At other times, external consultants may also be hired ad hoc to meet different challenges.

Data Gateway

Helping reporting teams to set up and maintain a *data gateway* is one of the most direct methods to add value by the CoE/support team. This session is covered in more detail in Chapter 2. Gateway is a critical part of the data refresh automation process, as shown in Figure 5-28.

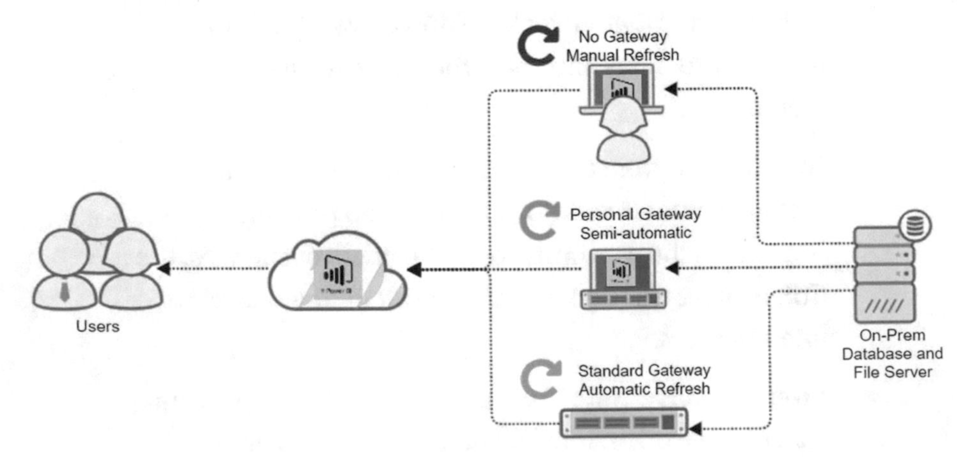

Figure 5-28. *Report gateway and data refresh automation*

Reports developed on the Power BI gateway are published to Power BI services in the cloud. Gateway is a mechanism to allow the cloud service to connect to the data sources that the report depends on. Once the connection is properly configured, the report can be refreshed according to refresh schedules.

There are two types of gateways, personal gateway, and standard gateway. They both allow Power BI services in the cloud to connect to the data sources.

- A **personal gateway** is typically installed on a developer's laptop and relies on their credentials to connect to the Power BI cloud service and the data sources.
- A **standard gateway** is typically installed on a server and relies on functional accounts to connect to Power BI cloud service and data sources.

The differences between standard and personal gateways are shown in Figure 5-29.

	On-premises data gateway (standard mode)	On-premises data gateway (personal mode)
Cloud services it works with	Power BI, Power Apps, Azure Logic Apps, Power Automate, Azure Analysis Services, Dataflows	Power BI
Serves multiple users with access control per data source	Included	
Runs as an app for users who aren't administrators		Included
Runs as a single user with your credentials		Included
Import data and set up scheduled refresh	Included	Included
DirectQuery support	Included	
Support for a live connection to Analysis Services	Included	

Figure 5-29. *Personal and standard gateway comparison*

A *functional account* is a system account with special privileges. It is often used with gateways to make the auto-refresh process more stable. This is because the functional account is not linked to any users, so not impacted by people's movements. Also, a password change is much less frequent than normal user password policies.

Once the gateway and functional account is configured properly, you can schedule the refresh as follows (see Figure 5-30).

1. Go to the workspace, find the dataset, and choose the settings.
2. Select the gateway (personal or standard).
3. Turn on **Scheduled refresh**. Select the frequency and time.

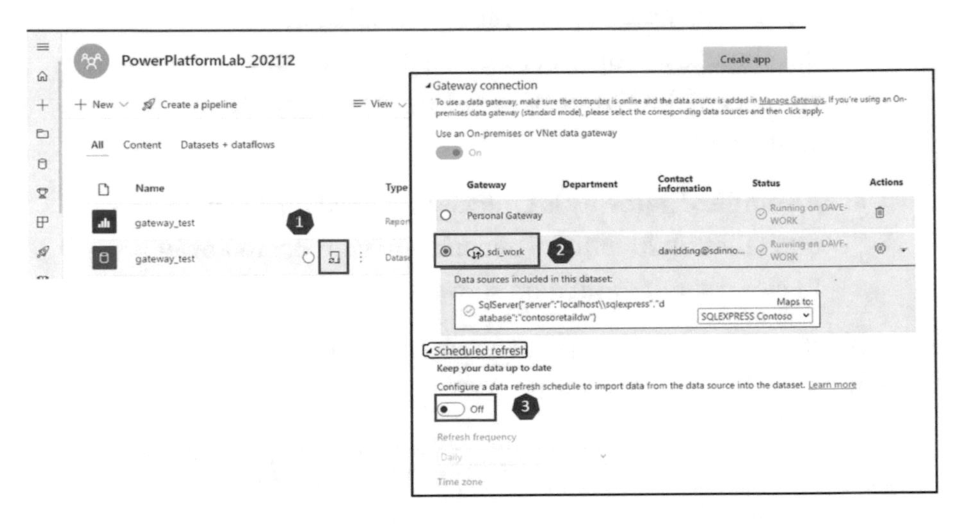

***Figure 5-30.** Schedule a data refresh with the gateway*

Once configured, the dataset is regularly refreshed automatically. The gateway has a setting to send a notification email when the data fails to refresh.

Data Security

Power BI Data security is a complex topic involving functionalities and processes. A secure report requires many layers of controls.

Access Points Analysis

Figure 5-31 shows 15 critical data access control points. As you can imagine, enforcing data security policies in a self-service environment can be challenging.

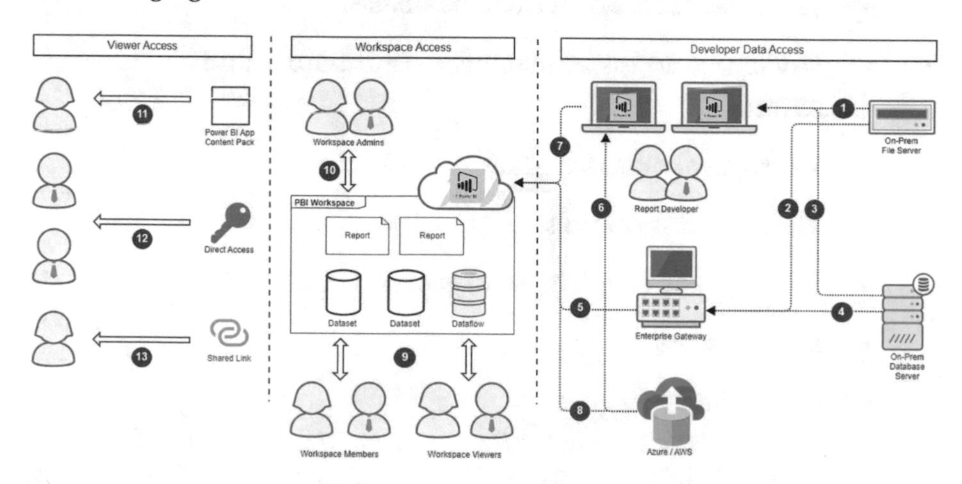

Figure 5-31. Power BI data access control points

The control points can be separated into standard controls, gateway controls, workspace controls, and viewer access controls.

The following are standard controls.

- Developer access to on-premises file server
- Developer access to on-premises database server
- Developer access to a cloud database

The following are gateway controls (if applicable).

- Gateway functional login access to on-premises file server
- Gateway functional login access to on-premises database server

The following are workspace controls.

- Workspace access to the enterprise gateway
- Workspace access to developer credentials
- Workspace access to a cloud database
- Workspace access for members, contributors, and so forth
- Workspace access for admins
- Viewer access controls
- Power BI app (Content pack) access
- Direct Access
- Shared link

Data Source Connection Process

As you can imagine, the multiple options at each area can result in multiple paths from the end user to the same data source. Restricting a single path for any given data source is a good idea. This requires clear rules, user training, and regular monitoring. This is in addition to the data connection standards discussed earlier in this chapter.

For example, many Excel/CSV flat data tables must be loaded into Power BI.

- **Rule**: All flat files (Excel, CSV, text) must be migrated into SharePoint sites before you can connect to it via Power BI SharePoint connector.
- **Training**: Develop online and regular lunch and learn sessions to communicate the rule and its implementation to the developers.

- **Monitoring**: Use the Get Datasource API or manual method to check the file path for all Excel files.

User Access Management

User Access Management (UAM) and the User Access Review (UAR) processes are hidden tasks in the self-service reporting package that no one likes to discuss.

UAM and UAR are used to answer the question at any given time about who has access to what report and data. This is one of the processes that can't be fully automated. The workspace access and report access registers are at the core of the processes. The following lists the key fields for each register.

Workspace Access Register

- Workspace name and GUID
- Username and Salary ID
- User access type (admin, member, contributor, viewer)
- Access start and end date
- Access the approver name and date

Report Access Register

- The report name and GUID
- Username and salary ID or group email (e.g., a distribution list)
- User access type (shared link, direct access, content pack app)

- Access start and end date
- Access the approver name and date

The activity event command described in the earlier chapter is also a good way to determine sharing activities. This can provide some automation toward new sharing events.

The UAR process is normally conducted every three to six months. This requires the approver to re-approve existing user access. People's movement in HR data can also trigger the UAR process.

Conclusion

This chapter contains less technical content but more process and framework. Hopefully, you have gained more insights about managing a Power BI development team, the importance of governance, and the many different areas of process improvement. When you decide to push any governance-related changes across multiple teams and developers, keep in mind working with people is always harder than working with technology. Success in the governance program requires the leader to be patient, constantly quantify all benefits, and build a strong relationship with people.

CHAPTER 6

SQL for Power BI

SQL Server was launched in 1988. A few decades later, it is still the most common enterprise data storage for structured data. There are many variants of SQL Server, including Microsoft SQL (T-SQL), MySQL, MongoDB, Teradata, PostgreSQL, and AWS Redshift. From an analyst's perspective, they are mostly the same. If you understand one, you understand the rest. In this chapter, Microsoft Azure SQL Server is used for demonstration.

For the following reasons, knowing SQL (Structured Query Language) is important for Power BI developers.

- SQL is in every organization.
- SQL is efficient with large datasets with millions of records.
- SQL pushes a lot of data processing to the server side.
- SQL is excellent for automation.

This chapter touches on the most common knowledge of SQL that is helpful to a Power BI developer. It focuses on the following topics.

- Setting up a SQL Server environment for practice
- Loading SQL queries into Power BI
- Basic SQL queries

Many Excel users believe SQL require advanced coding skill. The beauty of SQL lies in its simplicity.

D. Ding, *Transitioning to Microsoft Power Platform*,
https://doi.org/10.1007/978-1-4842-9239-6_6

Set up a SQL Server for Practice

There are two ways to set up the practice server: install SQL Express on your local machine or an Azure SQL Server. For simplicity, you learn how to set up a free Azure SQL and use Online Query Editor to write queries.

Set up a Free Azure Account

If you google *Azure SQL*, it should take you to the page shown in Figure 6-1.

Figure 6-1. *Azure SQL website*

You can follow the steps to set up a free trial account for 12 months, including some free credits. You can set it up with a personal email address, but you are asked to provide a valid credit card.

Set up a Sample SQL Database

Once your account has been successfully created. You can create a new SQL database from the create button in Figure 6-2.

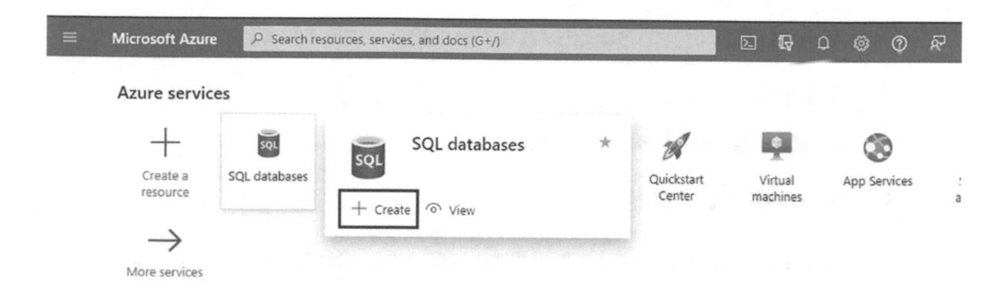

Figure 6-2. *Create a new Azure SQL database*

You can spin up a basic Azure SQL database for practice by going through the following steps.

Create a resource group using the following steps, as shown in Figure 6-3.

1. Create a resource group and create a name.

Create SQL Database ...
Microsoft

Basics Networking Security Additional settings Tags Review + create

Create a SQL database with your preferred configurations. Complete the Basics tab then go to Review + Create to provision with smart defaults, or visit each tab to customize. Learn more

Project details

Select the subscription to manage deployed resources and costs. Use resource g
manage all your resources.

Subscription * Azure subscription 1

Resource group * (New) SQL_Practice

1 Create new

A resource group is a container that holds related resources for an Azure solution.

Name *
SQL_Practice

OK Cancel

Figure 6-3. *Create a new resource group*

Create a server to host the database. The following steps explain how to do it, as shown in Figure 6-4.

2. Select Create New Server to create a server.

3. Name the server.

4. Select the region/location where the server is in.
5. Select Use SQL authentication and choose a username and password.
6. Name the database.
7. Select the Development workload environment.
8. Check the price.

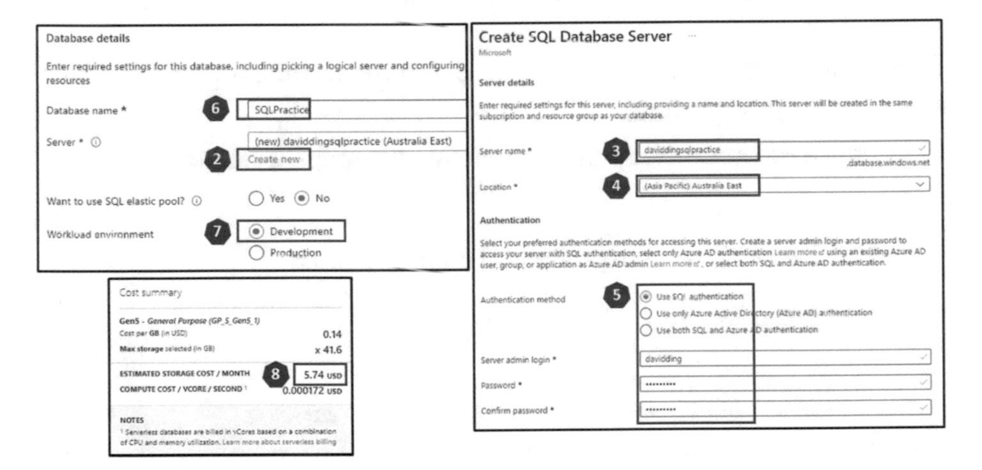

Figure 6-4. *Create a server to host the database*

Note The location is the location for the server. The nearest region provides better speed, but some regions are cheaper than others.

Configure network settings as shown in Figure 6-5.

9. Add current IP to Firewall rules by choosing Yes.

Basics Networking Security Additional settings Tags Review + create

Firewall rules

The settings displayed below are read-only. They can be modified from the "Firewalls and virtual networks" blade for the selected server after database creation. Learn more

Allow Azure services and resources to access this server No Yes

9 Add current client IP address * No Yes

Figure 6-5. *Add the current IP*

Choose the Sample database, as shown in Figure 6-6.

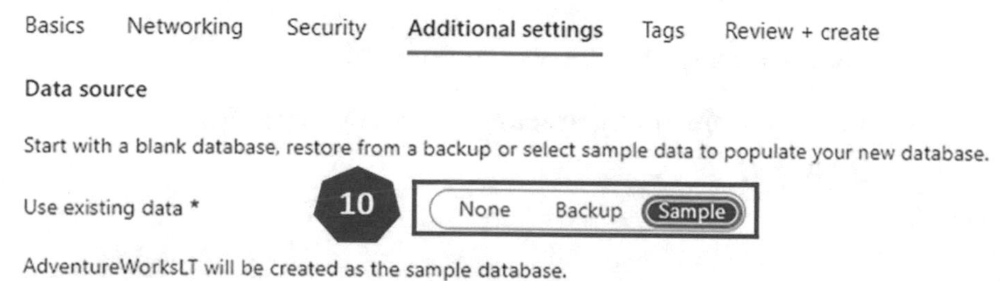

Figure 6-6. *Use existing sample database*

10. Review and create.

After a few minutes, the Azure SQL database server is ready with a sample Microsoft AdventureWorks data for you to consume.

Note Some parts of the process may have changed slightly but look for options with similar names.

Log in to Azure Query Editor

After successfully creating the Azure database, you need a query editor to connect to and query the database. There are many excellent tools to query databases, such as Visual Studio Code and Azure Data Studio. You can download any of these tools and establish the connection to the Azure SQL database you created earlier.

Obtain the connection strings to connect to the database. The following steps explain how to do it, as shown in Figure 6-7.

1. Select the newly created database from the Azure portal.
2. Choose **Connection strings** under Settings.
3. To simplify the query process, choose **Query editor (preview)**.

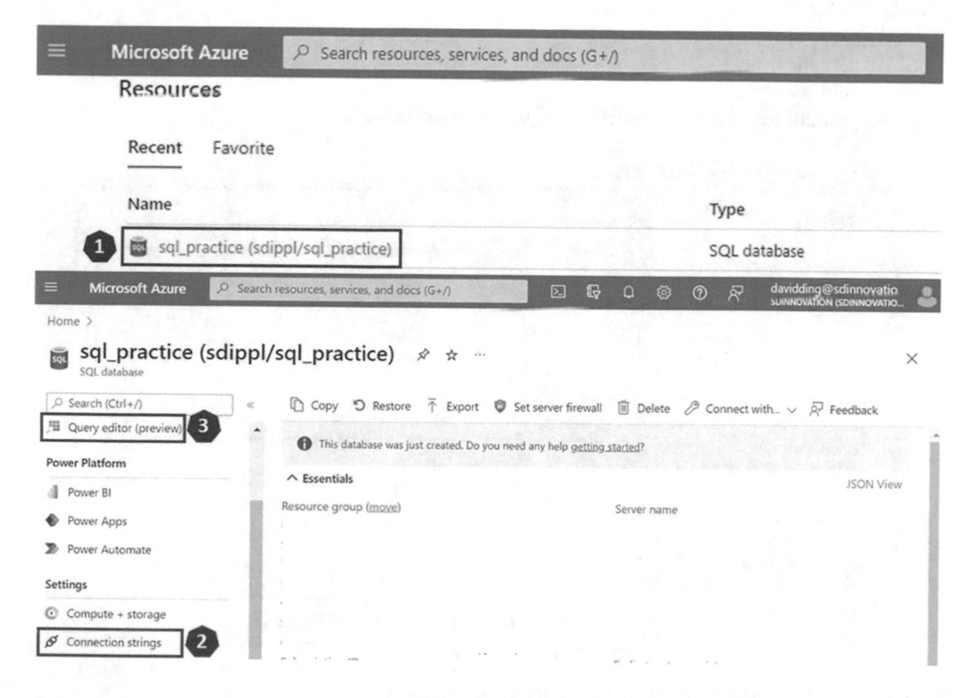

Figure 6-7. *Steps to connect to the database*

The query editor utilizes the cloud environment to generate an application that allows you to query the Azure database directly. Figure 6-8 shows that the query editor contains three functional areas.

1. The database navigation pane
2. A query editor
3. The result or message display area

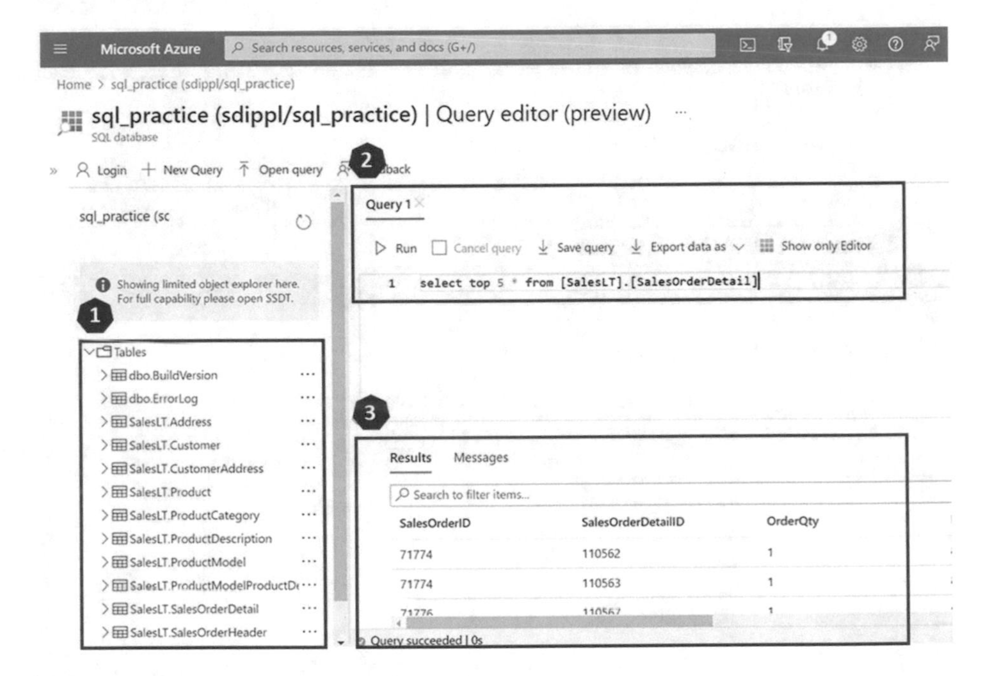

Figure 6-8. *Query Editor functional areas*

Load SQL Query in Power BI

With the Azure SQL database ready, you can configure a database connection in Power BI. There are two main methods for connecting to the SQL database: DirectQuery and Import.

The Import method allows Power BI to import data from a database into Power BI, and all the steps following the import are done at the local dataset. DirectQuery, in contrast, is a direct connection between Power BI and the database. There are three key practical differences between direct and import methods.

- DirectQuery does not load any data into Power BI.
- DirectQuery reflects the latest changes in the source tables, while the import method only updates when the Power BI dataset is refreshed.

- DirectQuery contains many limitations in data transformation, modeling, and measures.

To utilize the full functionalities in Power BI, your default choice should be the import method. The following steps connect to the Azure SQL database, as shown in Figure 6-9.

1. From Get data choose **More options**.
2. Choose **Azure SQL database**.

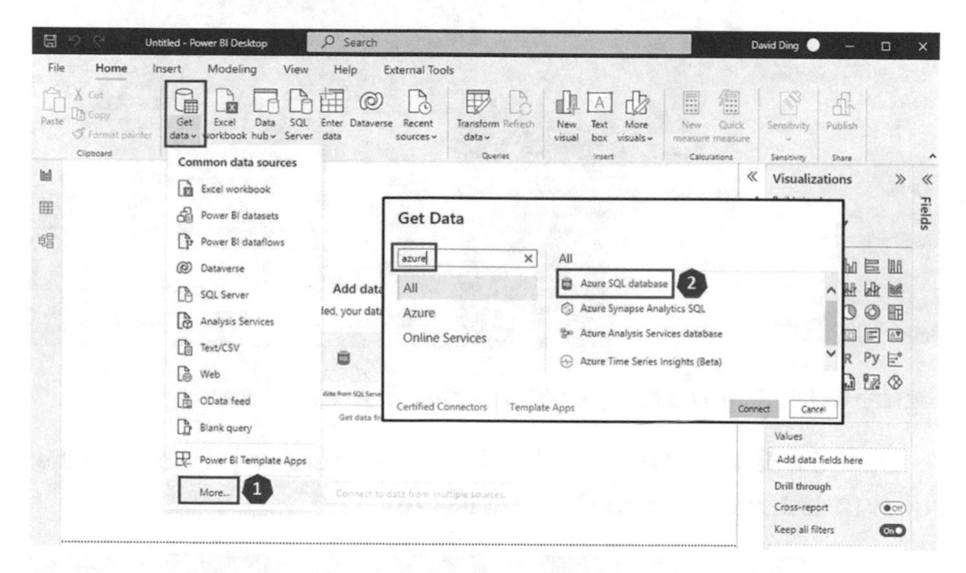

Figure 6-9. *Connect to Azure SQL database*

You can find the server and database detail from the connection strings session in the Azure portal under the selected database. The following steps finalize the connection, as illustrated in Figure 6-10.

1. Copy and paste the server name into Power BI.
2. Copy and paste the database name into Power BI.
3. Select Import.

4. Expand Advanced settings and enter this query:

   select top 5 * from SalesLT.SalesOrderDetail.

5. Select Database and enter database credentials.

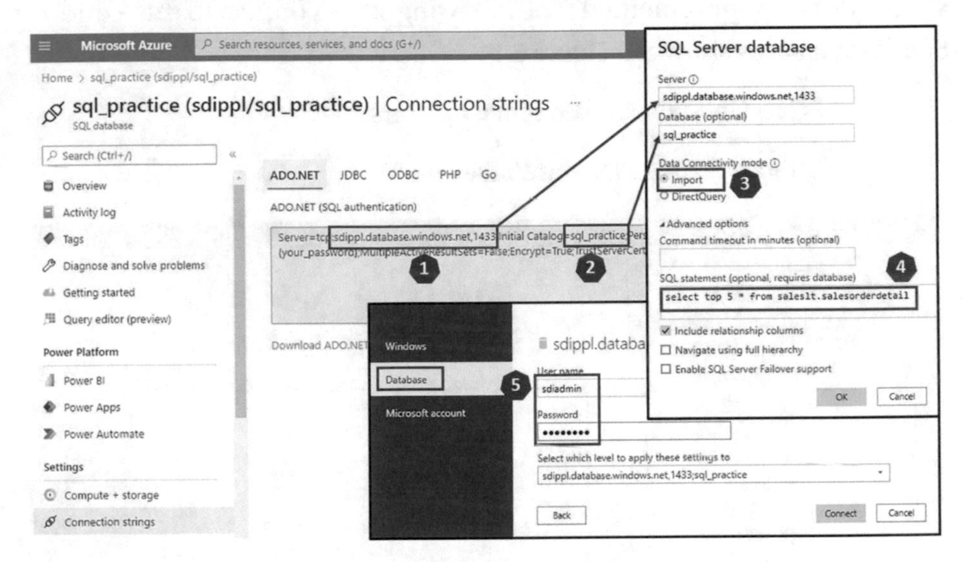

***Figure 6-10.** Connect Power BI to Azure SQL database*

Once the data has been successfully loaded, the query appears on the Power BI desktop screen, as shown in Figure 6-11.

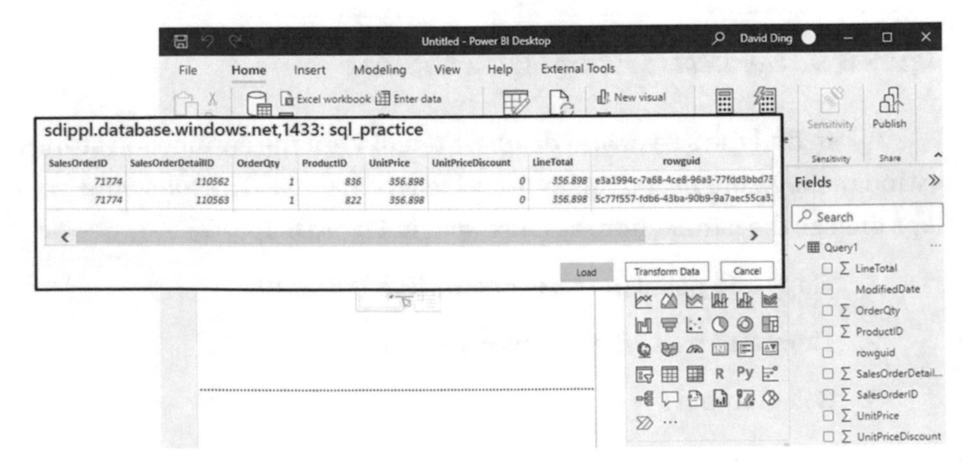

SalesOrderID	SalesOrderDetailID	OrderQty	ProductID	UnitPrice	UnitPriceDiscount	LineTotal	rowguid
71774	110562	1	836	356.898	0	356.898	e3a1994c-7a68-4ce8-96a3-77fdd3bbd73
71774	110563	1	822	356.898	0	356.898	5c77f557-fdb6-43ba-90b9-9a7aec55ca3

***Figure 6-11.** Successful connection*

SQL Query Basics

The following is one of the most common queries in most workplaces. The query returns the offline product sales in June 2008.

```
SELECT od.SalesOrderID
  ,od.ProductID
  ,pd.Name as ProductName
  ,od.OrderQty
  ,od.OrderQty
  ,od.UnitPrice
  ,od.LineTotal
  ,oh.CustomerID
  ,oh.ShipMethod
  ,pd.StandardCost

FROM [SalesLT].[SalesOrderDetail] as od
  left join [SalesLT].[SalesOrderHeader] as oh
    on od.SalesOrderID = oh.SalesOrderID
  left join [SalesLT].[Product] as pd
    on od.ProductID = pd.ProductID

WHERE oh.OrderDate >= '2008-06-01'
  and oh.OrderDate < '2008-07-01'
  and oh.OnlineOrderFlag = 'False'
```

This query consists of three parts.

- Select columns (SELECT)
- Tables Joints (FROM)
- Filter logic (WHERE)

Query Construction Process

The following summarizes the process of constructing the query following a different path.

1. Identify the core table (SalesOrderDetail) that meets the requirement.
2. Review the columns in the core table.
3. Search for additional tables.
4. Apply filter logic.

The elements of the SQL query are explained later. This session aims to understand the process and logic before diving into the syntax.

Identify the Core Table(s)

The requirements for each query are normally derived from the reporting requirements. For example, the business asks for a sales revenue report at the product level. This means you need a sales dataset that contains product-level detail. You already know that the Sales Order Detail table meets the requirements because it contains everything you need.

Review the Table Columns

You can view the columns in the table using the query in Figure 6-12. The sales order date and the product description are missing from this dataset. You must use the SalesOrderID and the ProductID keys to link the other tables.

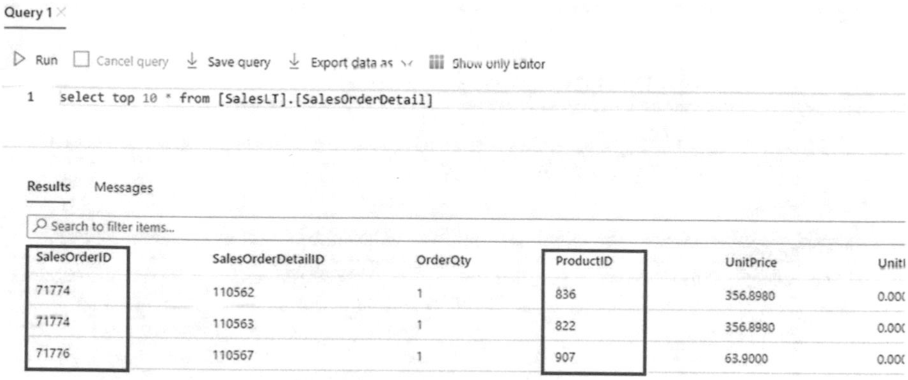

Query 1

Run Cancel query Save query Export data as Show only Editor

```
1 select top 10 * from [SalesLT].[SalesOrderDetail]
```

Results Messages

Search to filter items...

SalesOrderID	SalesOrderDetailID	OrderQty	ProductID	UnitPrice	Unitl
71774	110562	1	836	356.8980	0.00(
71774	110563	1	822	356.8980	0.00(
71776	110567	1	907	63.9000	0.00(

***Figure 6-12.** Columns sales order*

Search for Additional Tables

This sample database only contains a few tables, and it is easy to identify the additional tables by looking at the table description and column names inside the tables. The database can contain hundreds of fields in the real world, making it much harder to find relevant tables.

Information_Schema allows you to search for table and column names. The two most useful tables in information schema are *columns* and *tables*. Because you already have the column names (ProductID and SalesOrderID), you can search the tables that contain the relevant fields, as shown in Figure 6-13.

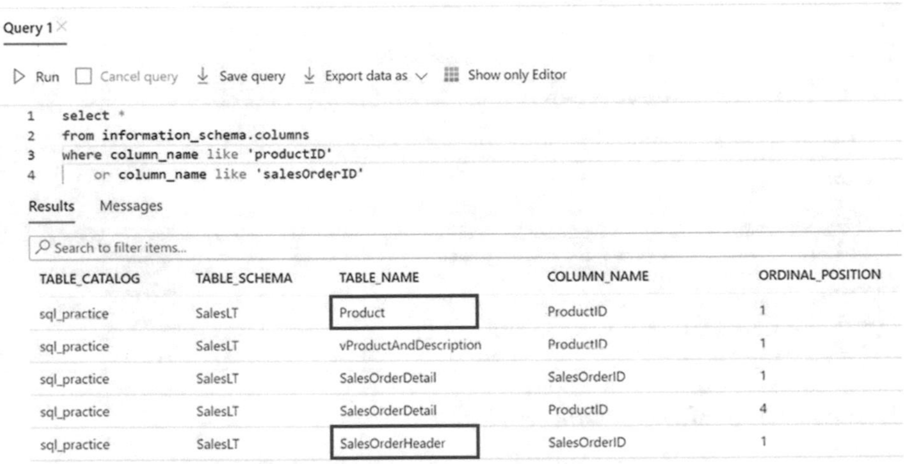

Query 1

Run Cancel query Save query Export data as Show only Editor

```
select *
from information_schema.columns
where column_name like 'productID'
    or column_name like 'salesOrderID'
```

Results Messages

Search to filter items...

TABLE_CATALOG	TABLE_SCHEMA	TABLE_NAME	COLUMN_NAME	ORDINAL_POSITION
sql_practice	SalesLT	Product	ProductID	1
sql_practice	SalesLT	vProductAndDescription	ProductID	1
sql_practice	SalesLT	SalesOrderDetail	SalesOrderID	1
sql_practice	SalesLT	SalesOrderDetail	ProductID	4
sql_practice	SalesLT	SalesOrderHeader	SalesOrderID	1

Figure 6-13. *Search for tables using information schema*

Once the tables have been identified, follow the steps to select the relevant columns from the table, as shown in Figure 6-14.

1. Use **select top 5** to bring all columns from the table.
2. Choose the required columns.
3. Change the **select** query to only include the required columns.

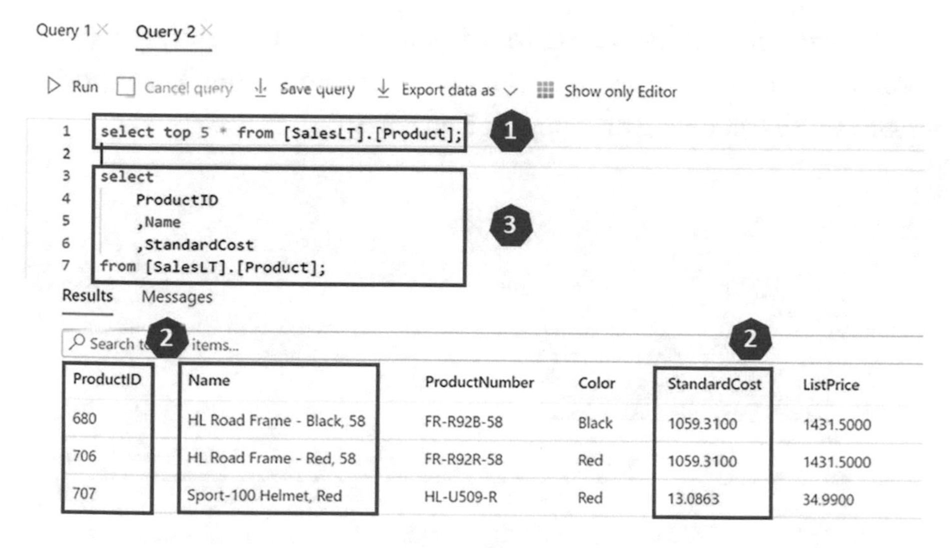

***Figure 6-14.** Query the product table*

You can repeat the exercise for the SalesOrderHeader table, as shown in Figure 6-15. Only the SalesOrderID and the OrderDate are required.

Query 1 Query 2 Query 3

Run | Cancel query | Save query | Export data as | Show only Editor

```
1  select top 5 * from [SalesLT].[SalesOrderHeader];
2
3  select
4      SalesOrderID
5      ,OrderDate
6  from [SalesLT].[SalesOrderHeader]
```

Results Messages

Search to filter items...

SalesOrderID	RevisionNumber	OrderDate	DueDate	ShipDate
71774	2	2008-06-01T00:00:00.0000000	2008-06-13T00:00:00.0000000	2008-06-08T00:(
71776	2	2008-06-01T00:00:00.0000000	2008-06-13T00:00:00.0000000	2008-06-08T00:(
71780	2	2008-06-01T00:00:00.0000000	2008-06-13T00:00:00.0000000	2008-06-08T00:(

***Figure 6-15.** Query the SalesOrderHeader table*

Once the new tables are confirmed, you can create a new query to combine the additional columns (Product Name and OrderDate) into a single table by following the query in Figure 6-16.

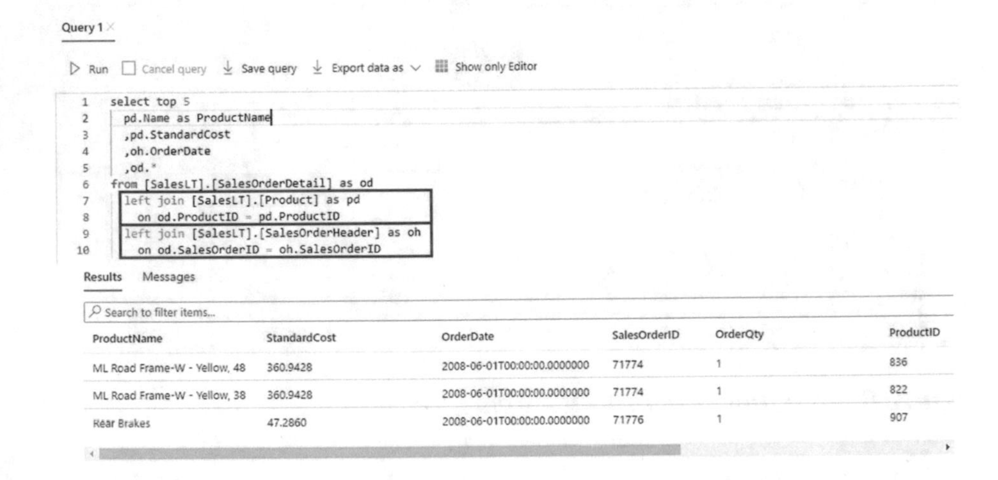

***Figure 6-16.** Combined table*

Apply Filter Logic

The last step in the query is to apply some filters for only June of 2008, as shown in Figure 6-17.

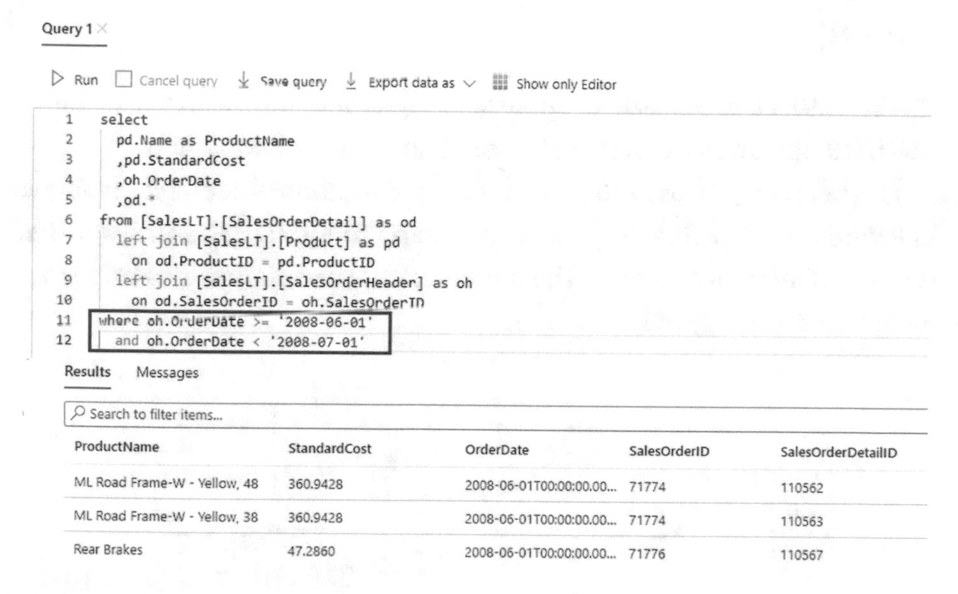

Figure 6-17. *Filtered table*

SQL Query Syntax

The sample database contains many tables. When used within Power BI, SQL query is used to access data in the database. The database contains tables, and the tables contain fields. Because Power BI also contains multiple ways to implement data logic, it allows developers to keep the SQL statement simple.

SELECT

The most basic SQL query is the SELECT statement. It asks the database to return the fields in a table. FROM always accompanies the SELECT statement. You may see the * following the SELECT statement. The * means all fields. The following is a simple example.

```
SELECT * FROM [SalesLT].[SalesOrderDetail]
```

WHERE

The WHERE clause is used to apply filter logic. You may need to use the AND/OR operators when there is more than one filter condition.

Sometimes you also need to use the ( ) to separate filter logic, as shown in Figure 6-18. The filter logic is to only keep ProductID 836 and 822, where the OrderQty is more than 1. The incorrect implementation also brings a row of data with Qty of 1.

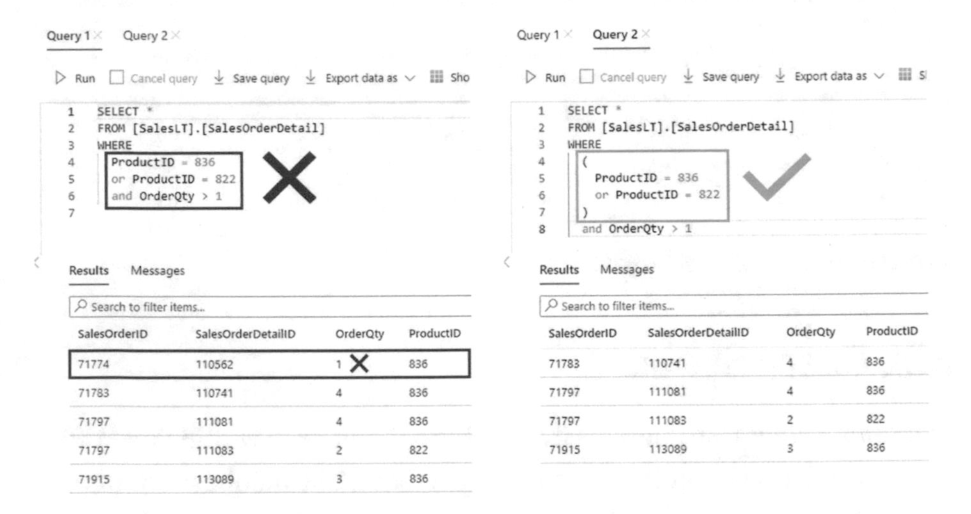

Figure 6-18. *WHERE operators with and without brackets*

TOP and ORDER BY

The TOP clause is a handy add-on to the Select statement. It helps to limit the data being brought into the query output. For large tables, this can save a lot of execution time, making it useful during development and investigations.

You can use the ORDER BY clause to determine which top N rows to display. As shown in Figure 6-19, there is the ASC (ascending default) order or DESC (descending) order.

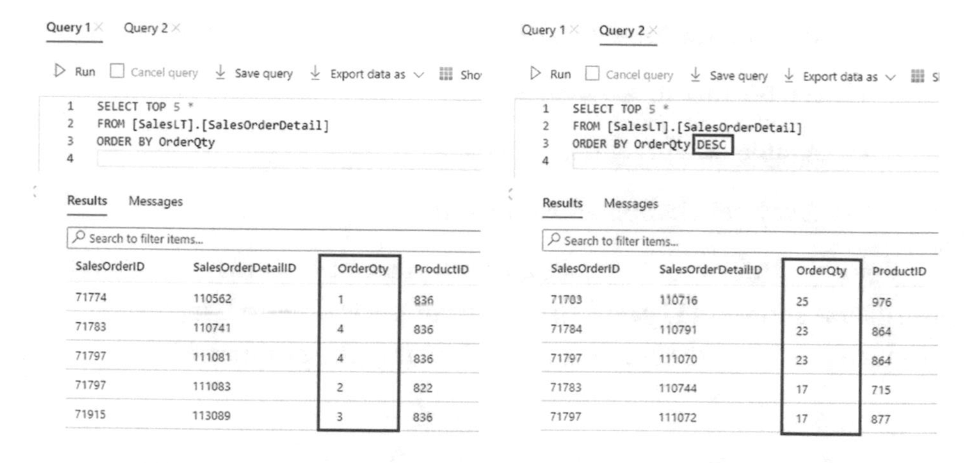

Figure 6-19. *Use Order By with TOP N in a different order*

JOIN

In the earlier chapter on Power Query, there is a session about table joints. Table join types in SQL work the same way as Power Query. Figure 6-20 shows that the yellow and blue circles represent the two joined tables. The area in red represents the output scope.

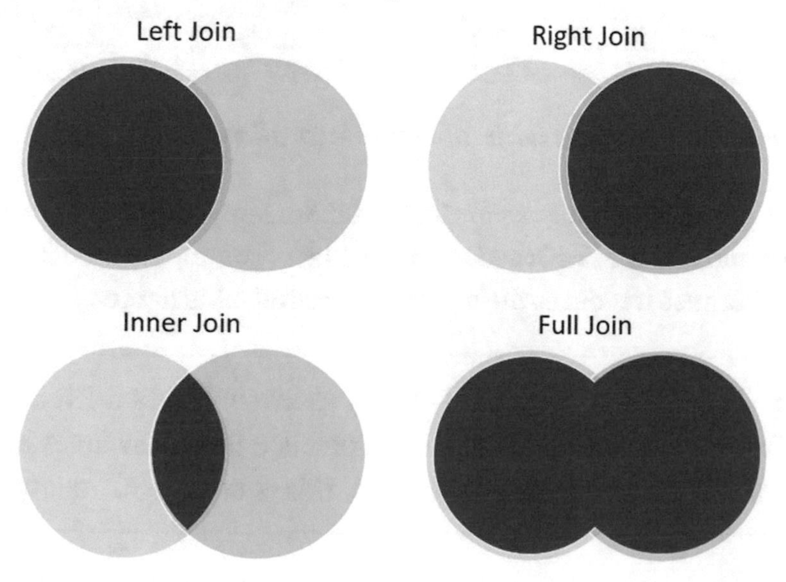

Figure 6-20. *SQL table joins*

The requirements for the SQL join require the following elements.

- A table on the left-hand side
- A table on the right-hand side
- A key field in both tables to join on

Once the three requirements are met, you can perform the required join clause shown in Figure 6-21. This example is doing a left join on ProductID to bring the Product Name field into the output table.

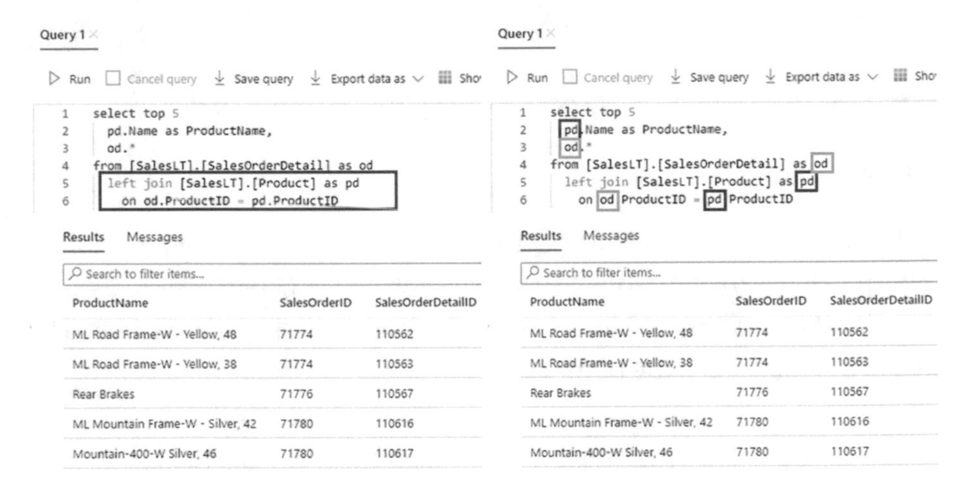

Figure 6-21. ***Left joins with table name as alias***

The example uses aliases to give the table a temporary name: *pd* for Product and *od* for SalesOrderDetail. An alias is commonly used in join clauses because it reduces the need to repeat full table names.

Note The table on the left is typically shown in Figure 6-21. It can be confusing, but you must try to practice this a few times to understand the left vs. right difference. This is core to SQL query.

You may have noticed in the example that the SELECT statement used pd.Name and od.* to bring in only the Name field in the Product table but all the fields from the SalesOrderDetail table.

As shown in earlier examples, the join clause can occur multiple times to join multiple tables.

One common mistake with table joins is the unintended duplication of records. The key on the right-hand table often causes this is not unique, as shown in Figure 6-22. The join key is the ProductModelID. There are six different records in the right-hand side table. This resulted in the same row in the left-hand side table being replicated six times.

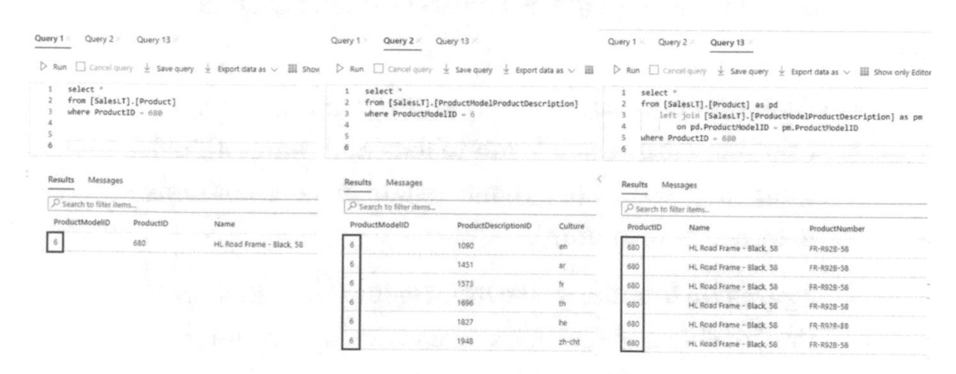

Figure 6-22. *Duplication of records when joining*

Note Duplication of records is not always bad. Sometimes you can use this behavior to replicate records to meet your data manipulation need.

SQL Functions

In the previous sessions, you have seen some common SQL queries for connecting, selecting, and filtering data in a database. Adding new columns is also a common need. This can be achieved using SQL functions.

Most SQL functions contain three components.

- **Input** (parameters) could be a column or a value.
- **Transformation** is the standardized steps performed using the input.
- Output is the output of the transformation.

For example, 1 + 3 can be generalized into an *ADD(1, 3)* function that takes two input parameters. In this case, 1 and 3 are the inputs; ADD() is a standardized step to perform 1 + 3 calculation; the function's output is 4.

The following describes the two most common SQL function categories.

- **Scalar functions** take input values from one or more columns within a single row of data and return a single value for each row (e.g., concatenate two text columns to form a new column).
- **Aggregate functions** take input values from multiple rows of data and return a single value (e.g., count the number of rows in a table).

The aggregation functions are discussed in more detail in the Group By session.

Data Types

Before getting into more functions, it is important to understand data types in SQL.

Each cell in the worksheet can have a different data type in Excel, which means the different cells in a column can have different data types. This is not the case for the table columns in SQL, where each column can only be a single data type.

The following describes some of the most common data types.

- **int** (integer) is a whole number without any decimal places.
- **decimal** is a fixed-digit decimal number.
- **varchar** is the variant length of characters/string/text.
- **date** is a date.
- **datetime** is a date and time.
- **timestamp** is similar to datetime but stored in UTC format (i.e., number of seconds since 1970-01-01 00:00:00).

Note There is no boolean data type in some SQL databases. Boolean values can be represented using bit (0/1) or varchar (True/False).

The data type of the column is defined when the table is created. It cannot be changed afterward. To check the column type on existing tables, consult the INFORMATION_SCHEMA shown in Figure 6-23.

Query 1

Run | Cancel query | Save query | Export data as

```
SELECT COLUMN_NAME, DATA_TYPE
FROM INFORMATION_SCHEMA.COLUMNS
WHERE TABLE_NAME = 'SalesOrderHeader'
```

Results | Messages

COLUMN_NAME	DATA_TYPE
SalesOrderID	int
RevisionNumber	tinyint
OrderDate	datetime
DueDate	datetime
ShipMethod	nvarchar
Status	tinyint
OnlineOrderFlag	bit
SalesOrderNumber	nvarchar

***Figure 6-23.** Check data type of an existing table*

In this example, the SalesOrderID column is *int* and SalesOrderNumber is *nvarchar*. When used in various SQL functions, this may result in data type errors. As shown in the example in Figure 6-24, mixing different data types (varchar and int) results in a conversion failure error.

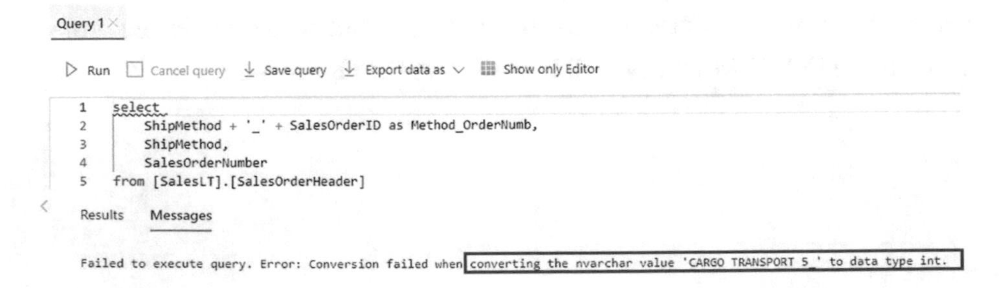

***Figure 6-24.** Conversion error when mixing different data types*

If you change to later to SalesOrderNumber, it displays the result successfully because both fields are of the same varchar data type. This is shown in Figure 6-25.

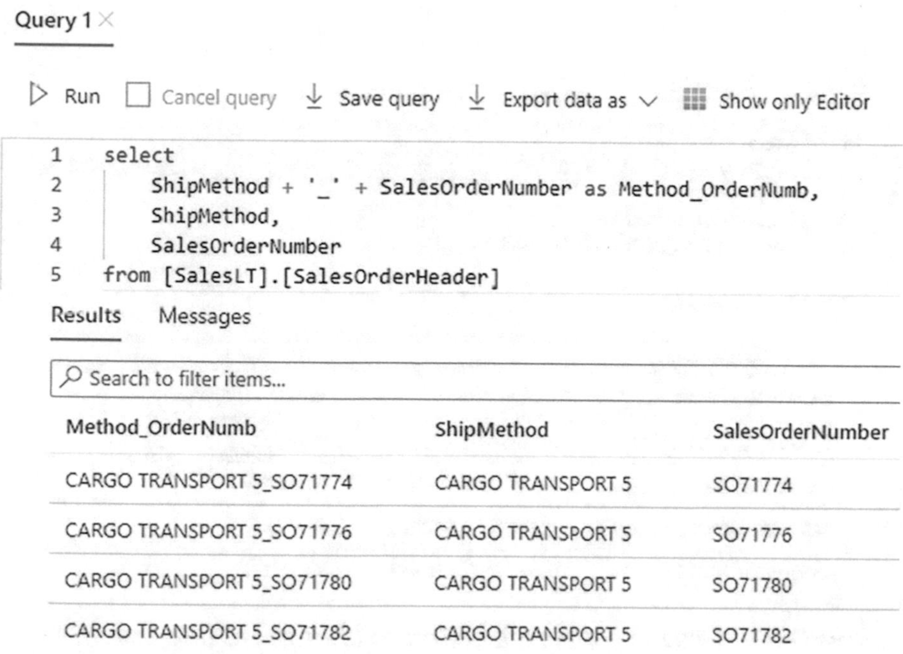

Query 1 ×

▷ Run ☐ Cancel query ↓ Save query ↓ Export data as ∨ Show only Editor

```
1   select
2       ShipMethod + '_' + SalesOrderNumber as Method_OrderNumb,
3       ShipMethod,
4       SalesOrderNumber
5   from [SalesLT].[SalesOrderHeader]
```

Results Messages

Search to filter items...

Method_OrderNumb	ShipMethod	SalesOrderNumber
CARGO TRANSPORT 5_SO71774	CARGO TRANSPORT 5	SO71774
CARGO TRANSPORT 5_SO71776	CARGO TRANSPORT 5	SO71776
CARGO TRANSPORT 5_SO71780	CARGO TRANSPORT 5	SO71780
CARGO TRANSPORT 5_SO71782	CARGO TRANSPORT 5	SO71782

Figure 6-25. *Combine fields of the same type*

The more common method of fixing the problem is to use a SQL function to convert the int data field into a varchar as shown in Figure 6-26.

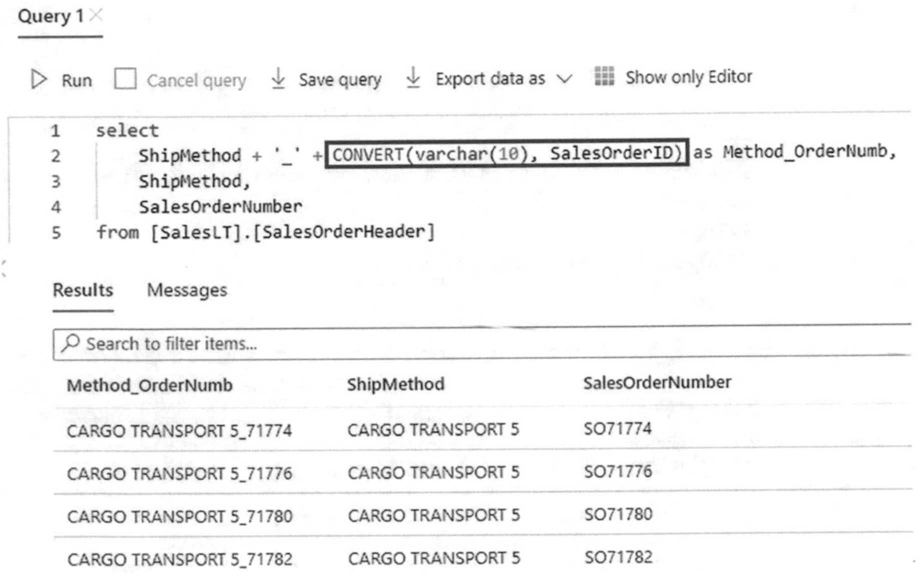

Query 1

Run | Cancel query | Save query | Export data as | Show only Editor

```
1  select
2      ShipMethod + '_' + CONVERT(varchar(10), SalesOrderID) as Method_OrderNumb,
3      ShipMethod,
4      SalesOrderNumber
5  from [SalesLT].[SalesOrderHeader]
```

Results | Messages

Search to filter items...

Method_OrderNumb	ShipMethod	SalesOrderNumber
CARGO TRANSPORT 5_71774	CARGO TRANSPORT 5	SO71774
CARGO TRANSPORT 5_71776	CARGO TRANSPORT 5	SO71776
CARGO TRANSPORT 5_71780	CARGO TRANSPORT 5	SO71780
CARGO TRANSPORT 5_71782	CARGO TRANSPORT 5	SO71782

***Figure 6-26.** Use the convert function to change int into varchar*

While it is easy to convert an integer into a varchar, it is often harder to convert a varchar into an integer. You may get the error message shown in Figure 6-27 because SQL cannot ignore the first two characters (SO) preceding the number.

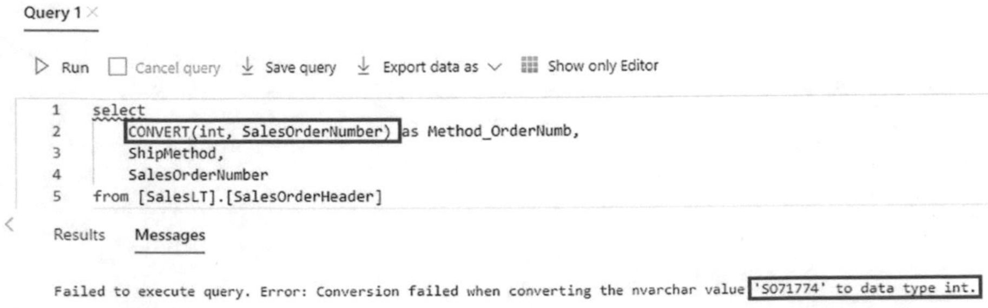

Query 1

Run | Cancel query | Save query | Export data as | Show only Editor

```
1  select
2      CONVERT(int, SalesOrderNumber) as Method_OrderNumb,
3      ShipMethod,
4      SalesOrderNumber
5  from [SalesLT].[SalesOrderHeader]
```

Results | Messages

```
Failed to execute query. Error: Conversion failed when converting the nvarchar value 'SO71774' to data type int.
```

***Figure 6-27.** Conversion error when converting varchar to integer*

Converting the date to a varchar is also difficult because of the different date format standards around the world. Figure 6-28 shows that you can use different codes to convert the date/datetime field into varchar.

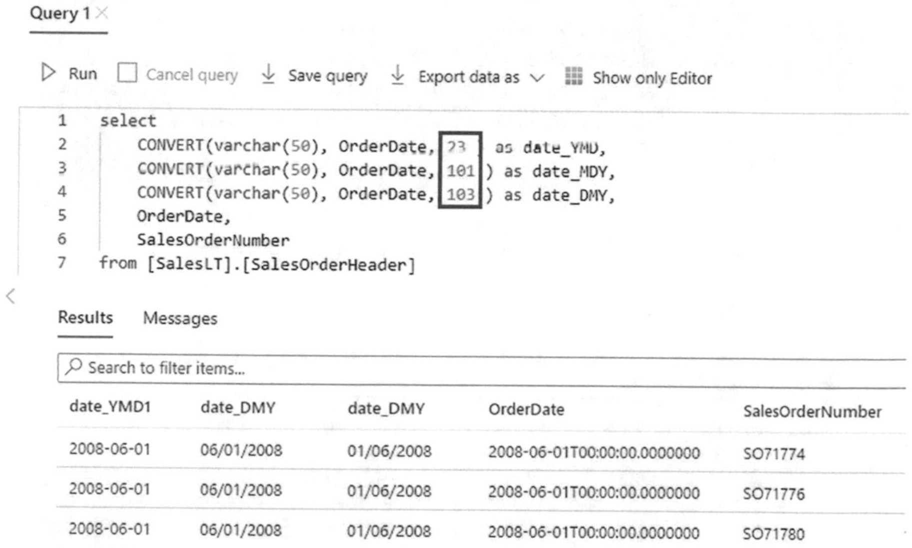

Query 1

Run | Cancel query | Save query | Export data as | Show only Editor

```
1  select
2      CONVERT(varchar(50), OrderDate, 23 ) as date_YMD,
3      CONVERT(varchar(50), OrderDate, 101) as date_MDY,
4      CONVERT(varchar(50), OrderDate, 103) as date_DMY,
5      OrderDate,
6      SalesOrderNumber
7  from [SalesLT].[SalesOrderHeader]
```

Results | Messages

Search to filter items...

date_YMD1	date_DMY	date_DMY	OrderDate	SalesOrderNumber
2008-06-01	06/01/2008	01/06/2008	2008-06-01T00:00:00.0000000	SO71774
2008-06-01	06/01/2008	01/06/2008	2008-06-01T00:00:00.0000000	SO71776
2008-06-01	06/01/2008	01/06/2008	2008-06-01T00:00:00.0000000	SO71780

Figure 6-28. *Example codes for different date formats*

You can get the full date format by googling *CAST and CONVERT T-SQL* and selecting the Microsoft documentation.

Common Scalar Functions

Figure 6-29 shows three new columns created in a SQL script.

1. Total_Amount (OrderQty multiplied by UnitPrice)
2. rowID (first four characters of the rowguid column)
3. ModDate (a date text in the format of YYYY-MM-DD)

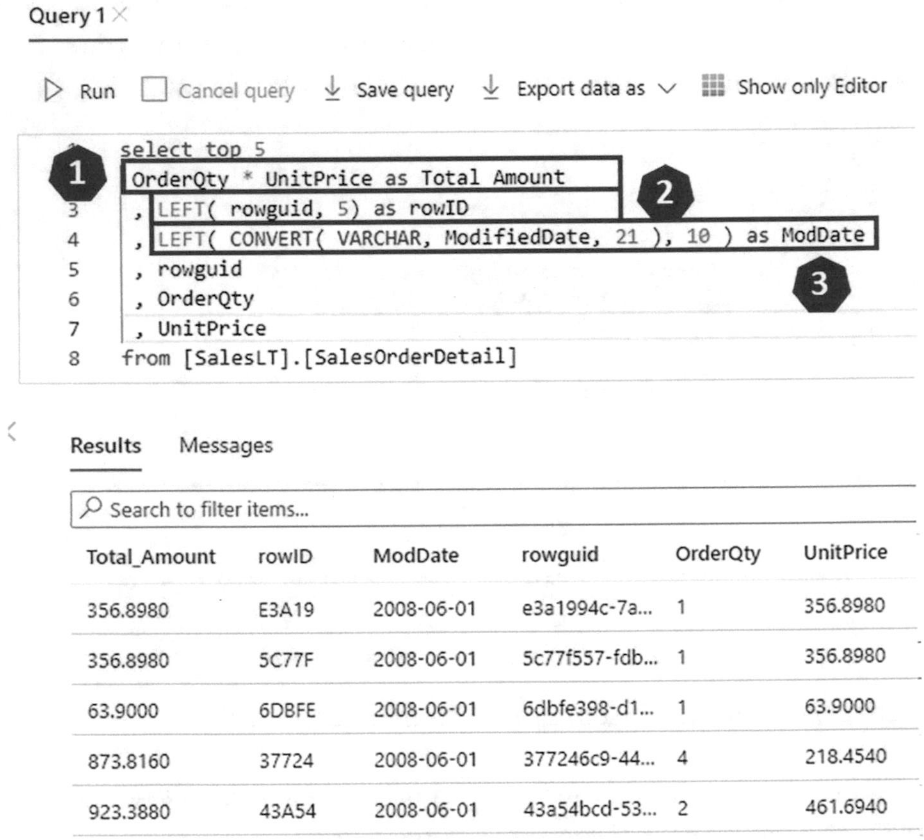

Query 1

Run | Cancel query | Save query | Export data as | Show only Editor

```
1  select top 5
2  OrderQty * UnitPrice as Total_Amount
3    , LEFT( rowguid, 5) as rowID
4    , LEFT( CONVERT( VARCHAR, ModifiedDate, 21 ), 10 ) as ModDate
5    , rowguid
6    , OrderQty
7    , UnitPrice
8  from [SalesLT].[SalesOrderDetail]
```

Results | Messages

Search to filter items...

Total_Amount	rowID	ModDate	rowguid	OrderQty	UnitPrice
356.8980	E3A19	2008-06-01	e3a1994c-7a...	1	356.8980
356.8980	5C77F	2008-06-01	5c77f557-fdb...	1	356.8980
63.9000	6DBFE	2008-06-01	6dbfe398-d1...	1	63.9000
873.8160	37724	2008-06-01	377246c9-44...	4	218.4540
923.3880	43A54	2008-06-01	43a54bcd-53...	2	461.6940

Figure 6-29. *Adding new columns in SQL*

In the preceding examples, you first call the functions, then use *AS* to provide the name of the new columns.

The first calculated column utilized the *multiplication* function. This function is used to multiply two values, in this case, OrderQty, and UnitPrice.

The second calculated column utilized the LEFT function. This function takes two parameters as input. The first parameter is a text column, and the second is the number of characters to keep. You used the rowguid as the column and the first five characters in the example.

The third calculated column combines the LEFT function with the **CONVERT** function. The CONVERT function's output becomes the LEFT function's input in this case. The CONVERT function converts the date into a text with a specific display format (YYYY-MM-DD hh:mm:ss.000). The LEFT function then takes the format and only keeps the year-month-date part of the text, which is the first ten characters (YYYY-MM-DD).

Group By and Aggregate Function

Group By is another common statement in SQL. The aggregate functions normally accompany it. The following session breaks down the Group By statements.

Aggregate Function

Before you start to explore the Group By statement, you first need to understand two simple but useful aggregate functions, count(), sum(), and average().Figure 6-30 shows the use of these aggregate functions to calculate the number of transactions, total revenue, and average order quantity of all the products.

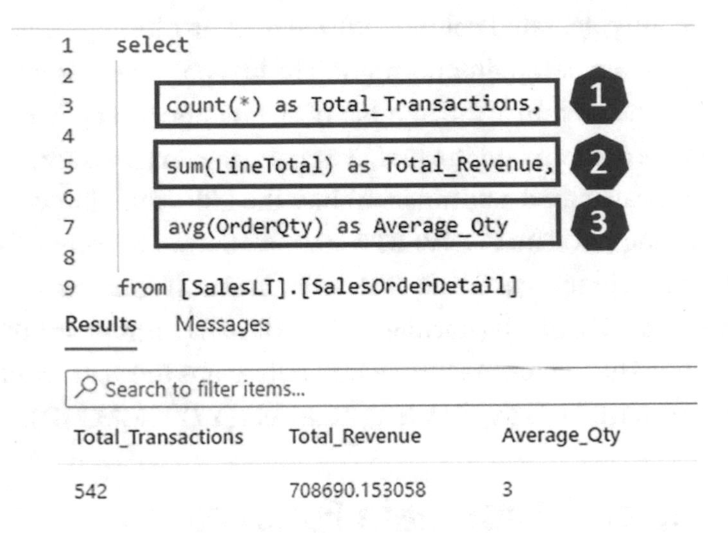
1 select
2
3 count(*) as Total_Transactions,
4
5 sum(LineTotal) as Total_Revenue,
6
7 avg(OrderQty) as Average_Qty
8
9 from [SalesLT].[SalesOrderDetail]
1
2
3
Results
Messages
Search to filter items...
Total_Transactions
Total_Revenue
Average_Qty
542
708690.153058
3

Figure 6-30. *Simple aggregate functions*

The resulting table contains a single row with three columns, summarizing each of the metrics for the entire SalesOrderDetail table.

Note count(*) returns the total number of rows in the SalesOrderDetail table. Each row in the table is interpreted as a transaction.

Group By

Group By allows the metrics to be further broken down by other columns, such as ProductID. This is the same as the Pivot table function in Excel. The following steps implement the Group By at the ProductID level, as shown in Figure 6-31.

1. Apply GROUP BY after the FROM statement.
2. Add ProductID in the SELECT statement.
3. Add the aggregation functions in the SELECT statement.

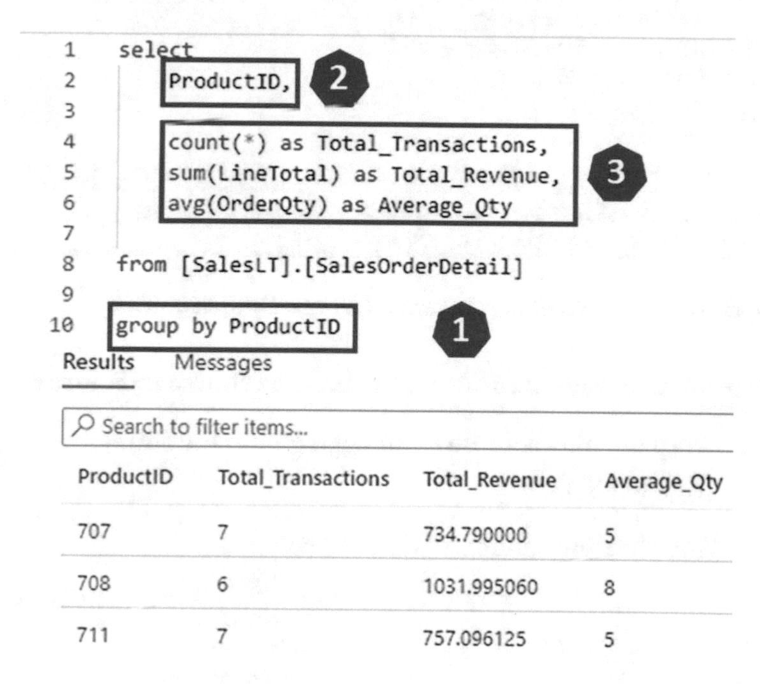

***Figure 6-31.** Simple Group By statement*

One of the most common errors when using Group By is shown in Figure 6-32. The error message was caused by a new column being included. Once Group By is used, all columns in the SELECT statement must fit one of the two scenarios.

- The column is listed in Group By.
- The column is used in an aggregation function.

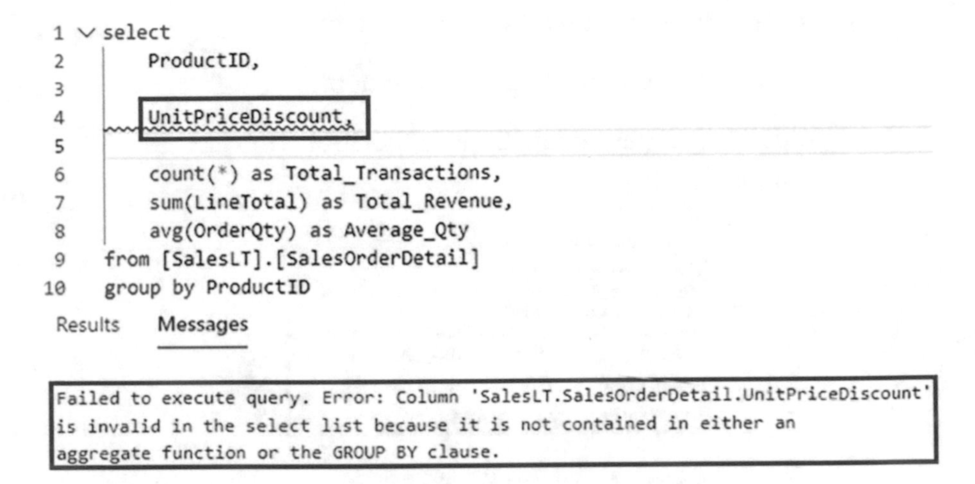

1 select
2 ProductID,
3
4 UnitPriceDiscount,
5
6 count(*) as Total_Transactions,
7 sum(LineTotal) as Total_Revenue,
8 avg(OrderQty) as Average_Qty
9 from [SalesLT].[SalesOrderDetail]
10 group by ProductID
Results
Messages
Failed to execute query. Error: Column 'SalesLT.SalesOrderDetail.UnitPriceDiscount' is invalid in the select list because it is not contained in either an aggregate function or the GROUP BY clause.

Figure 6-32. *Common error using Group By statement*

There are two ways of resolving the issue as shown in Figure 6-33.

1. wrap the new column in an aggregation function, (e.g., MAX()).
2. Add the new column to GROUP BY.

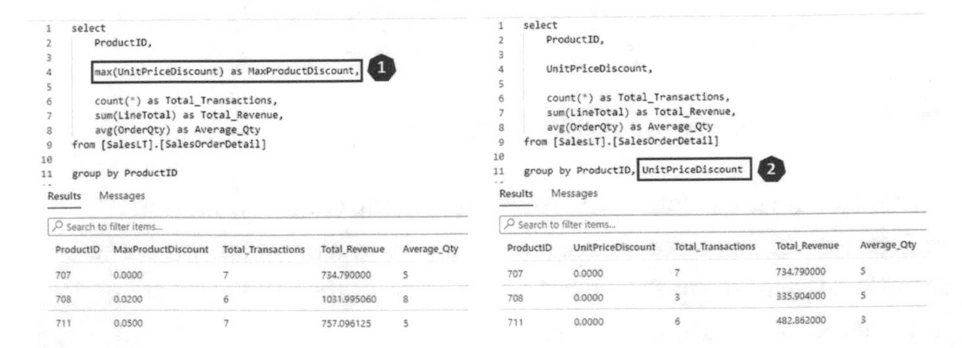

Figure 6-33. *Two methods to fix the common Group By error*

With Group By there are two filter options, as shown in Figure 6-34.

1. **WHERE** clause to filter original table *before* GROUP BY operation. It follows the FROM clause.
2. **HAVING** clause to filter transformed table *after* GROUP BY. It follows the Group By statement.

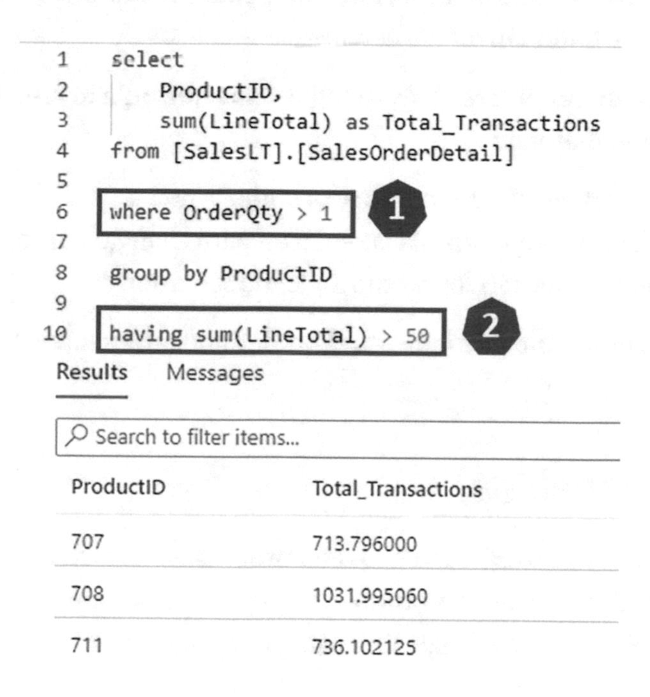

Figure 6-34. *Filter options in Group By*

SQL Basics Summary

If you haven't used SQL before, remember that SQL is easy to start. you have seen some of the most common SQL operations in practice.

Predefined Variables and Tables

You can enter a value or calculation in Excel in a cell. You can then name the cell, which allows you to refer to the value or calculation repeated in other calculations. This approach provides three main benefits.

- **Saves time**. You don't need to repeat the same calculation in different places.
- **Reduces errors**. Only a single value/formula to create and maintain.
- **Improves the readability of code**. When a complicated formula is replaced with a single name, the reader can follow the logic much easier.

In SQL, predefined variables and predefined tables have the same benefits.

Create Variables

Creating a variable in SQL is a two-step process. As shown in Figure 6-35, first you need to declare the variable, then you can set the value. SET can also be combined into a single line using the = sign after DECLARE.

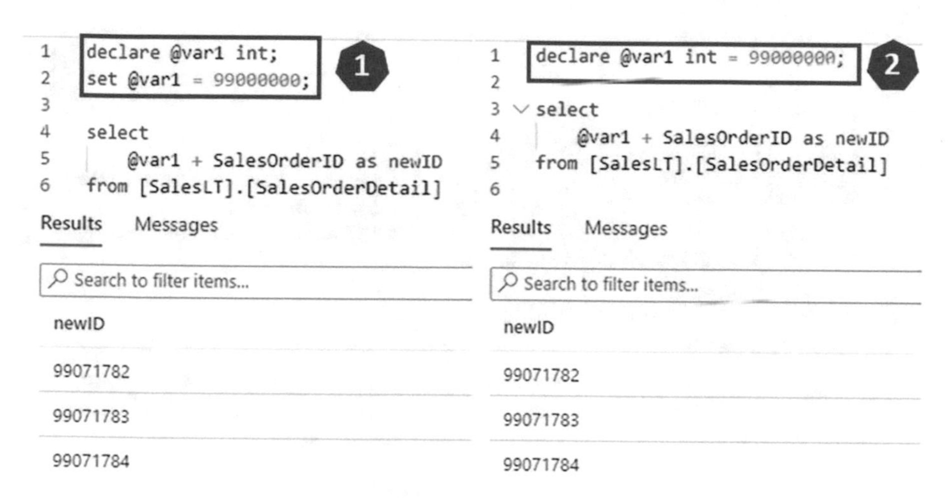
1 declare @var1 int;
2 set @var1 = 99000000;
3
4 select
5 @var1 + SalesOrderID as newID
6 from [SalesLT].[SalesOrderDetail]
1
Results Messages
Search to filter items...
newID
99071782
99071783
99071784
1 declare @var1 int = 99000000;
2
3 select
4 @var1 + SalesOrderID as newID
5 from [SalesLT].[SalesOrderDetail]
6
2
Results Messages
Search to filter items...
newID
99071782
99071783
99071784

Figure 6-35. *Create variables in SQL*

Caution All variables must start with the @ symbol. You would choose an easily understandable name. *var1* is a bad example.

This process also works for multiple variables, as shown in Figure 6-36.

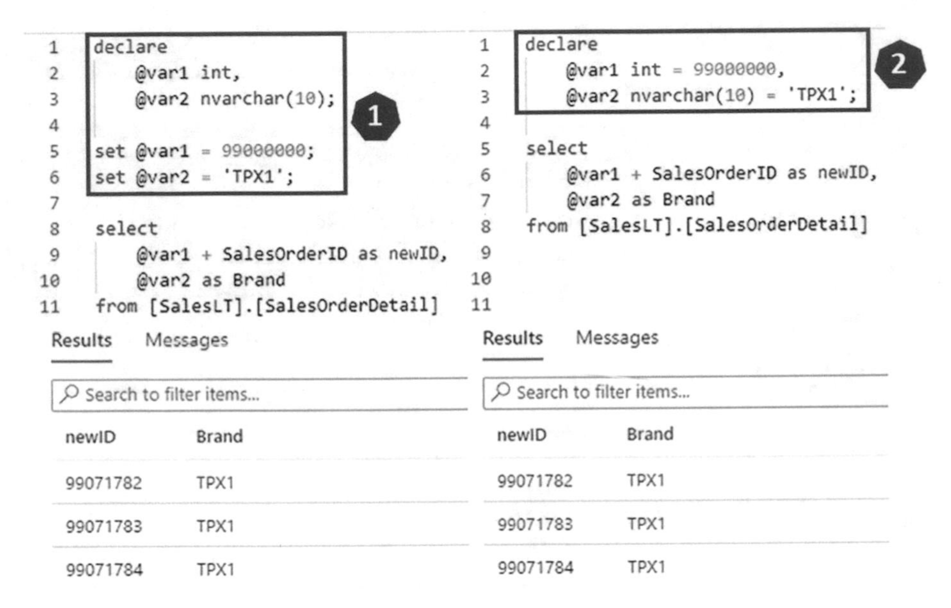

Figure 6-36. *Create multiple variables*

Create Variables from Data

In the prior examples, the variable values are independent of the data. In practice, you may need to create a variable based on existing data. Figure 6-37 shows an example where the variable is set as the max DueDate of the order. The variable only filters the orders due on the max DueDate; it is done using the following steps.

1. Declare the @maxDueDate variable.
2. Set the value of the variable from a SELECT query.
3. Use the variable in the WHERE clause.

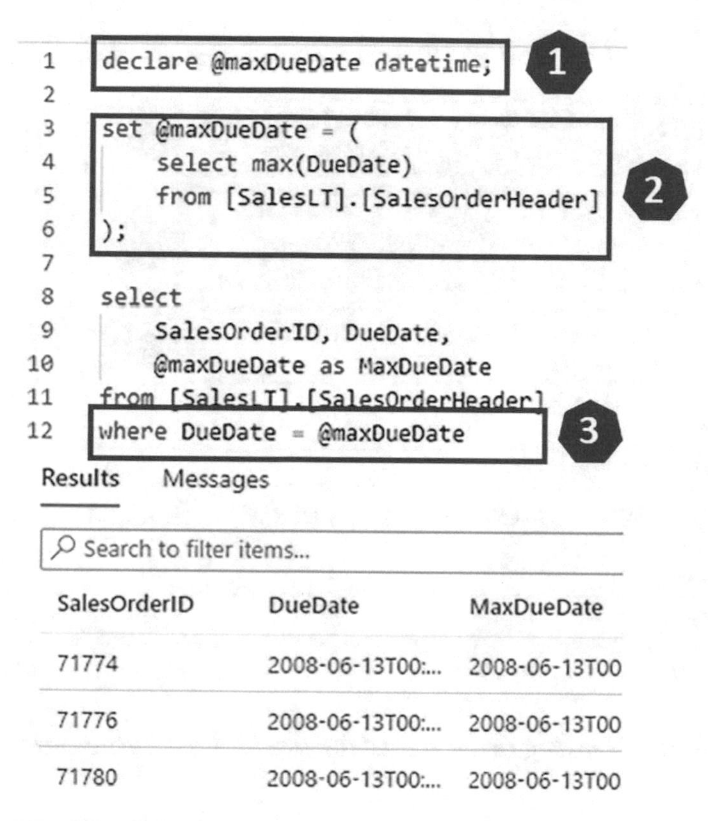

Figure 6-37. *MaxDueDate variable created from data*

The newly created variables can also be used to create other variables. The following steps identify the order detail for the order with the highest value, as shown in Figure 6-38.

1. Create the TopSubTotal variable to store the max order value from the OrderHeader table.

2. Create a TopOrder variable to store the order number for the max value order from the OrderHeader table.

3. Use TopOrder to SELECT the detail from the OrderDetail table.

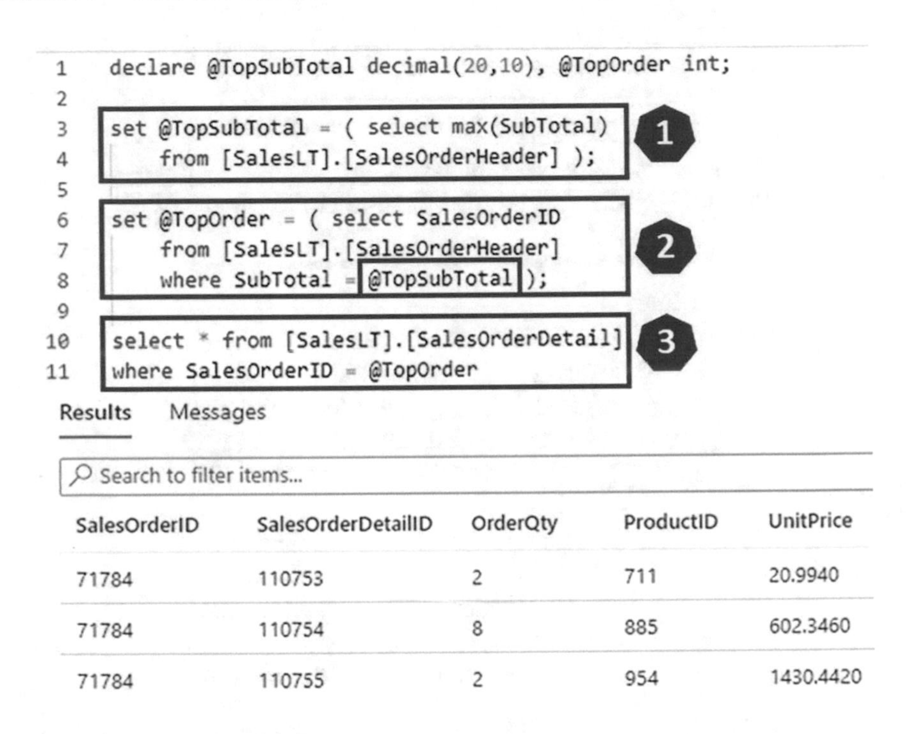
1 declare @TopSubTotal decimal(20,10), @TopOrder int;
2
3 set @TopSubTotal = (select max(SubTotal)
4 from [SalesLT].[SalesOrderHeader]);
5
6 set @TopOrder = (select SalesOrderID
7 from [SalesLT].[SalesOrderHeader]
8 where SubTotal = @TopSubTotal);
9
10 select * from [SalesLT].[SalesOrderDetail]
11 where SalesOrderID = @TopOrder
1
2
3
Results
Messages
Search to filter items...
SalesOrderID SalesOrderDetailID OrderQty ProductID UnitPrice
71784 110753 2 711 20.9940
71784 110754 8 885 602.3460
71784 110755 2 954 1430.4420

***Figure 6-38.** Finding order detail for the highest value order*

Create Table from Data (WITH)

The DECLARE and SET methods can only create a variable with a single value. In practice, the data logic often requires storing a transformed table. Figure 6-39 shows a different way to get the TopOrder sales detail using the WITH clause.

1. Create a new variable to store the top value.
2. Create a new table to store the top value row from the OrderHeader table.
3. Use INNER JOIN to select only the rows in OrderDetail table that matches the SalesOrderID in the new table.

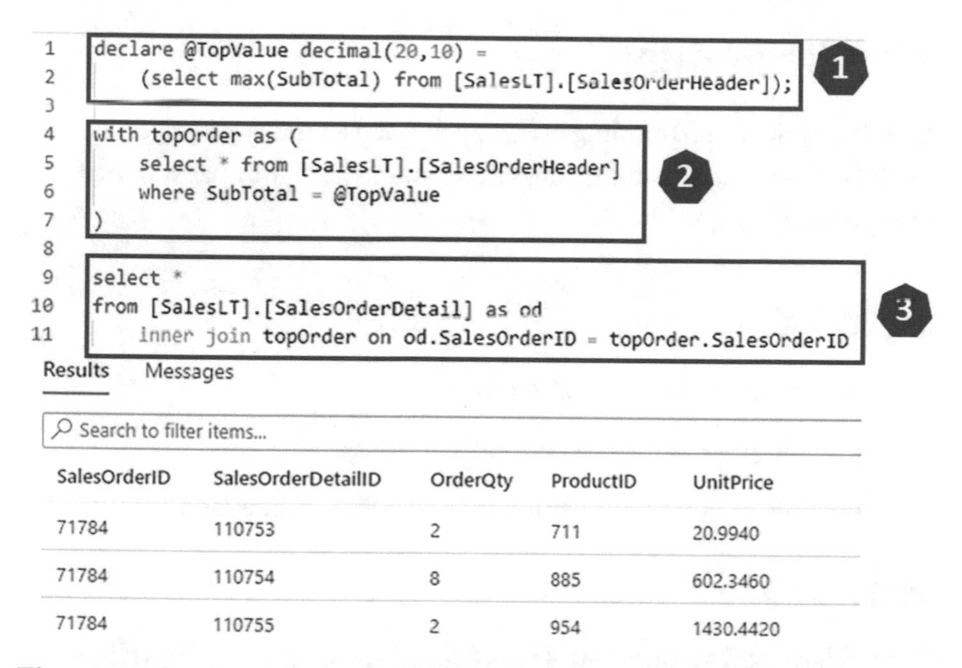

declare @TopValue decimal(20,10) =
(select max(SubTotal) from [SalesLT].[SalesOrderHeader]);
1
with topOrder as (
select * from [SalesLT].[SalesOrderHeader]
where SubTotal = @TopValue
)
2
select *
from [SalesLT].[SalesOrderDetail] as od
inner join topOrder on od.SalesOrderID = topOrder.SalesOrderID
3
Results
Messages
Search to filter items...
SalesOrderID SalesOrderDetailID OrderQty ProductID UnitPrice
71784 110753 2 711 20.9940
71784 110754 8 885 602.3460
71784 110755 2 954 1430.4420

***Figure 6-39.** Reproduce top value order detail using WITH clause*

Even though the method using the WITH clause produces the same result as the earlier method. The WITH approach is more robust because it also covers the case if there are multiple orders with the same top value.

Note While you can store multiple transformed tables with the WITH clause, you can't refer to one stored table when creating another stored table.

Conclusion

SQL is easy to start with. The sessions in this chapter cover 80% of the reporting requirements. Power BI lets you keep the SQL queries simple by allowing additional transformation and modeling layers in Power Query and with data models.

Mini-Hackathon

With this basic understanding of SQL, you need as many practices as possible. HackerRank is a site that allows you to gain more experience through practice.

Objective

There are two objectives in this Hackathon.

- Complete some HackerRank challenges.
- Build a Power BI Dashboard with a SQL connection.

Objective 1

Clear all easy and medium-level SQL (Basic) challenges, as shown in Figure 6-40.

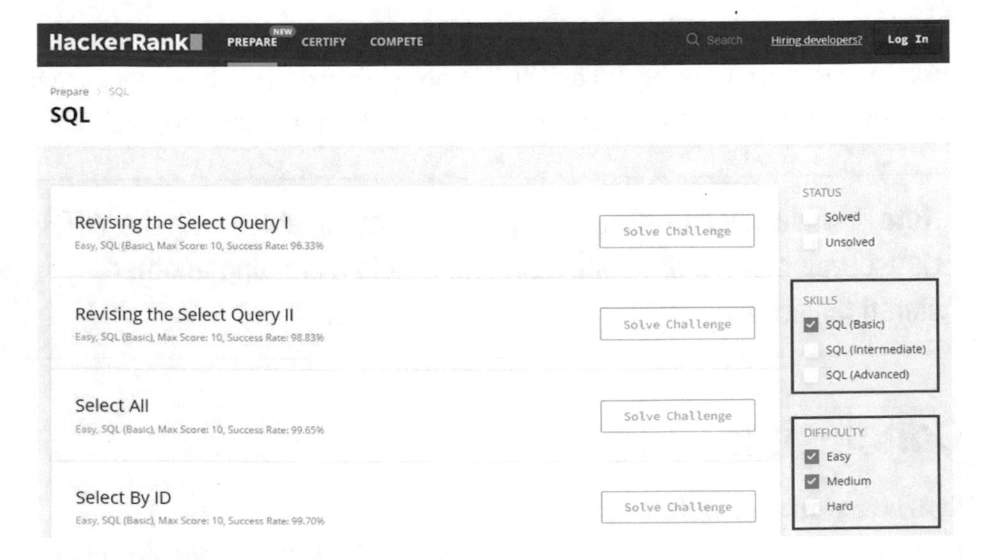

***Figure 6-40.** SQL (Basic) challenges in HackerRank*

Objective 2

Try to create a dashboard based on data from SQL Server using the Import method. If a workspace is available, try to publish the report into the workspace and configure auto refresh.

Time Limit

Objective 1 should take three to five hours. Objective 2 should take three hours.

CHAPTER 7

SharePoint Lists

This chapter may come as a bit of a surprise after SQL Query. It is mostly true that SQL Server is a better data storage solution than SharePoint list. The issue with SQL Server is that most analysts cannot add or edit the data in a SQL table. This ability to edit data records is fundamental for the following chapters on automation and building business applications.

This chapter is an important bridging chapter between business intelligence and business applications. A SharePoint list is preferred over other solutions due to its native integration with Power Automate and Power Apps. Even better, a SharePoint list is free in most organizations.

While SharePoint and SharePoint lists have many different use cases, this chapter only focuses on the following areas.

- Creating a SharePoint list
- Editing a SharePoint list
- User access and security
- Connection to Power BI

Business Scenario

The operation and technology team has sought a new incident management ticketing system for internal system issues. There are a few well-established ticketing systems in the market. The team found the cost higher than expected and required additional approval from the finance team.

D. Ding, *Transitioning to Microsoft Power Platform*,
https://doi.org/10.1007/978-1-4842-9239-6_7

The technology team asks Kim to help to build a business case. After some initial analysis, Kim found the solution cost to be three times more than the approved funding.

After an initial discussion with the technology team, Kim extracted the following fundamental ticketing system requirements.

- The system allows **many users** to **create** new incidents.
- The system allows users to upload screenshots.
- The system allows admins to **make changes** to the incident.
- The system contains incident management **reports**.

In the past 12 months, the team has been using Microsoft Excel to report on this. After a few system releases and changes, the Excel method was overwhelming, resulting in many collaboration errors.

Kim has experience with SharePoint. She knows that SharePoint list is a better collaboration tool than Excel. She proposed to build a prototype using a SharePoint list as an interim solution while she helped build the business case.

Create a SharePoint List

Microsoft created SharePoint as a collaboration, knowledge, and content hub for teams and organizations. It is generally included with any modern Microsoft Office license arrangements. This book focuses on SharePoint from Microsoft 365 (Online).

Create a SharePoint Site

SharePoint lists exist on the SharePoint site. This means that to create a list, you first need a site. In most organizations, you do not have the privilege to create a SharePoint site. You can submit a request for IT to create one for you. A SharePoint site is easy to create. Use the following steps, as shown in Figure 7-1.

1. Sign in at `www.SharePoint.com` and click **Create site**.
2. Create a team site.
3. Name the site.
4. Add users to the site.

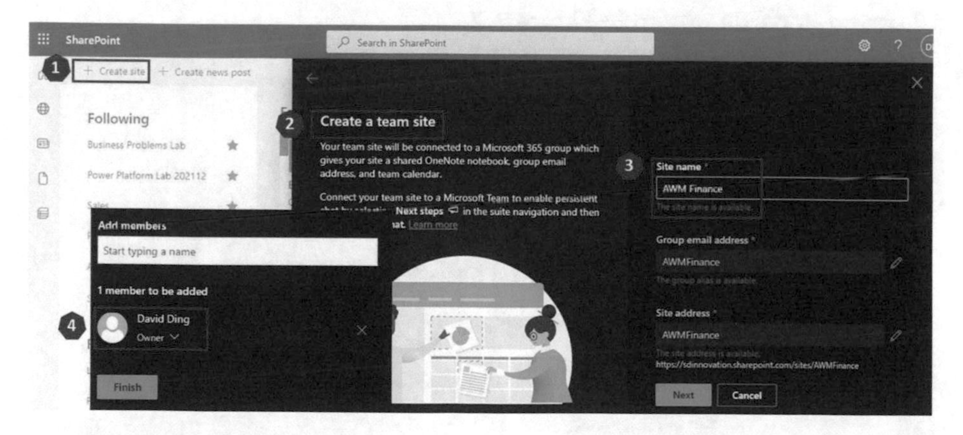

Figure 7-1. *Create a SharePoint site*

The SharePoint site is a powerful tool that contains many useful contents. It can be used as a website for information sharing. It contains folders that can be used as a team share drive. It also contains a flexible data storage format, the SharePoint list.

Create a SharePoint List

A SharePoint list is a special type of data storage with a flexible structure. For example, you can store structured data (e.g., texts and numbers) and unstructured data (images or files up to 2 GB) in different columns of the same table. Each row of data is referred to as an item with different attributes.

Theoretically, a SharePoint list can contain up to 30 million items (or rows). In practice, unless the list is carefully designed, you should limit it to 100,000 items. This is mostly because performance-related problems start to be noticeable around 100,000 items.

Create a New List

You can create the SharePoint list from the SharePoint site using the following steps, as shown in Figure 7-2.

1. Select New ➤ List.
2. Create a **Blank list**.
3. Name the list.

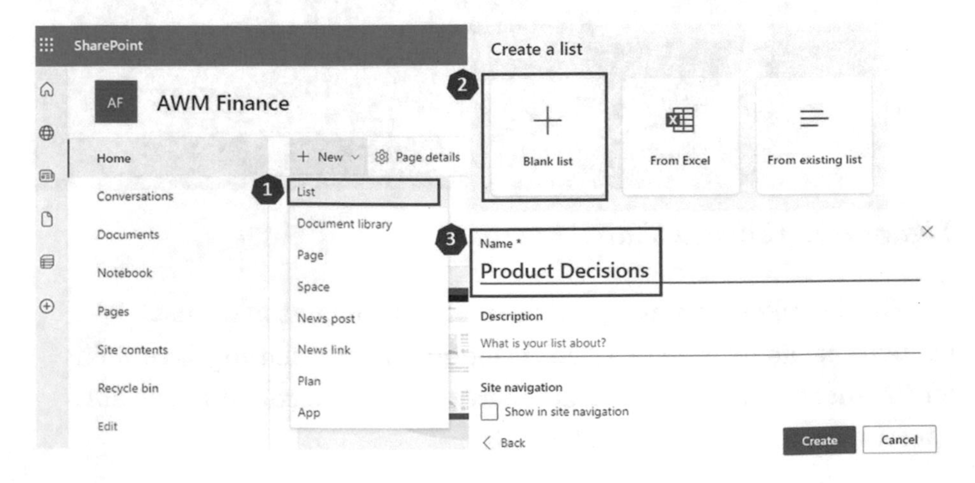

***Figure 7-2.** Create a new SharePoint list*

Add Columns

A SharePoint list comes with one mandatory default column called Title. You can add more columns. The following steps explain how to do this, as shown in Figure 7-3.

1. Click **Add column.**
2. Choose the column type.
3. Name the new column.
4. Choose **More options.**
5. Make the column a mandatory column (optional).
6. You can also choose to Enforce the unique values (optional). This means the issue ID cannot be duplicated.

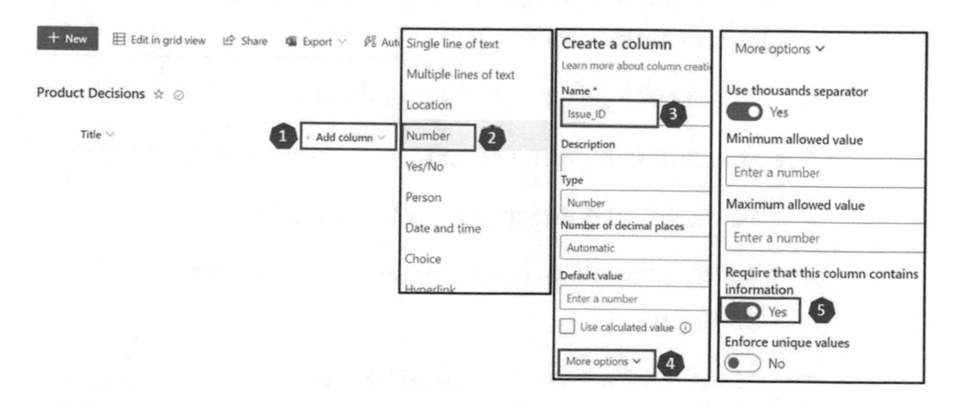

Figure 7-3. *Create a new column*

Tip Try to avoid using spaces in the column names. Space in names can sometimes be displayed as *_x0020_*. This may require some additional steps in data cleansing.

Column Property Guideline

For this book, a SharePoint list is being used as an easily accessible data storage. Although a SharePoint list accepts many data types in columns, let's focus on the following guidelines.

- Only use three data types.
    - A single line of text
    - Multiple lines of text
    - Numbers
- Avoid any settings that may result in an error.
    - Do not create mandatory fields. (Select No for Require that this column contains information.).
    - Do not enforce unique values. (Select No for Enforce unique values.)
    - Do not configure column validation.
- Keep dates and datetime as text.
    - Use standard date format (*YYYY-MM-DD*).
    - Use standard date time format (*YYYY-MM-DD hh:mm:ss*).
    - If the time zone is required, store it in a separate text column with the format (*UTC +/-h.mm (UTC -10:00)*).

The guideline is to help you to keep things simple. Integration errors are common and random between SharePoint list and Power Platform tools (Power BI, Power Automate, and Power Apps).

Caution A SharePoint list contains useful formats such as dates, choice, and person. When integrating with Power Apps, it is better to avoid these data types. They can result in unpredictable failures in the app. Even though it doesn't often happen, when it does happen, it causes frustration.

Add Columns for a Ticketing System

You can now create data storage for a ticketing system. You can add the following columns to the SharePoint list.

- Title (keep as is)
- Issue_ID (number)
- Issue_Name (single line of text)
- Issue_Description (multiple lines of text)
- System (single line of text)
- Priority (single line of text)
- Impact_Value (number)
- Impact_Group (single line of text)
- Status (single line of text)
- Created_Date (single line of text)
- Resolution_Date (single line of text)

The empty list shown in Figure 7-4 is used in the upcoming exercises.

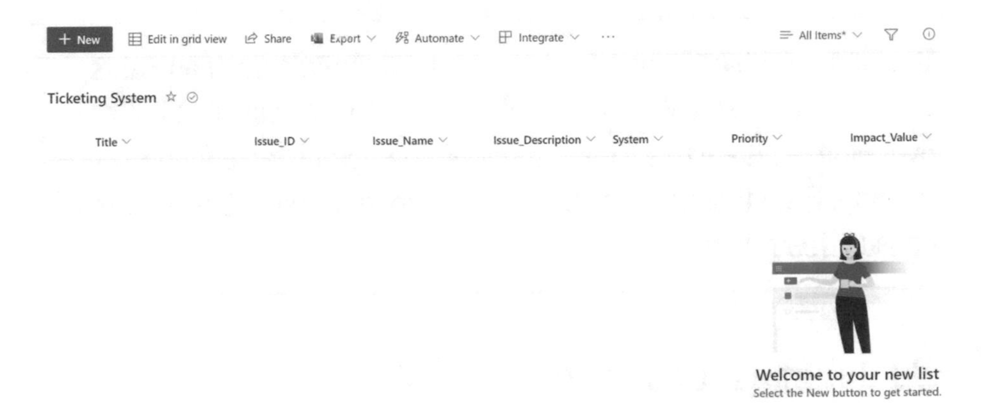

Figure 7-4. *New SharePoint list*

Note The Title column is a mandatory field in all SharePoint lists. It can't be removed but can be renamed for mandatory text purposes.

Edit a SharePoint List

Once you create a SharePoint list, you may want to add some data.

Standard Edit

The following steps add a new item to the ticketing system list, as shown in Figure 7-5.

1. Select + **New**.
2. Populate the content for each field (enter **1** in the title).
3. After all the information has been entered, click Save.

4. Repeat steps 1–3 *three* times.

   a. Make the ticket 3 status = Not Assigned and the resolution date blank.

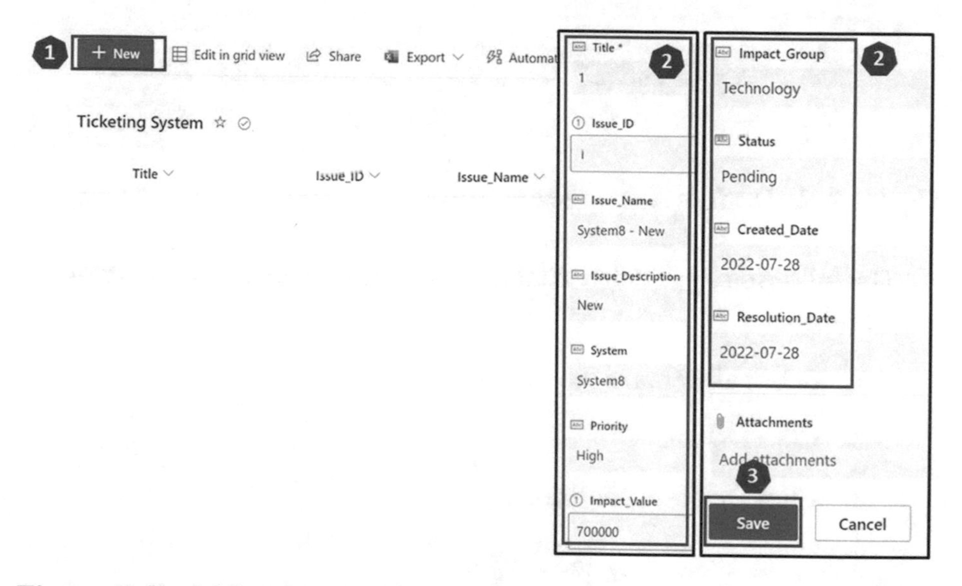

Figure 7-5. *Add a new issue*

The following process changes the status for ticket 2 from Not Assigned to Resolved and adds a resolution date to 10 days after the creation date (see Figure 7-6).

1. Double left-click ticket 3.
2. Change the status from Not Assigned to Resolved.
3. Add a Resolution_Date of 2022-02-09.

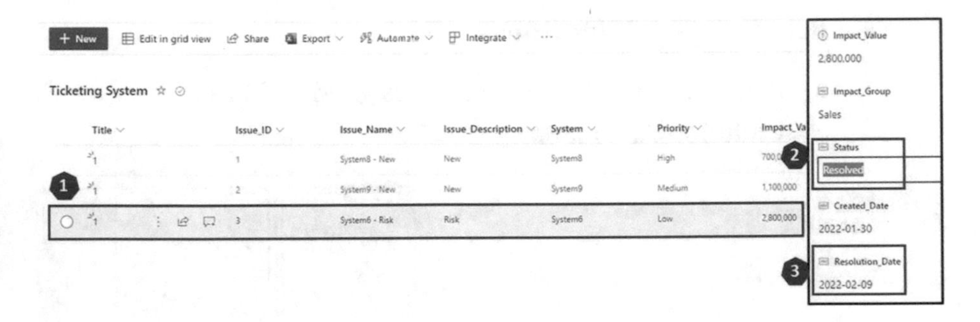

Figure 7-6. Update ticket status

The following steps delete all items using the standard step, as shown in Figure 7-7.

1. Select all items in the current view.

2. Click Delete.

3. Click Delete to confirm.

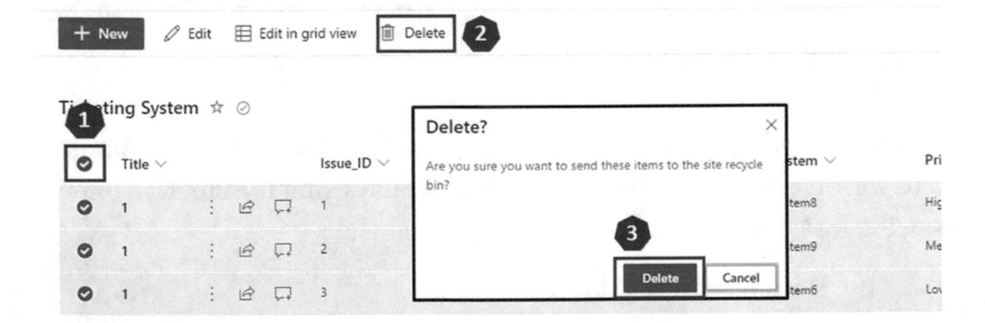

Figure 7-7. Delete items with standard method

Bulk Edit

The standard edit process is the preferred method of editing a SharePoint list. It is okay during the operational process because tickets are added/updated one at a time. The standard edit process is too time-consuming when you need to bulk add/edit/delete items (e.g., during list migration).

There are two bulk edit methods. Both are often used with Microsoft Excel.

- Use **Edit in grid view** (Quick edit) for lists/views with less than 500 items.
- Use **Microsoft Access** to bulk update large lists with more than 500 items.

Edit in Grid View

To bulk edit a small number of items, **Edit in grid view** is likely the best option. In the Chapter 7 folder, there is an Issues Lists flat.xlsx file. The Small tab contains 50 issues. The following steps add all items to the SharePoint list, as shown in Figure 7-8.

1. Select **Edit in grid view** from the SharePoint list.
2. Go to Excel, and copy all 50 rows of data (Ctrl+C).
3. Go to the first field in the SharePoint list and paste (Ctrl+V).
4. Click **Exit grid view**.

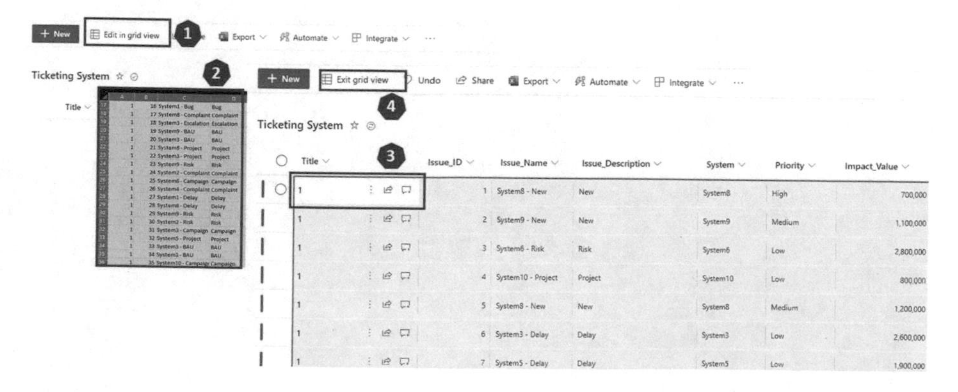

***Figure 7-8.** Copy and paste from Excel*

The *grid view* provides an Excel-like experience to edit items in the SharePoint list. When you paste the 50 items, you would have noticed there is a validation process for each item.

This validation process limits the speed of bulk edit in a SharePoint list. This leads to one of the key properties of a SharePoint list. Unlike SQL databases, a SharePoint list is **not suitable for bulk operations**. With this important property in mind, you need to design the interaction of SharePoint differently from Excel and other databases.

There is no bulk delete function in the grid view. You can follow the previous step to bulk select all 50 items and delete them.

Edit Using Microsoft Access

If there are more than 200 items, even the grid view method for editing becomes too slow and unreliable. You can use Microsoft Access to integrate and edit the SharePoint list in such cases.

MS Access is normally known as the desktop database tool. It also allows you to establish database connections. The following steps connect to the SharePoint list in this case, as shown in Figure 7-9.

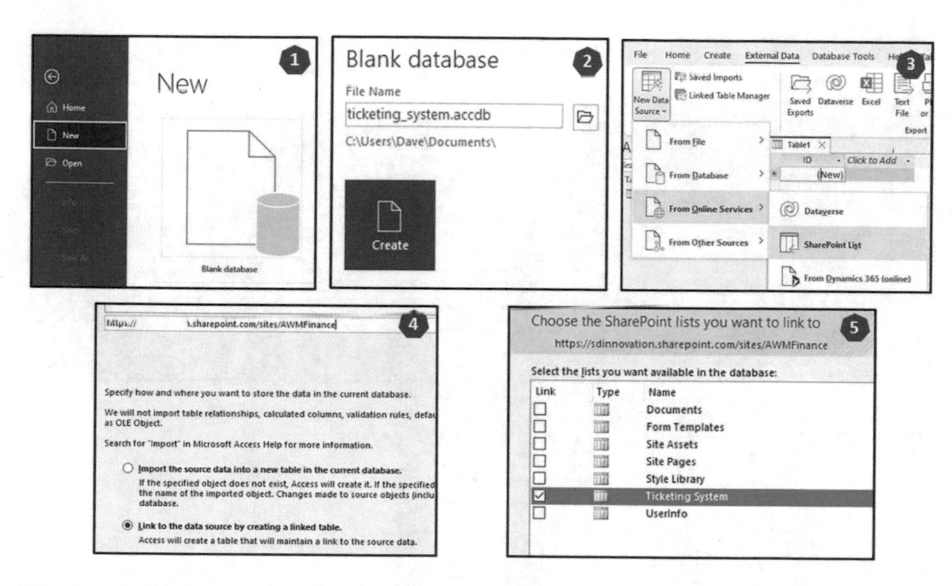

***Figure 7-9.** Create a database in Access with connection to SharePoint list*

The Medium dataset in Issues Lists flat.xlsx contains 2,943 rows of data. This contains all the existing tickets from the Excel file. This data is a bit too big using the grid view edit method. To load this data, you can use MS Access. Once the connection to the SharePoint list has been created, you can load the data into the SharePoint list. The following steps explain how to do this, as shown in Figure 7-10.

1. Select External Data ➤ From Files ➤ Excel.
2. Browse for the Excel file.
3. Append the data to an existing table, and choose the Ticketing Systems table (from the previous step).
4. Make sure the right Excel worksheet is selected (Medium).
5. Click Finish and start loading.

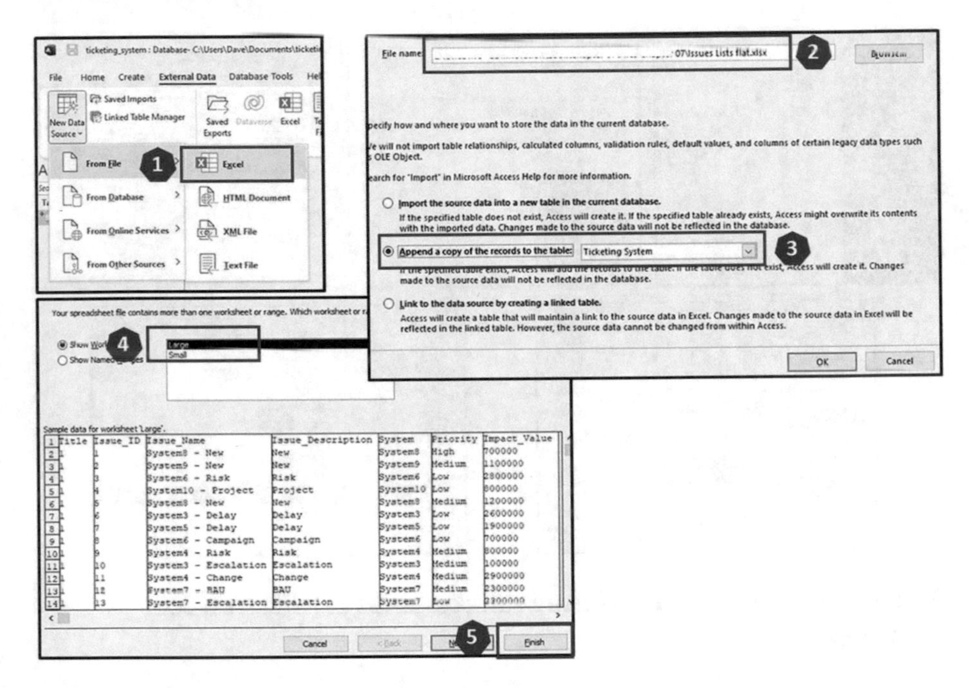

Figure 7-10. *Load data from Excel into SharePoint list*

It takes between 30 to 60 seconds to load all the items. Once finished, there are 2,943 items in the SharePoint list, as shown in Figure 7-11.

Ticketing System
Title Issue_ID Issue_Name Issue_Description System
1 9 System4 - Risk Risk System4
1 10 System3 - Escalation Escalation System3
1 11 System4 - Change Change System4
1 12 System7 - BAU BAU System7
Items Modified
0 7/30/2022 6:20 AM
0 8/12/2022 2:43 AM
2 8/12/2022 2:44 AM
0 7/30/2022 6:20 AM
Ticketing System List 2943 8/26/2022 12:10 AM

Figure 7-11. *Check loaded data in SharePoint list*

This bulk editing SharePoint list data method using Microsoft Access and Excel is much more reliable and efficient than grid view and standard edit methods.

User Access and Security

The ticketing system requires two different levels of access groups. You've been asked to ensure the standard user cannot delete existing tickets.

- **Standard User** can only view, add and edit tickets.
- **System Admin** can delete and bulk update the list.

This session explores three concepts to set up permissions.

- **Permission Level** determines what the user can and cannot do.
- **Groups** is a collection of users.
- **Assign Permissions** means the permission level can be assigned to users or groups for documents, folders, and lists.

Permission Level

Permission levels are set up for the entire site. Permission level can be created on the Permission Settings page. You can navigate to the permission settings page using the following steps, as shown in Figure 7-12.

1. Click the site settings icon.
2. Select **Site permissions**.
3. Select **Advanced permissions settings**.

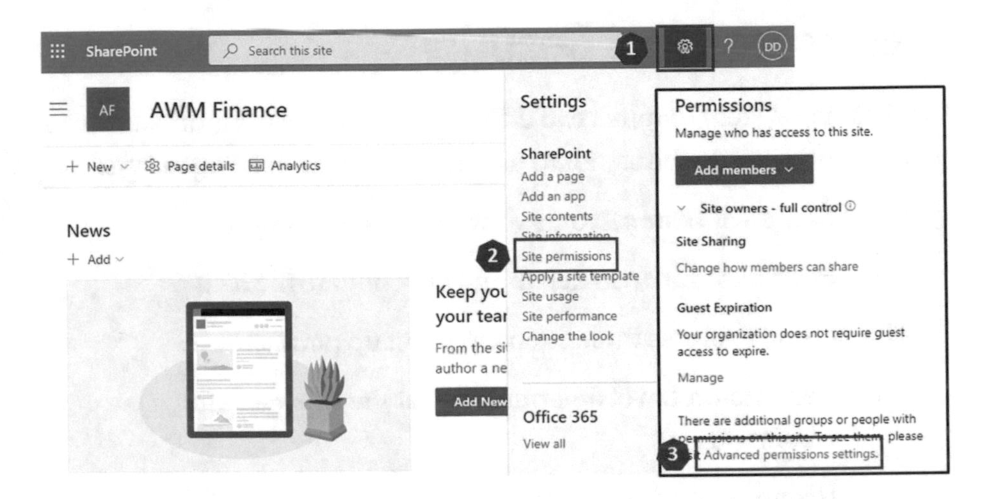

Figure 7-12. *Navigate to SharePoint site advanced permission settings*

The Permissions page normally contains the three default groups, as shown in Figure 7-13.

1. Members with Edit permission
2. Owners with Full Control permission
3. Visitors with read permission

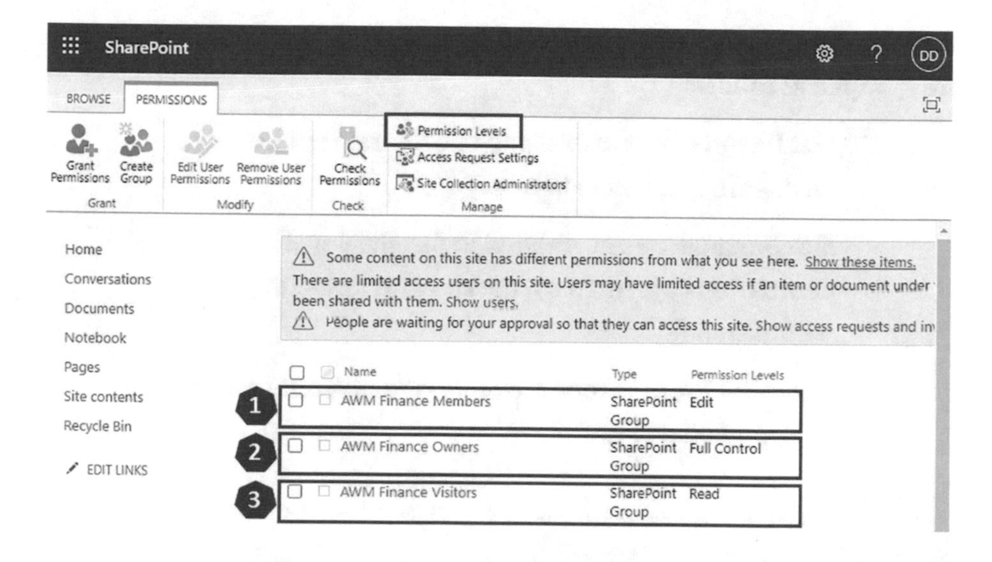

***Figure 7-13.** Default groups and permission levels*

Select the permission levels at the top of the permission page to learn more about the actual permission. As shown in Figure 7-14, there are five default permission levels.

AF

EDIT LINKS

Permissions ▸ Permission Levels

Home
Conversations
Documents
Notebook
Pages
Site contents
Recycle Bin
EDIT LINKS

Add a Permission Level | ✕ Delete Selected Permission Levels

Permission Level	Description
Full Control	Has full control.
Design	Can view, add, update, delete, approve, and customize.
Edit	Can add, edit and delete lists; can view, add, update and delete list items and documents.
Contribute	Can view, add, update, and delete list items and documents.
Read	Can view pages and list items and download documents.

***Figure 7-14.** Default permission levels*

Figure 7-15 shows the permissions for the *read* level. The permissions are grouped as follows.

- **List Permissions** are related to a SharePoint list (e.g., read, edit, and delete items).
- **Site Permissions** are related to the site-level permission (e.g., manage the site and manage permissions).
- **Personal Permissions** are related to personal settings (e.g., personal views).

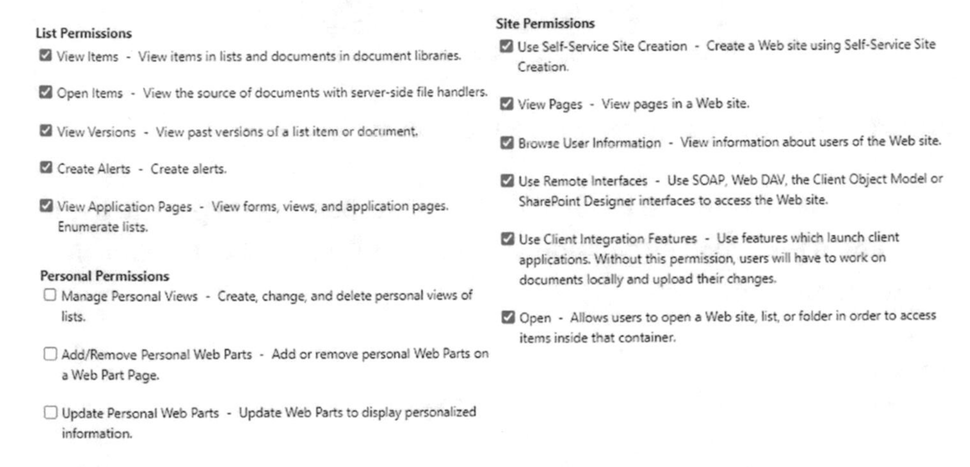

List Permissions
View Items - View items in lists and documents in document libraries.
Open Items - View the source of documents with server-side file handlers.
View Versions - View past versions of a list item or document.
Create Alerts - Create alerts.
View Application Pages - View forms, views, and application pages. Enumerate lists.
Personal Permissions
Manage Personal Views - Create, change, and delete personal views of lists.
Add/Remove Personal Web Parts - Add or remove personal Web Parts on a Web Part Page.
Update Personal Web Parts - Update Web Parts to display personalized information.
Site Permissions
Use Self-Service Site Creation - Create a Web site using Self-Service Site Creation.
View Pages - View pages in a Web site.
Browse User Information - View information about users of the Web site.
Use Remote Interfaces - Use SOAP, Web DAV, the Client Object Model or SharePoint Designer interfaces to access the Web site.
Use Client Integration Features - Use features which launch client applications. Without this permission, users will have to work on documents locally and upload their changes.
Open - Allows users to open a Web site, list, or folder in order to access items inside that container.

***Figure 7-15.** Permissions settings for read level permission*

Even though there is a default Edit permission level, it provides too many permissions for the standard user. One such permission is for the user to delete the entire list, which is not a permission you want the user to have.

The following steps create a new permission level, as shown in Figure 7-16.

1. From the Permissions page, select Add a Permission Level.
2. Provide the name and description (optional) for the new permission level.
3. Select the four list permissions.
4. Select the two site permissions.

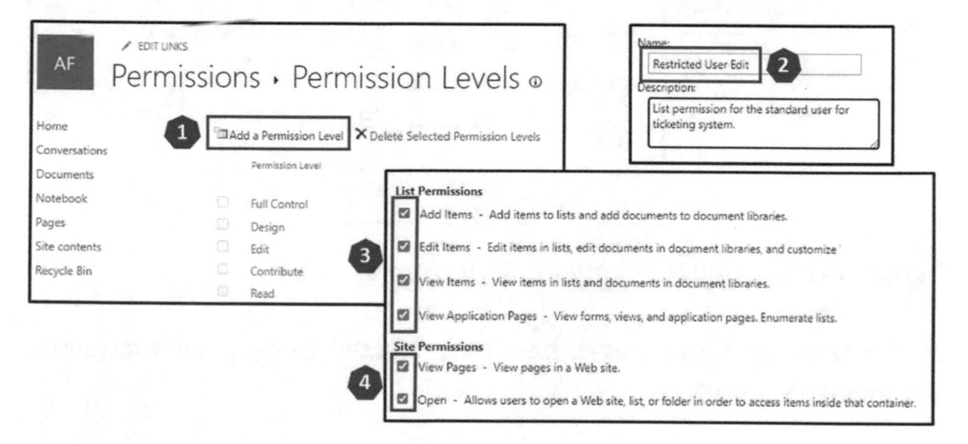

Figure 7-16. *Permission level for a standard user*

This permission level grants the minimum permission for the user to view, add and edit the SharePoint list. It also prevents the user from deleting items.

The system admin can have the contribute permission. The following steps copy the permission level, as shown in Figure 7-17.

1. From the Permissions page, select Contribute.
2. At the bottom of the Edit page, select Copy Permission Level.
3. Give the permission level a name and description.

4. Modify the permissions (optional).

5. Click the Create button.

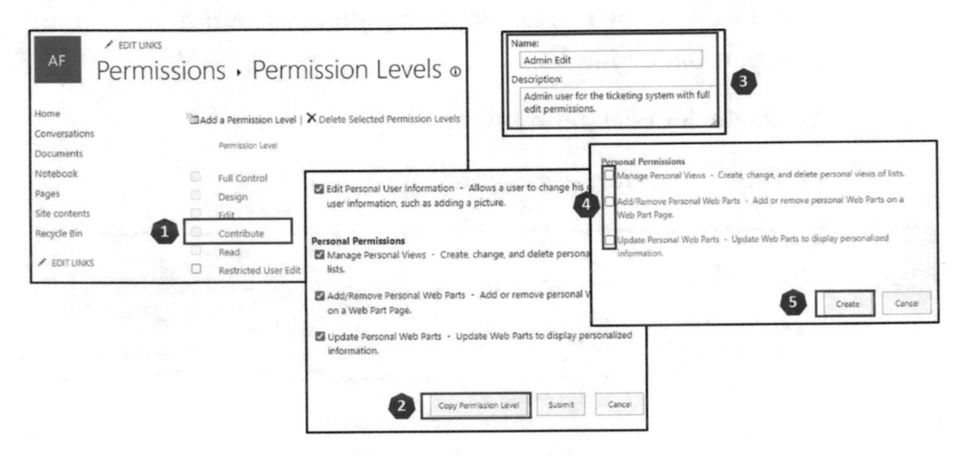

Figure 7-17. *Copy the Contribute permission level*

Two new permission levels have been created for the ticketing system, as shown in Figure 7-18.

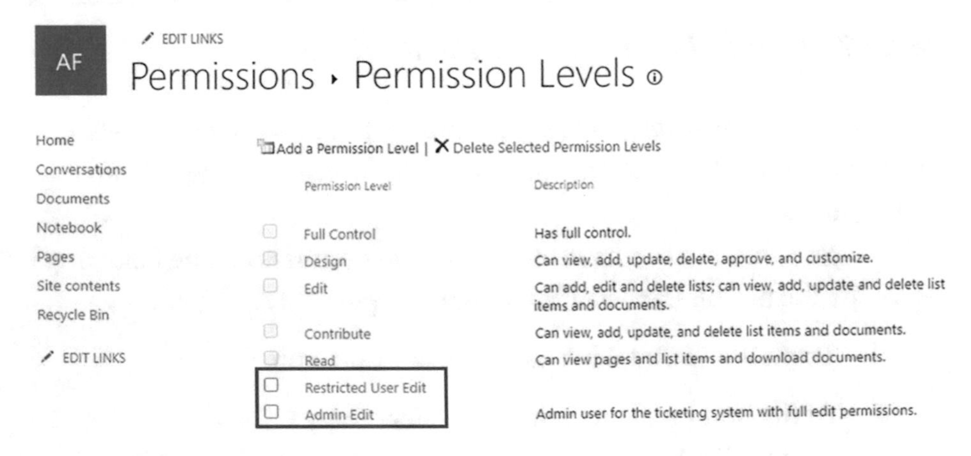

Figure 7-18. *New permission levels*

Groups

Once the permission levels have been determined, you can navigate to the create new SharePoint groups page. The following steps explain how to do this, as shown in Figure 7-19.

1. Click the Settings icon.
2. Go to **Site permissions**.
3. Select **Advanced permissions settings**.
4. Select **Create Group**.

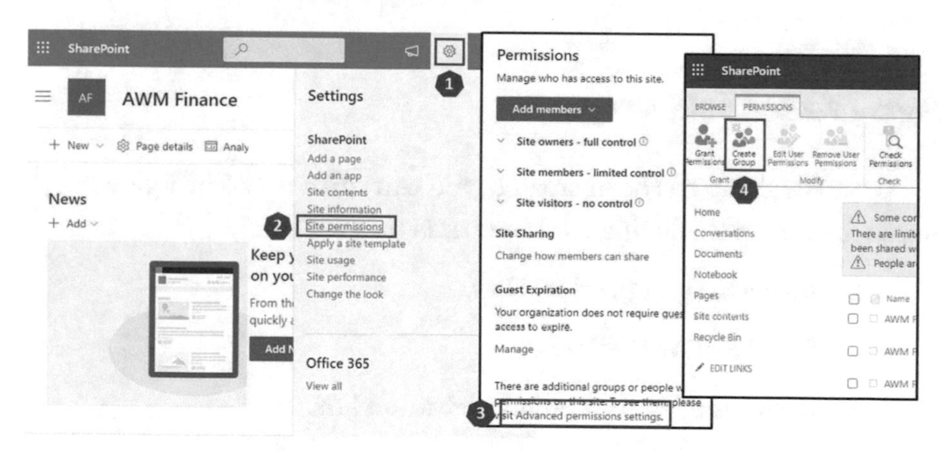

Figure 7-19. Navigate to create a new group page

The following steps create the new group, as shown in Figure 7-20.

1. Give the group a name and description.
2. Make sure all of the site permission settings are not ticked.
3. Repeat the same with AWM Ticket System Admins.

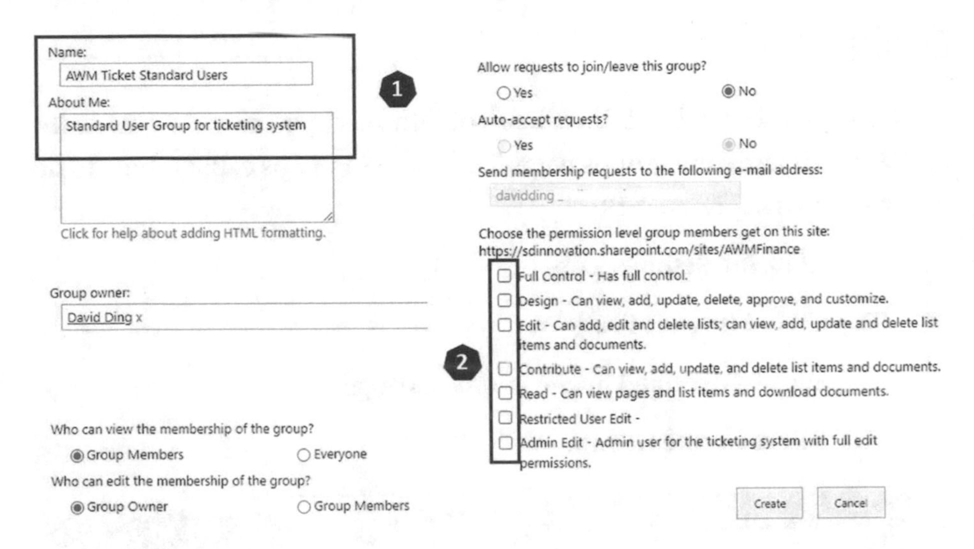

Figure 7-20. *Create new group*

Once the group has been created, you can add a user. The following steps explain how to do this, as shown in Figure 7-21.

1. Inside the group, select New.

2. Select SHOW OPTIONS.

3. Untick the **Send an email invitation** box.

4. Add the user and click the Share button.

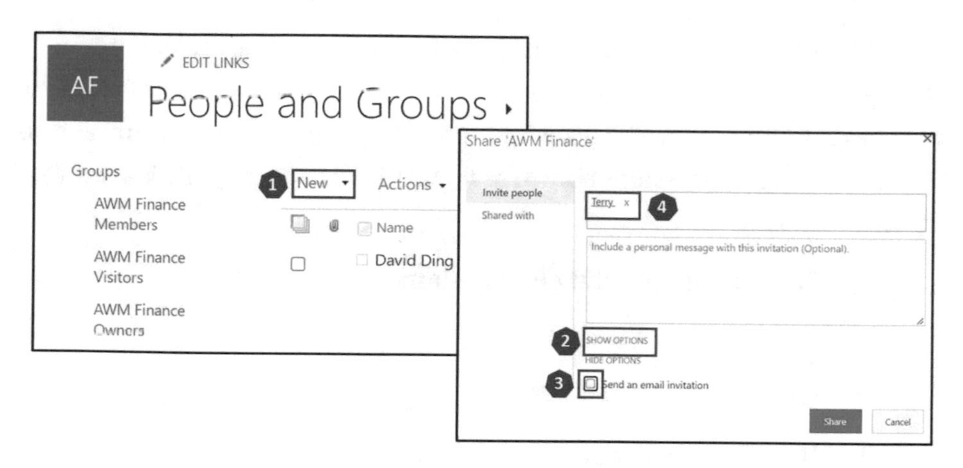

Figure 7-21. *Adding a new user to group*

List Permissions

Once you have created both the permission level and the groups, they must be applied to the contents. In this case, the content is the ticketing system SharePoint list. The following steps navigate to the list Permissions page, as shown in Figure 7-22.

1. Select List ➤ Settings.
2. Select **Permissions for this list**.

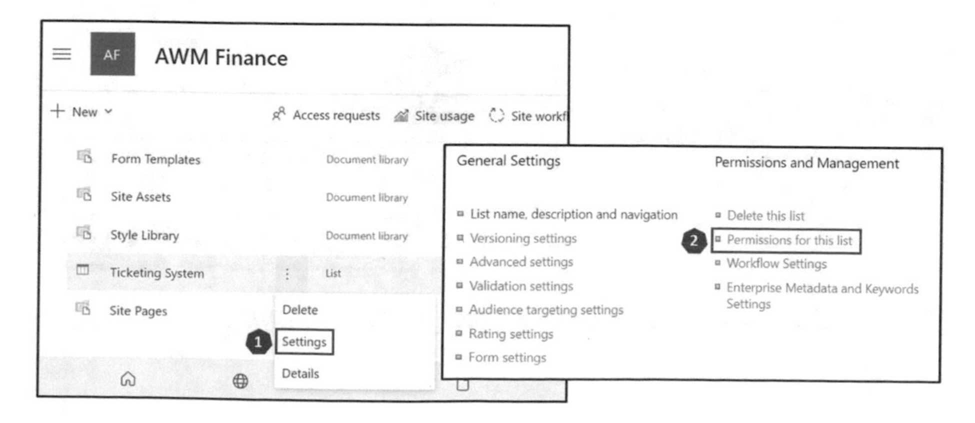

Figure 7-22. *Navigate to the list settings page*

If you're on the list Permissions page for the first time, the default setting is to inherit permission from the site. In this case, you want users to only have permission to the list without them being able to have any access to the sites. Follow the steps shown in Figure 7-23 to grant permissions to the list.

1. Select Stop Inheriting Permissions.
2. Choose Grant Permissions.
3. Select SHOW OPTIONS.
4. Untick **Send an email invitation**.
5. Add the AWM Ticket Standard Users group.
6. Choose the Restricted User Edit permission created earlier.
7. Repeat for AWM Ticket System Admins with Admin Edit.

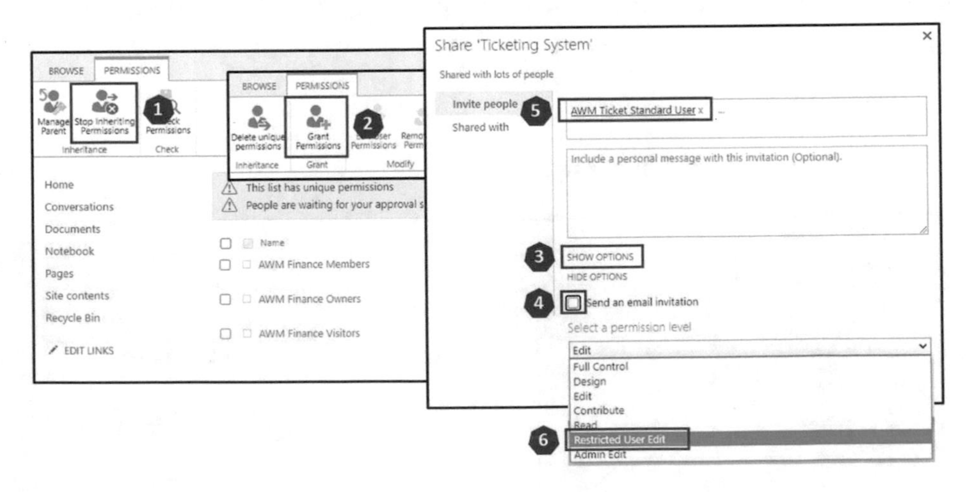

***Figure 7-23.** Grant group permissions for SharePoint list*

Share the List with Users

Once the access permission has been granted properly, you can share the list by copying its URL address. This URL address contains four parts, as shown in Figure 7-24.

1. Company SharePoint address
2. Site name (AWM Finance)
3. List name (Product Decisions)
4. List view name (All Items)

Figure 7-24. *SharePoint list URL*

Standard users and system admins can use the name URL link. They have different levels of permissions as per your setup in earlier sessions.

List Views

You can also customize what the user sees by configuring the list views. Figure 7.25 shows the All Items view and the Edit current view option.

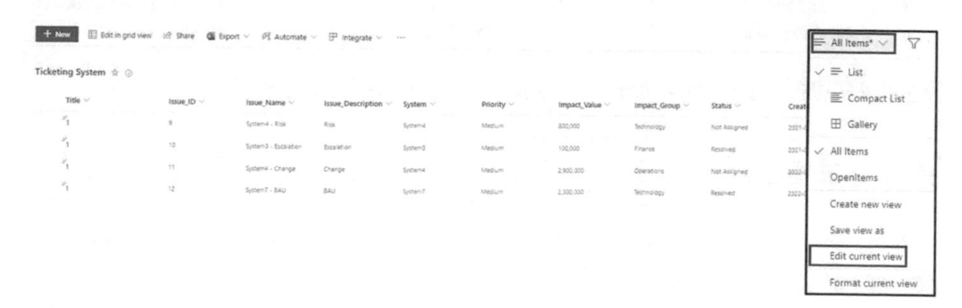

Figure 7-25. *SharePoint list view*

If you edit the view, you see the following configuration groups.

- Name of the view
- Columns selector
- Sort options
- Filter options
- Group By options
- Other options

Columns Selection

A SharePoint list creates about 20 columns of metadata for every item, such as created_by, creation date, and so forth. You can display these columns in the user view by ticketing the display box next to the column, as shown in Figure 7-26.

Display	Column Name	Display	Column Name	Display	Column Name
☑	Title (linked to item with edit menu)	☐	App Modified By	☐	Item is a Record
☑	Issue_ID	☐	Attachments	☐	Label applied by
☑	Issue_Name	☐	Compliance Asset Id	☐	Label setting
☑	Issue_Description	☐	Content Type	☐	Modified
☑	System	☐	Created	☐	Modified By
☑	Priority	☐	Created By	☐	Retention label
☑	Impact_Value	☐	Edit (link to edit item)	☐	Retention label Applied
☑	Impact_Group	☐	Folder Child Count	☐	Title
☑	Status	☐	ID	☐	Title (linked to item)
☑	Created_Date	☐	Item Child Count	☐	Type (icon linked to document)
☑	Resolution_Date	☐	Item is a Record	☐	Version

Figure 7-26. *Column selection for default All Items*

Sort Options

Sorting the display order could be helpful. You can sort the columns from latest to oldest, as shown in Figure 7-27.

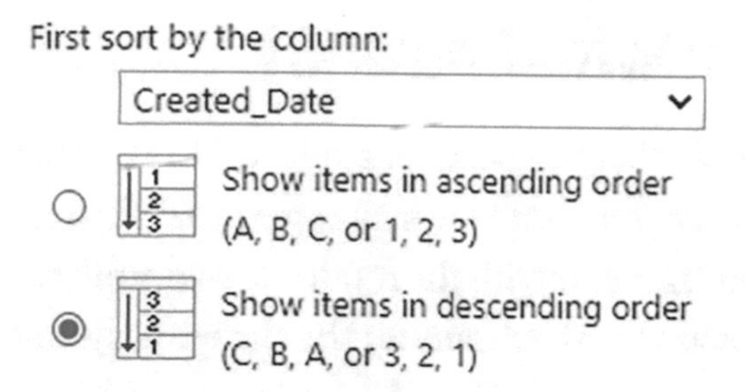

Figure 7-27. *Sorted view for SharePoint list*

Filter Options

Creating separate views using the Filter option can be very helpful in limiting the information to what's relevant for the user. For example, the standard user may only want to see the outstanding open items that belong to the user. This can be done using the following steps, as shown in Figure 7-28.

1. Choose Show items only when the following is true.
2. Select created by [Me].
3. Choose **Status is not equal to Closed**.
4. Choose **Status is not equal to Resolved**.

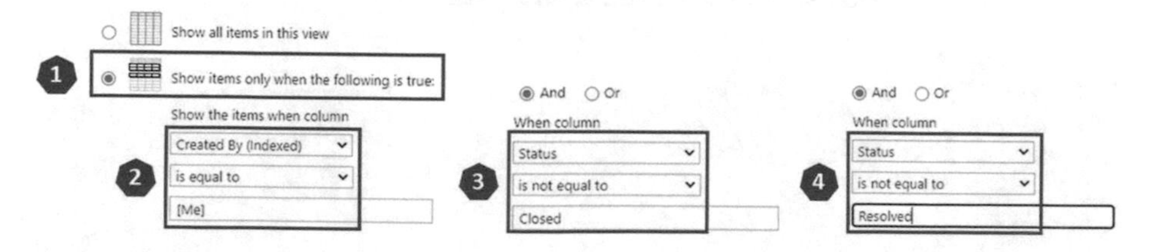

Figure 7-28. *Filter configuration for SharePoint list*

Connection with Power BI

Once the SharePoint list has been created, and the user access has been set up. The last step is to create a Power BI report to link to the SharePoint list. As you are already familiar with the report development, this session only covers the connections. The report for the ticketing system is covered in later chapters.

The following three options exist when connecting to the SharePoint list in Power BI.

- SharePoint List
- SharePoint Online List Version 1.0
- SharePoint Online List Version 2.0

Most organizations use SharePoint Online, for which the default connector is the Online List connector. You can only connect via the SharePoint list connection if you have an on-premises SharePoint server.

This session focuses on versions 1.0 and 2.0 of the SharePoint Online List. In summary, version 1.0 contains more information content, whereas version 2.0 is faster.

A Power BI report has been created to connect to the same list using two separate methods, as shown in Figure 7-29.

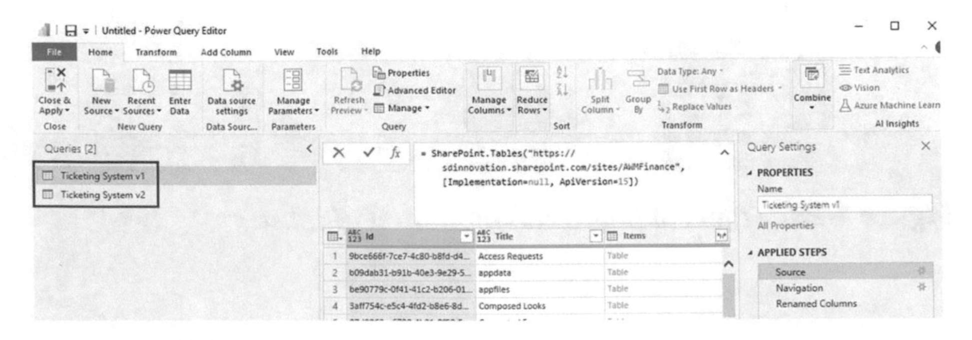

Figure 7-29. *SharePoint list connectors in Power BI*

The Query Editor shows that the difference between the two methods is only in the parameter selection.

The following is for version 1.0. Source =

```
SharePoint.Tables("https://cmp.sharepoint.com/sites/
AWMFinance", [Implementation= null, ViewMode="All"])
```

The following is for version 2.0. Source =

```
SharePoint.Tables("https://cmp.sharepoint.com/sites/
AWMFinance", [Implementation= "2.0", ViewMode="All"])
```

Difference in Content

Version 1.0 extracts more content from the SharePoint list. In this example, version 1.0 extracts 41 columns, and version 2.0 extracts 31 columns.

Even though you only created 11 columns, a SharePoint list contains additional metadata columns not shown in the default All Items view. Some items also contain additional information in the form of List and Record, as shown in the Created By column in Figure 7-30.

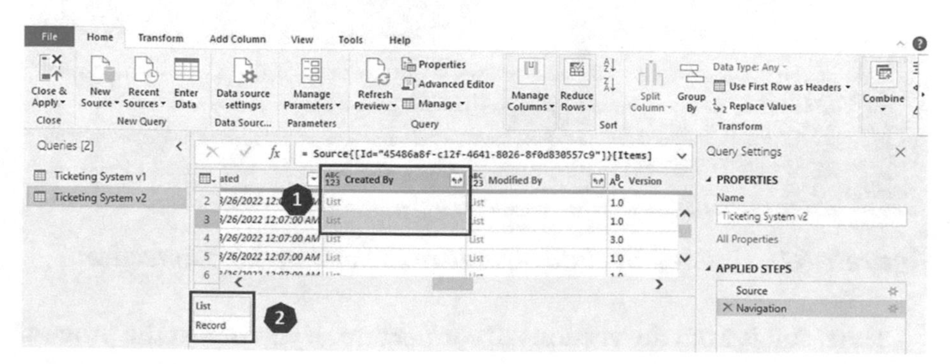

Figure 7-30. SharePoint list columns from Power BI

List and Record in Power Query are data structures that contain multiple values. This is also why they can't be displayed directly in a column. The following steps display the values, as shown in Figure 7-31.

1. Select the expand button next to the Created By column and choose Expand to New Rows.
2. Select any Record in the column.
3. Check the data structure inside the record with four fields.

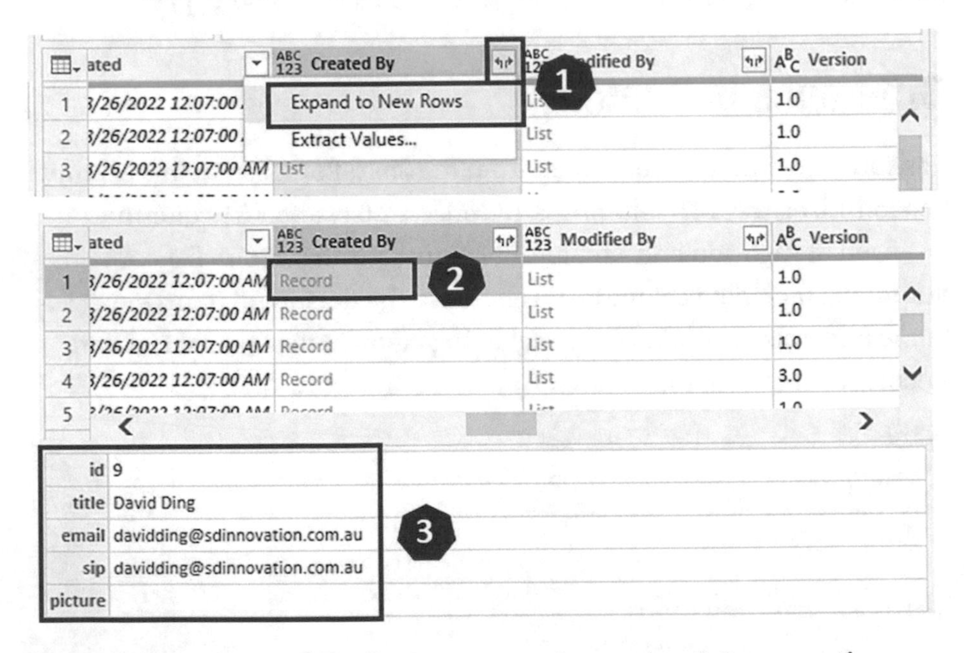

***Figure 7-31.** Created By data structure in version 2.0 connection*

The result is from the version 2.0 connection. If you repeat the process in the version 1.0 connection, you see that the Author column (equivalent to Created By) contains far more fields, as shown in Figure 7-32.

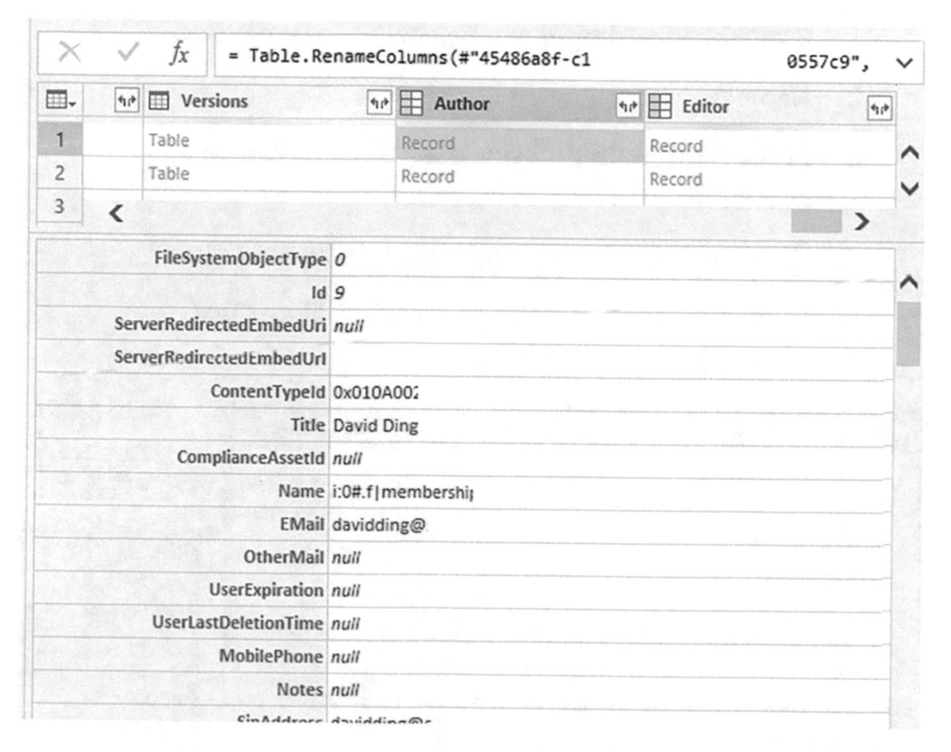

Figure 7-32. *Author data structure in version 1.0 connection*

This is also why version 1.0 is much slower than version 2.0 because it contains exponentially more information.

Improve Speed with Default View

Loading and transforming large SharePoint list data in Power BI can be a frustrating experience as Power Query has to constantly reload SharePoint data. One effective method of improving the list connection is via default views. Default views can use column selection and filter operations to limit the amount of transferred data.

The following steps configure a new view that filters issues created in 2022, as shown in Figure 7-33.

1. From the View drop-down, select **Create new view**.
2. Name the view.
3. From the View drop-down, select **Set current view as default**.
4. From the View drop-down, select **Edit current view**.
5. Configure the Filter setting to **Created_Date (Text column) begins with 2022**.

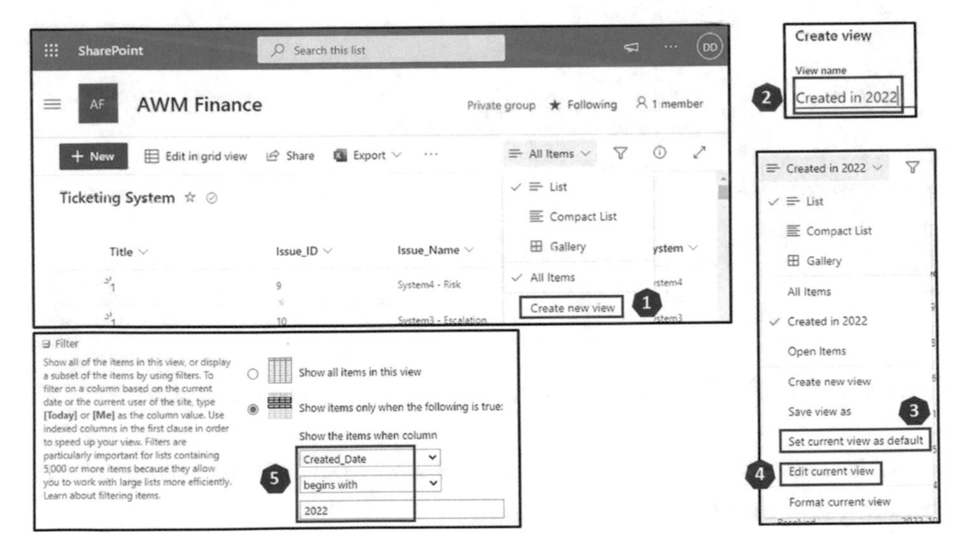

Figure 7-33. Create a new filtered view

Caution You must always use the site link in Power BI, not the full link to the list. This applies to connecting to a SharePoint folder and a SharePoint list.

In Power BI, you can configure the SharePoint Online list connector to extract data from the default view, as shown in Figure 7-34.

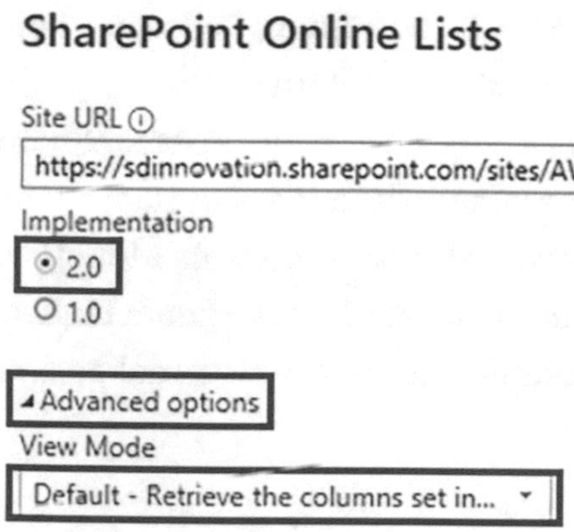

SharePoint Online Lists

Site URL ⓘ

https://sdinnovation.sharepoint.com/sites/AWMFinance

Implementation

⦿ 2.0

○ 1.0

◢ Advanced options

View Mode

Default - Retrieve the columns set in...

Figure 7-34. *Extract data from default view*

The default view can significantly reduce the data load time with the SharePoint list. Creating the SharePoint list requires some consideration of the required reporting fields.

Tip The default view is also useful in limiting data during development. You can change the amount of data to be loaded into Power BI by adjusting the default view filter.

Conclusion

Kim's temporary incident management system with SharePoint and Power BI is well received by the company. Although it lacks a good user interface for data input, it is highly flexible and customizable to suit additional needs. The technology team likes this solution because of the following reasons.

- It was developed in such a short period of time.
- There are no additional infrastructure or costs involved.
- It is very easy to maintain.

The next three chapters focus on the improvement based on the current solution using Power Apps and Power Automate.

This chapter highlights some of the key knowledge areas for SharePoint lists. The last few chapters used a SharePoint list as the key data storage.

One alternative is Dataverse (previously Common Data Model). While Dataverse is a better solution, the cost associated is also much higher. A SharePoint list would be more than sufficient for most internal business applications.

CHAPTER 8

First Power Apps

Power Apps allows data professionals to build and share business applications in a cloud environment without the knowledge of any programming languages. Power Apps is built in the Microsoft Azure Cloud infrastructure. It can be shared with thousands of users without any noticeable performance impacts.

The following are some of the most common Power Apps.

- Smart forms and online catalogs
- Asset management solutions
- Customer relationship management (CRM) solutions
- Project management solutions
- Ticketing system solutions

This chapter continues with the ticketing system requirement from Chapter 7. As you understand the following aspects of Power Apps, you build a proper front-end application for the ticketing system using canvas apps.

D. Ding, *Transitioning to Microsoft Power Platform*,
https://doi.org/10.1007/978-1-4842-9239-6_8

Business Scenario

Before Kim delivered the SharePoint list solution, she started exploring the other Power Platform tools. The tool that excited her the most was Power Apps. After delivering the temporary ticketing system solution using a SharePoint list, She asks the technology team for three more weeks to test an alternative Microsoft solution.

Richard is the technology manager in charge of the ticketing system. With over 15 years in the technology industry, he is experienced in system implementation and integration. Even though he likes the SharePoint list solution Kim delivered, he is skeptical about her being able to develop a proper solution. Since the decision on the final solution is at least two months away, Richard sees no harm in letting Kim try another solution. He asks Kim to keep him in the loop with progress while also ensuring she understands the top priority is securing the funding for the proper ticketing system solution.

Power Apps Licensing

When it comes to licensing, Power Apps can be confusing. The licensing rules are still evolving. The following licensing summary refers to the rules in Sep 2022.

A Power Apps license has two parts.

- Access to Power Apps
- Access to Premium and Custom connectors

There are a few licenses that give you access to Power Apps.

- Direct Power Apps licenses (per-app plan, per-user plan, or pay-as-you-go)
- Power Apps licenses as part of Microsoft Business/Enterprise subscriptions

Direct Power Apps licenses provide full access to Power Apps, including premium connectors. As part of the Business/Enterprise subscription, Power Apps only provide access to standard connectors.

Check Access to Power Apps

Kim decided to check on the licensing arrangement for her AWM account. She self-checked the existing licensing agreement using the following steps, as illustrated if Figure 8-1.

1. Log in to office.com.
2. Select the user logo.
3. Click **View account** in the drop-down.
4. Find Subscriptions and select VIEW.
5. Find the subscription that contains Power Apps.
6. Check if there is a separate license for Power Apps.

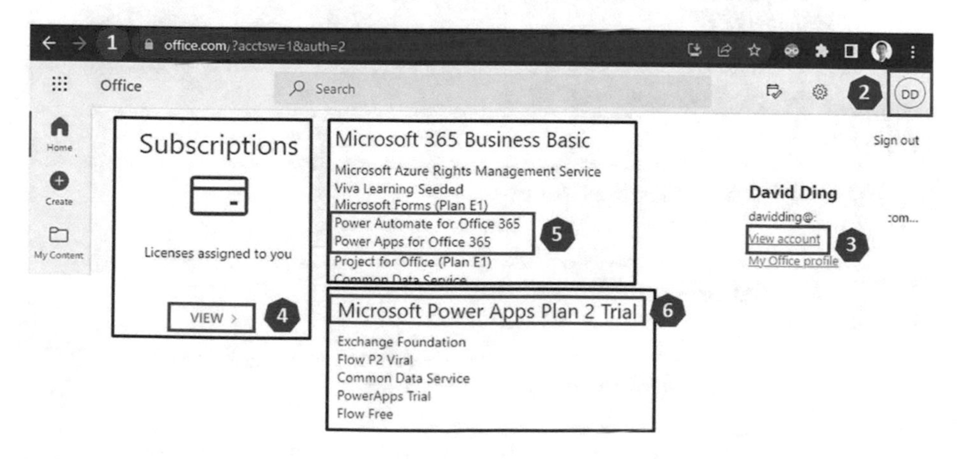

Figure 8-1. *Check existing Microsoft license agreement*

Such as is the case for Kim, many organizations already covered all users with Power Apps licenses through their Microsoft 365 or Office 365 subscriptions. If that is the case for your organization, you can deliver Power Apps using standard connectors.

If your organization doesn't already have Power Apps, you have the option to purchase a Power Apps license or to discuss with your IT department the possibility of choosing a different enterprise plan with Microsoft. You may find that IT is already planning the switch. But if all options fail, your last option is to obtain the Power Apps licenses directly.

Power Apps Connectors

The enterprise license covers Power Apps for the standard connectors. If you google *list of all Power Apps connectors*, you should find a link that gives you a long list of connectors for Power Apps, as shown in Figure 8-2.

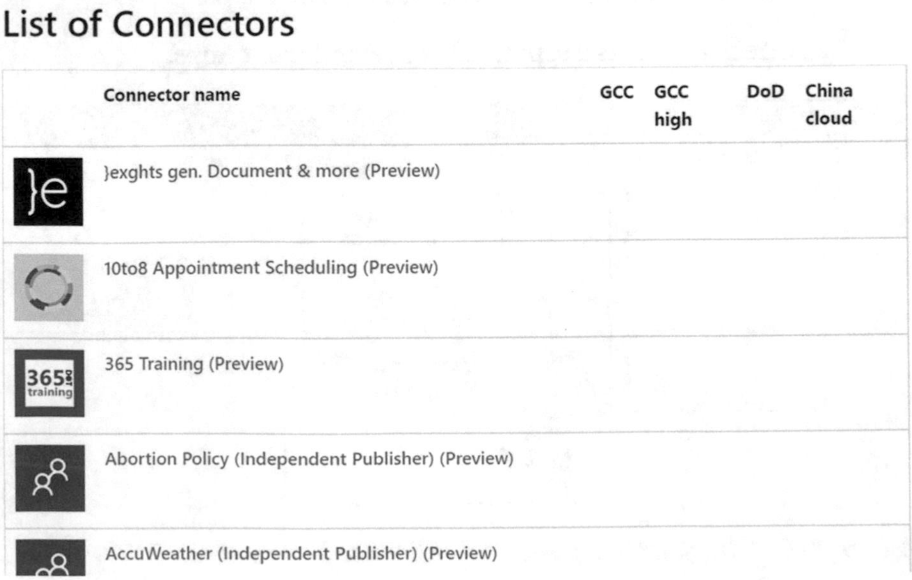

List of Connectors

	Connector name	GCC	GCC high	DoD	China cloud
}e	}exghts gen. Document & more (Preview)				
	10to8 Appointment Scheduling (Preview)				
365 training	365 Training (Preview)				
	Abortion Policy (Independent Publisher) (Preview)				
	AccuWeather (Independent Publisher) (Preview)				

Figure 8-2. *List of Power Apps connectors*

The connectors can be separated into two groups.

Application connectors allow developers to integrate Power Apps with various applications, such as Twitter, LinkedIn, or Microsoft Teams.

Data storage connectors allow developers to integrate Power Apps with various data storages, such as Microsoft SQL Database, MySQL Database, and SharePoint lists.

This book mainly focuses on data storage connectors. Application connectors are partly covered in the next chapter.

Premium Connectors

The enterprise license only covers some standard connectors. While you can develop many applications with SharePoint list and Excel, having a proper enterprise-grade database to support production applications is generally better. There are several reasons for this.

- **Scalability**: Enterprise databases can support billions of lines of records.
- **Speed**: Querying an enterprise database is much faster.
- **Reliability**: There are fewer integration issues associated with the most common database servers. Redundancy is also often built into the design of the database.

If you already have access to an enterprise-grade database, you can justify the cost of a Power Apps license via a strong business case. Access to enterprise databases via premium connectors always be preferred.

Cost and Licensing Strategies

There are three types of Power Apps licenses with premium connectors.

- **Per app**: US$5 per user per app per month. A license is required for all user/developer that accesses the app. If the same user requires access to three separate apps, the user needs three separate licenses (i.e., US$15 per month).
- **Per user**: US$20 per user per month. The per-user plan is more cost-effective if the same user/developer requires access to five or more apps.
- **Pay-as-you-go**: US$10 per user per app per month. Even though this license provides the same access but is more expensive than the per-app license, new users don't need to commit to the per-app license upfront.

The licensing options lead to three licensing strategies: controlled, organic, and hybrid.

- A **controlled strategy** requires the company to specify the users for the app upfront. It assumes that all the users with a license use the app. This strategy gives the team a lot of control over cost. The strategy also requires much more planning, monitoring, and governance effort.
- An **organic strategy** does not require any licensing control process to be put in place. The team only be charged for the month after the user uses the app. This approach encourages a more organic growth of the usage. The overall cost may exceed the controlled strategy for the same level of access.

- A **hybrid strategy** always starts with the organic strategy and overlays its usage monitoring. When a frequent user is identified, the user can be migrated to the per-app license to reduce cost. This is the recommended approach because it can significantly reduce the management effort while keeping the cost low.

Power Apps with a SharePoint List

This book focuses on using SharePoint lists with Power Apps for the following reasons.

- It comes as a standard connector (free) with Power Apps, which makes it ideal as a starting point for many applications.
- Compared to Excel, it comes with delegation allowing the app the load more data.
- It can be integrated with other databases via standard APIs. This allows automating the data transfer from a SharePoint list into an enterprise DB.
- It can integrate with Power Automate for further process automation.

Build the First Power Apps

Now Kim knows everyone in the organization can access Power Apps via standard connectors. She has decided to use Power Apps with a SharePoint list. She can now continue to replace the ticketing system developed using a SharePoint list from the previous chapter.

The good news is that she can continue to use the same SharePoint list and focus on developing the Power Apps as a smart form.

Create an App from a SharePoint List

The following steps create a new app based on the SharePoint list, as shown in Figure 8-3.

1. Go to make.powerapps.com and log in.
2. Choose SharePoint.
3. Choose SharePoint again.
4. Choose the site (AWM Finance).
5. Choose the SharePoint list (Ticketing System).
6. Click Connect.

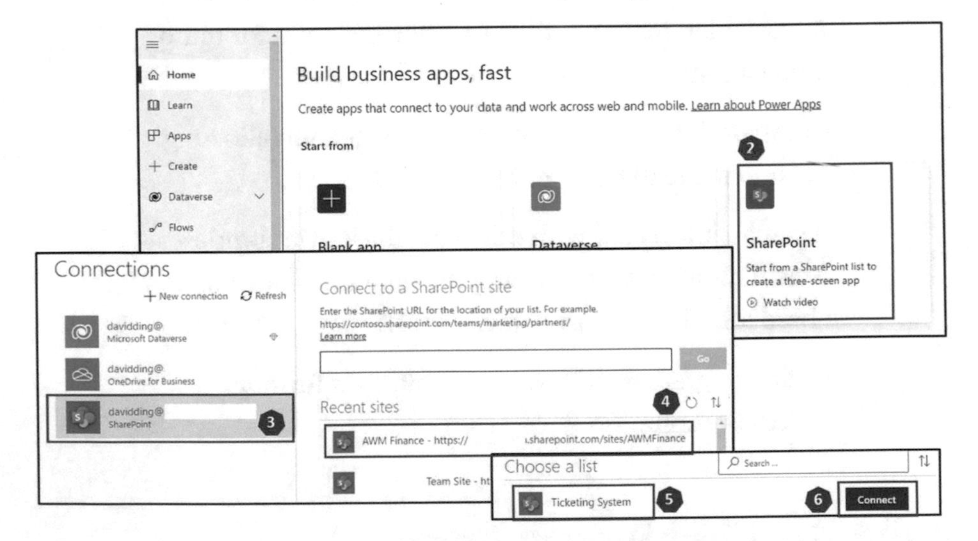

***Figure 8-3.** Create a new app based on SharePoint list*

A new app is created automatically based on the contents from the list. You can see the app if you press F5 or the Play button, as shown in Figure 8-4. This app is referred to as the *default app*.

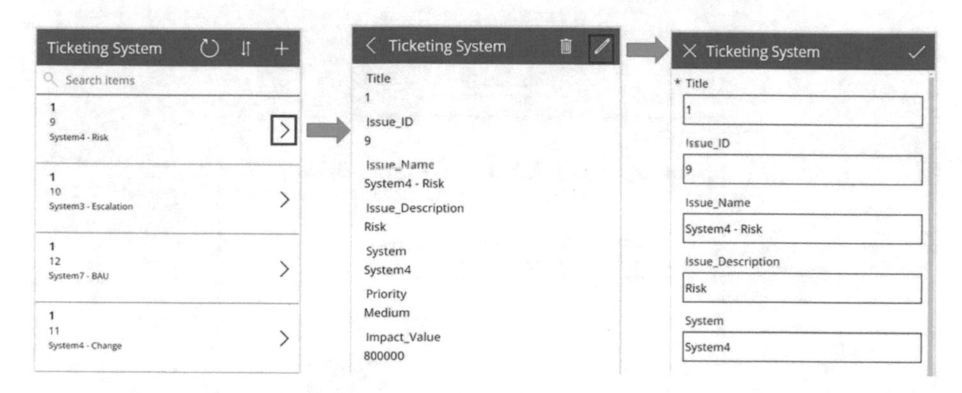

Figure 8-4. *The default app*

There are the following observations about the default app.

- It is a mobile app, given by the portrait layout.
- It contains multiple *screens* for different purposes.
- It allows users to add/remove/edit the SharePoint list items using the app.

You can keep the default app by clicking the Save option and naming it, as shown in Figure 8-5.

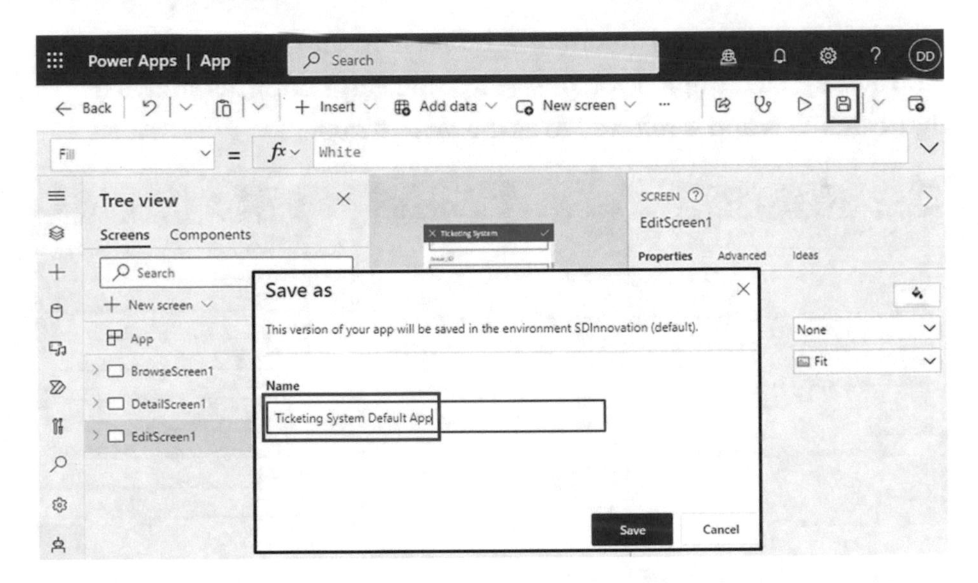

Figure 8-5. *Saving the app*

Understanding Power Apps Studio

Figure 8-6 shows the Power Apps Studio. The following are some of the key components.

1. The **canvas pane** can have different control items placed in it. The design layout of the app follows the WYSIWYG (What-You-See-Is-What-You-Get) approach to simplify the creation process. You can in the canvas pane.

2. The **authoring menu** allows users to switch between the key components of the app, including design elements, data sources, and automation.

3. **Authoring options**: based on the selection in the Authoring Menu, this pane shows the various detail options under the selection.

4. **Properties** is based on the item selected in the canvas. This pane shows the properties related to the selected control item.

5. The **properties drop-down list** is similar to the Properties pane. This drop-down menu shows a list view of the properties.

6. The **formula bar** is based on the property selection from the drop-down list. The formula bar allows you to assign one or multiple functions to the property.

7. The **command bar** is based on the elements or control items selected. The command bar shows a different set of commands.

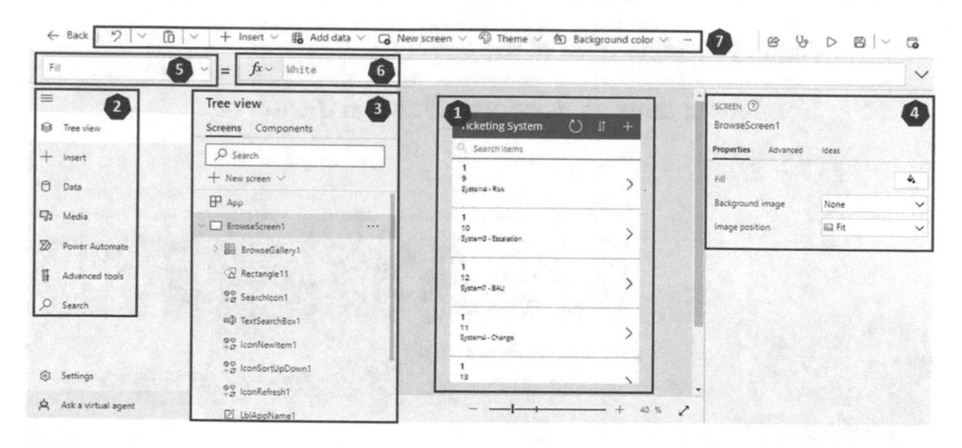

Figure 8-6. *Power Apps Studio overview*

Understanding the Default App

The default app consists of three separate screens.

- A Browse screen with a list of all tickets.
- A Detail screen that displays details of a single ticket.
- An Edit screen that allows the user to edit or add new tickets.

The following sessions break the screen down into different components. This helps you to understand the basic construct of the app.

Browse Screen

In the tree view shown in Figure 8-7, the navigation screen has two parts.

1. **Gallery** displays a list of the tickets.
2. **Header** contains a few control items at the top of the screen.

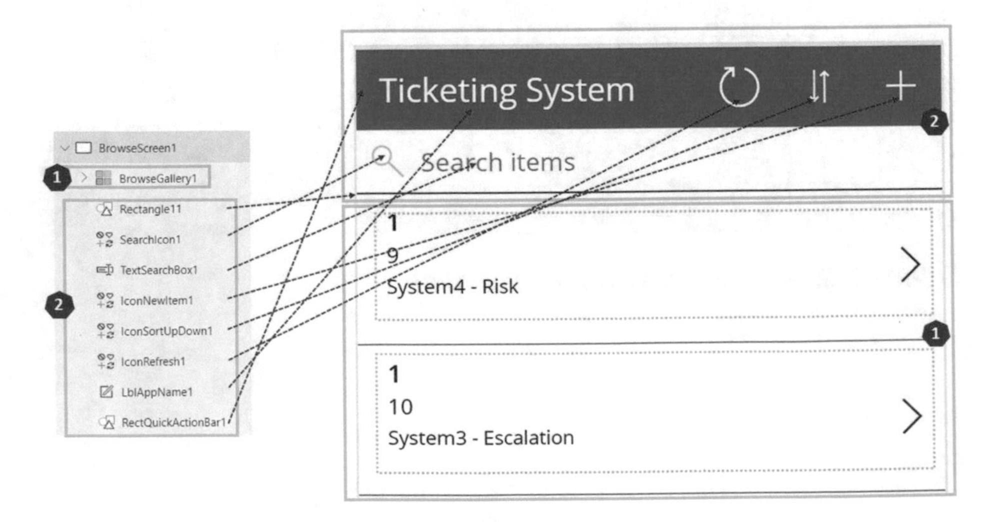

Figure 8-7. *Navigation screen tree view*

Figure 8-7 breaks down the details of the various control items. The gallery is an important feature. It is used to display the *selected fields* from each row of data in the SharePoint list. The gallery display repeats the same display format for all rows of data in the SharePoint list. Figure 8-8 shows that the following fields are displayed automatically in the default app: Title, Issue_ID, and Issue_Name.

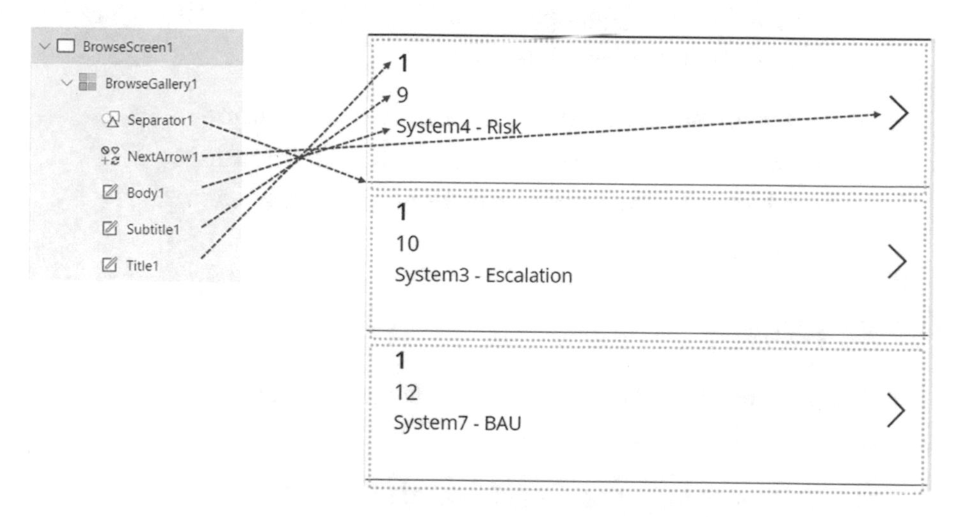

***Figure 8-8.** Controls in the gallery of the navigation screen*

The BrowseGallery1 takes the following data source, as shown in Figure 8-9. The data source is provided in the Items property of the gallery.

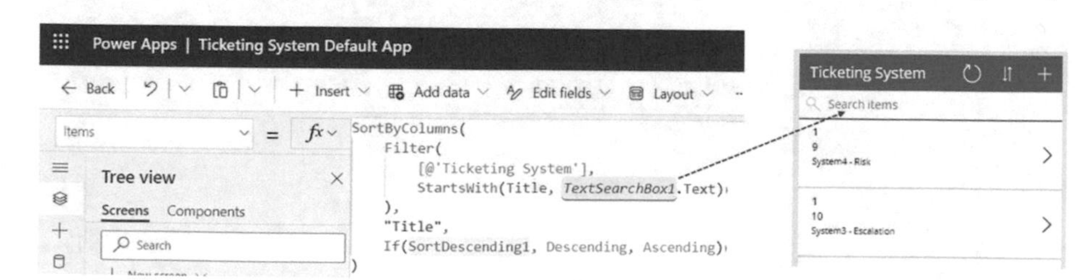

***Figure 8-9.** BrowseGallery1 data source*

The items formula has three parts.

- **Dataset**: The dataset used in this gallery is the Ticketing System SharePoint List.
- **Transformation Filtering**: The Gallery only displays the items with initial letters in *Title* that matches the search bar text (TextSearchBox1). If the search bar is blank, then display all items.
- Transformation Sorting: The sorting order is done according to the Title column. The SortDescending1 is a local context variable that determines the sorting order.

Detail Screen

From the tree view in Figure 8-10, the Detail screen is also made of two parts.

1. **Forms** displays the detail fields of the ticket and
2. **Header** contains a few control items at the top of the screen.

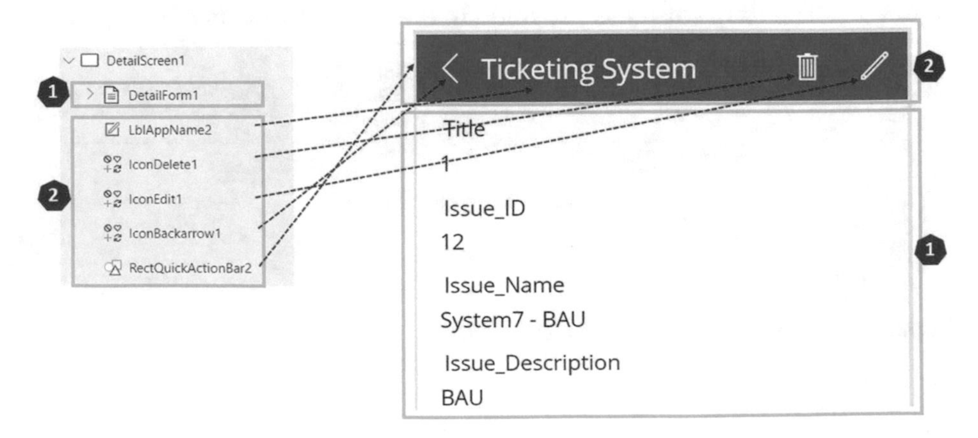

Figure 8-10. Detail screen tree view

Similar to the gallery in the navigation screen, the detail screen uses the Display form to list the different fields, as shown in Figure 8-11. The screen displays the key fields from the same ticket. The form contains multiple cards; each card contains two controls for the field title and value.

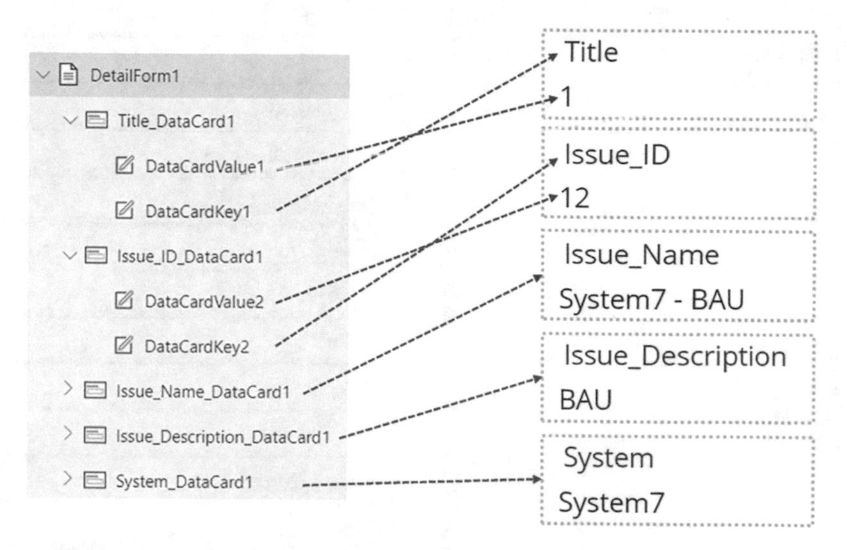

***Figure 8-11.** Forms and cards in the Detail screen*

Note The main difference between gallery and form is that the gallery displays multiple rows of data, but the form displays multiple columns from the same row of data.

Edit Screen

The Edit screen in Figure 8-12 is similar to the Detail screen. It also uses a form. The main difference is that the form in the Edit screen is an Edit form, and the detail screen uses a Display form.

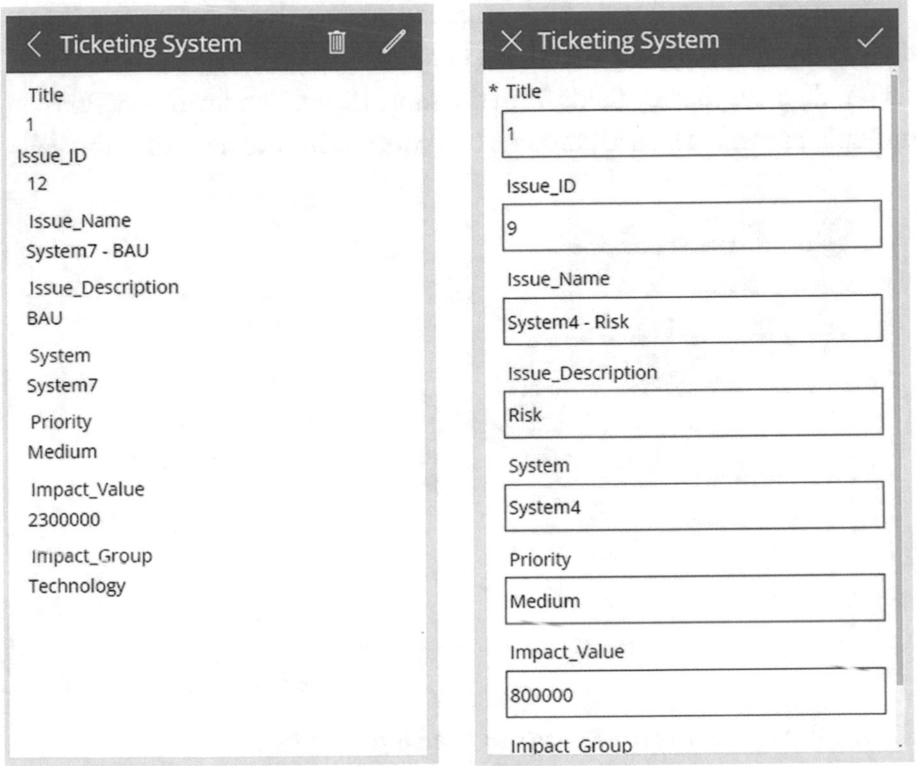

Figure 8-12. *The Display screen compared to the Edit screen*

With the three screens working together, the users can choose the relevant tickets, see the detail, add new tickets, and edit existing tickets on mobile and other devices.

Share the Default App

Even though the default app is far from perfect, Kim decided to share this app with a colleague to test the app. The following steps explain how to share the app, as illustrated in Figure 8-13.

1. Go to make.powerapps.com and select Apps.
2. Select the app and choose Share from the drop-down.
3. Enter the name, salary ID, or email of the person you want to share it with.
4. Decide if you want Power Apps to send an automatic email to notify the users. You can customize the message if you choose to do so.
5. Decide if you want the person have been able to edit the app as a co-owner.
6. Select Share.

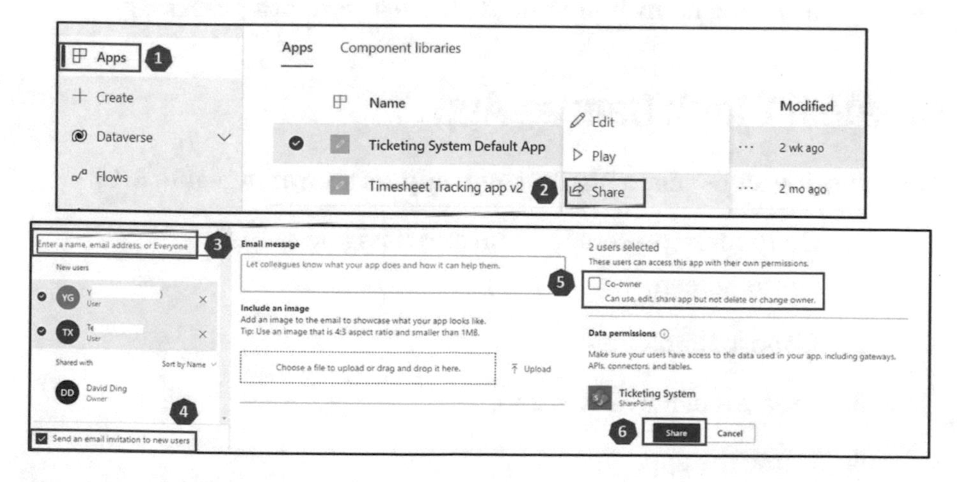

Figure 8-13. *Sharing a PowerApp*

Kim's colleague is very impressed with the app and how he can easily update the backend data. The main point of improvement is that he normally uses a computer to complete the data input, so he prefers a wide screen instead of a mobile screen. He also thinks the benefit of widescreen is that all input fields can be fit into a single page.

Build the App from a Blank Canvas

If you want to build a quick and simple proof-of-concept application, you can use existing templates. For anything else, starting from a blank canvas is recommended.

In addition, when writing this book, it was strongly recommended that you *avoid* the following.

- **Dataverse** usually involves additional licensing costs for both developers and users. While a SharePoint list contains many limitations, it is generally sufficient for internal applications.
- **Forms & Cards** is designed to hide complexity by inserting many limitations to force a standard process.

Create a Blank Canvas App

The following steps start a blank canvas app, as shown in Figure 8-14.

1. Go to make.powerapps.com and navigate to the Home screen.
2. Choose **Blank app**.
3. Create a blank canvas app.
4. Name the app.
5. Choose Tablet or widescreen format.
6. Select Create.

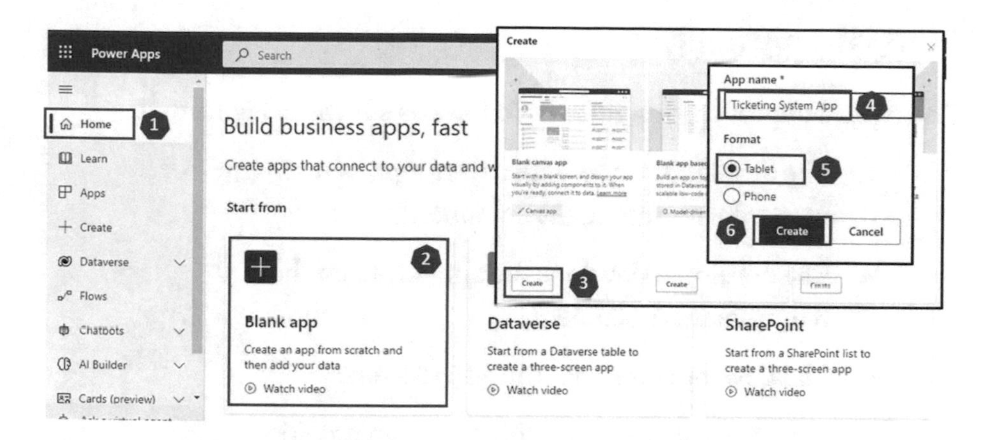

Figure 8-14. *Create a blank canvas app*

The result is a blank app with a single screen, Screen1, as shown in Figure 8-15. This is also the perfect start as it's full of opportunities.

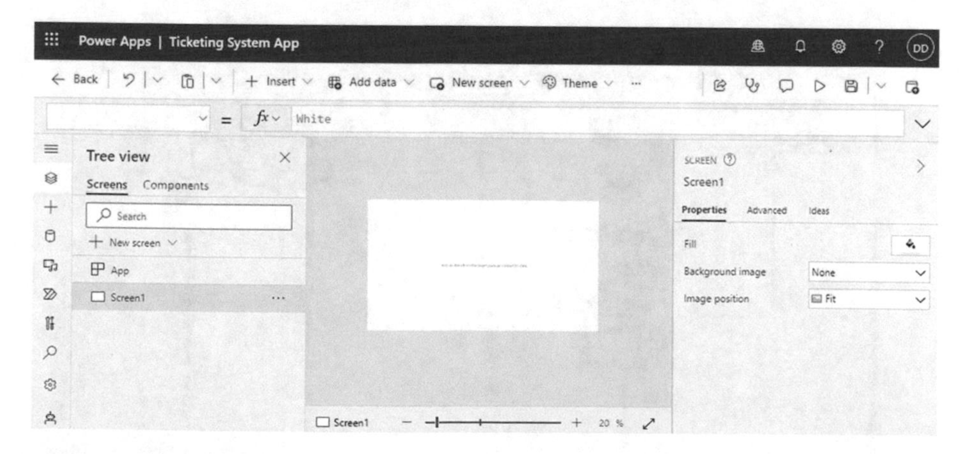

Figure 8-15. *A blank canvas app*

Agile Mindset for App Development

It is often time-consuming to establish the initial starting point. The main concern is that if you start with the wrong design, there will be a lot of work later to rebuild the app.

Typical Approach

The typical approach, described as follows, is shown in Figure 8-16.

1. The developer works with a user representative to create the requirements document.
2. The developer develops the app based on the requirements document.
3. The developer releases the app to users.
4. If changes are required, the users can provide feedback to the representative and feedback on the requirements.

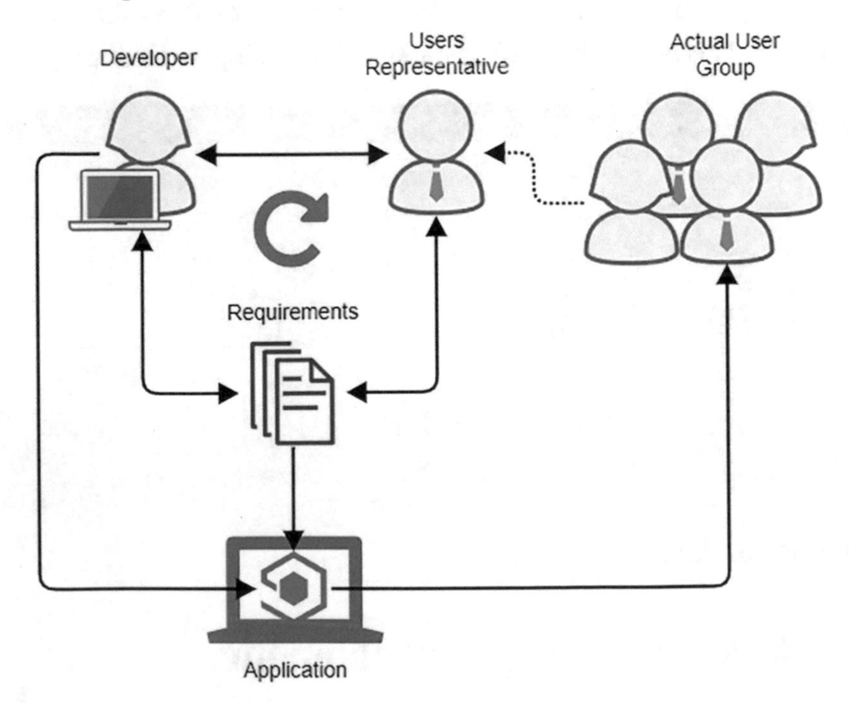

Figure 8-16. *Typical development approach*

Although this approach is the most popular project approach, it has *not* delivered the required outcome for many projects. The causes of project failures vary. Here are a few common issues.

- **Users don't know what they need** at the beginning of the project. This makes the requirements document to be misleading.
- **User requirements change** too often. Even if the requirements are captured correctly initially, there is no guarantee that the requirements will remain the same a few months later.
- Too much **time is wasted on the requirements process**. It is common for the requirements process to take 40% of the total project time. Users won't see any outcome for a ten-month project until six months later.

Agile Approach

When working with tools such as Power BI and Power Apps, the developer can switch to a much more agile mindset aiming at receiving user feedback as early as possible. This process is summarized in Figure 8-17.

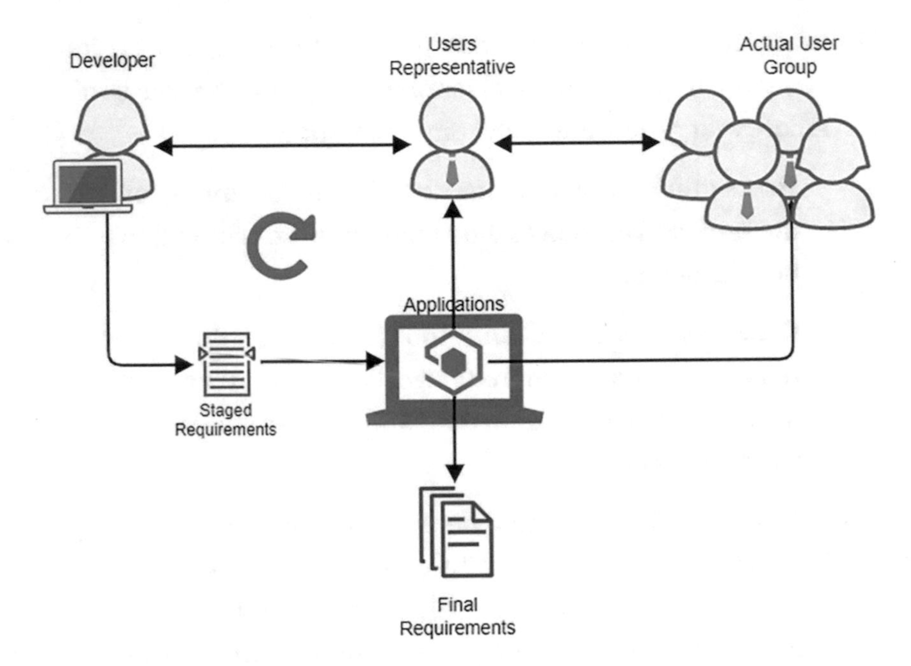

***Figure 8-17.** Development with an agile mindset*

The following are the main differences with the Agile approach.

- The **requirements process** is split into two separate processes.
  - **Staged requirements** are fully controlled by the developer (with help from the team) to guide the next round of implementation/sprint.
  - **Final requirements** are the documentation of the application after the development. This is not to guide further development but to summarize all existing staged requirements into a single document.

- Key **iteration** focuses on application improvement rather than requirements refinement.
- User representatives can use the developed application to **actively seek feedback** from actual users early in the development process.

This approach can be integrated into a flexible, agile process with weekly sprints and daily catchups. The development process can be flexible because Power Apps generally only requires a small team of one developer, one user representative, and various other subject matter experts as required.

Establish the Initial Requirements

The initial requirement can be extracted from the existing data sources, tools, and processes in the ticketing system scenario. This is often the case with solutions that must be migrated from one platform to another. The following requirements are gathered in Chapter 7.

1. The system allows **many users** to **create** new incidents.
2. The system allows users to upload screenshots.
3. The system allows admins to **make changes** to the incident.
4. The system contains incident management **reports**.

You also know that the data is already stored in a SharePoint list. This does not change as, in this case, the Power Apps aims to provide a better user experience for entering the data.

With the agile iteration approach discussed earlier, you only need a starting point for the initial sprint. You might feel this requirement to be abstract because it doesn't have any user experience requirements. Instead of having three separate screens, you can try to merge the browse, detail, and edit screens into a single screen, as shown in Figure 8-18.

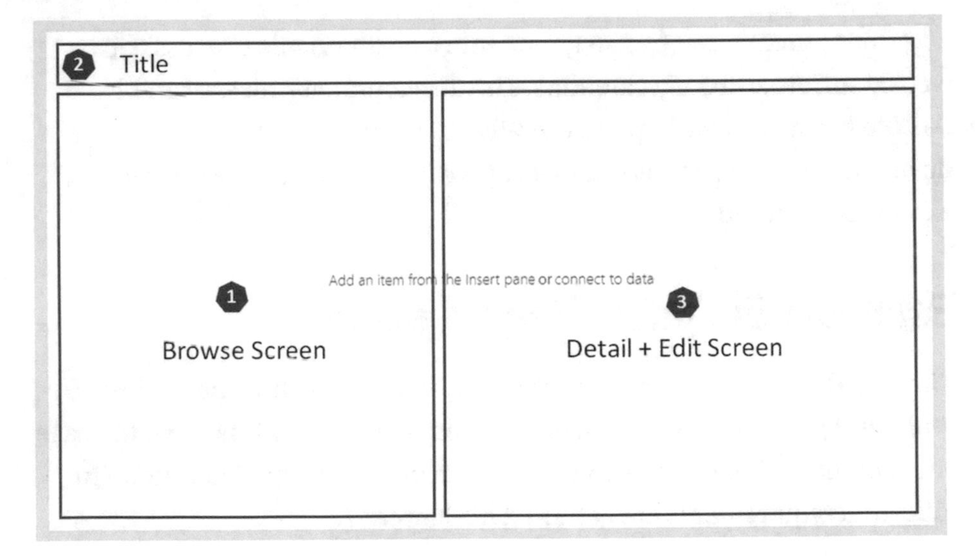

Figure 8-18. *Custom app design*

Tip The starting point doesn't matter. The key is to deliver a prototype way ahead of users' expected timeline. Speed is a quality in its own right. The purpose of any prototype is to facilitate the next round of improvement discussion.

Add Data

The first step in creating the app is to add the data source, in this case, the ticketing SharePoint list. The following steps add the list, as shown in Figure 8-19.

1. Go to the data view.
2. Choose **Add data**.
3. Search for "SharePoint" and choose SharePoint.
4. Choose the relevant SharePoint site.
5. Choose the list.
6. Click Connect.

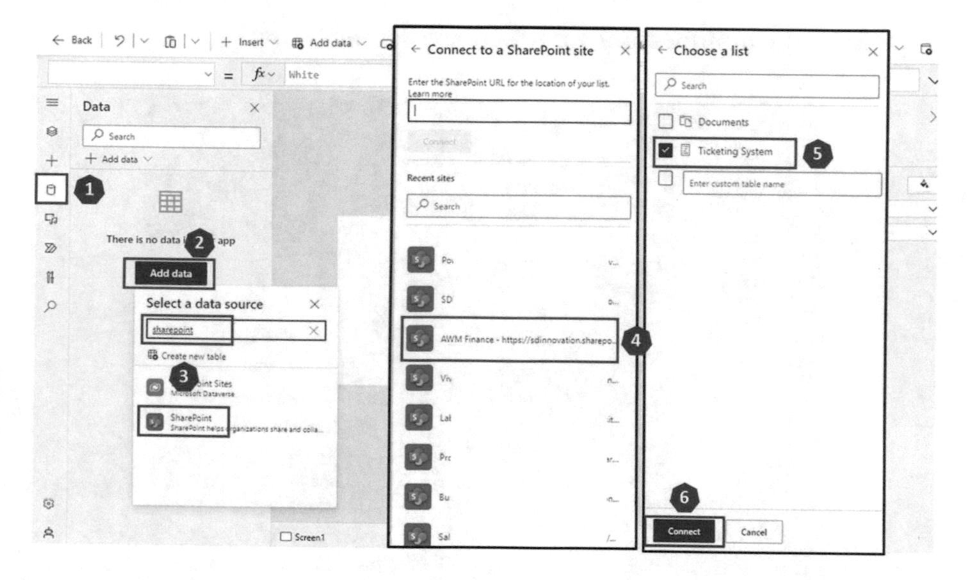

Figure 8-19. *Steps to add a data source*

With the data connection established, you can move on to screen development.

Note If you don't yet have data in a SharePoint list, the first step is creating a SharePoint list with some data by following Chapter 7. This ensures that the galleries and forms display the correct data during development.

Main Screen

Before adding controls to the screen, rename Screen1 to **BrowseScreen**. The following steps explain how to do so, as shown in Figure 8-20.

1. Choose **Tree view.**
2. Select more options next to Screen1.
3. Choose Rename.
4. Type in BrowseScreen to replace Screen1.

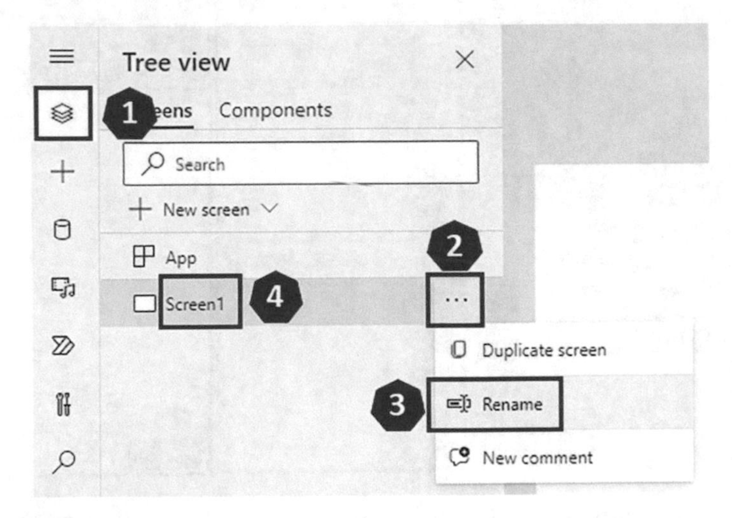

Figure 8-20. *Rename the screen*

Gallery

The gallery is one of the most important layout control items in Power Apps. It allows users to customize the display of a data table.

Create a Blank Vertical Gallery

You can now add a *blank vertical gallery* to the BrowseScreen. The following steps explain how to do it, as illustrated in Figure 8-21.

1. Choose Insert.
2. Expand Layout.
3. Choose Blank vertical gallery.
4. Choose Ticketing System.

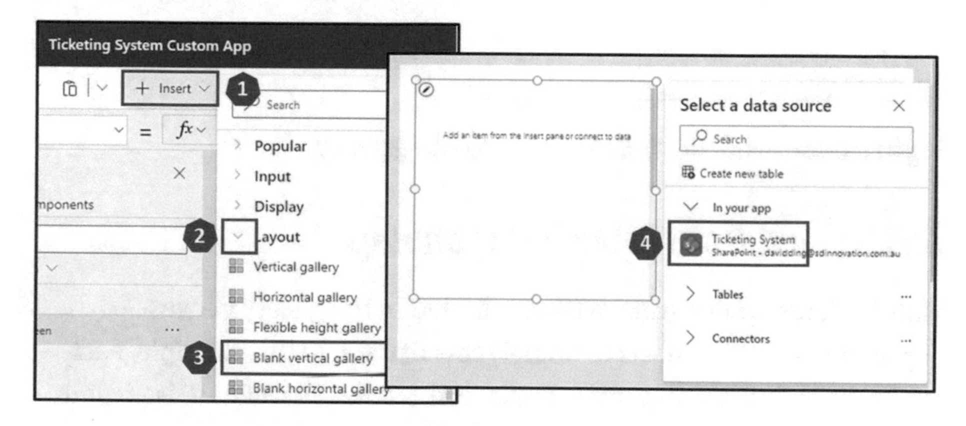

Figure 8-21. *Add a new gallery*

The next step is to add some control items to the gallery. The following steps explain how to do it, as shown in Figure 8-22.

1. Choose Insert View.
2. Drag **Text label** into the gallery and drop it inside the purple square.
3. Select **Text label** and change the text property from ThisItem.Title to **ThisItem.Issue_Name**.

4. The Issue_Name text was displayed twice in the gallery with two different display names.

5. Drag another text label into the gallery to display ThisItem.ID.

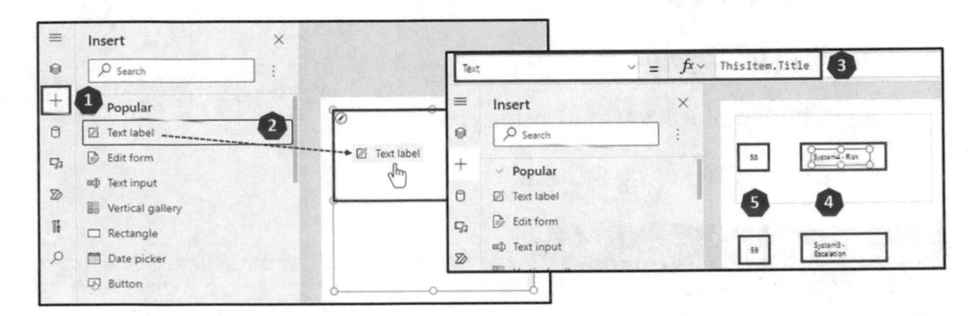

Figure 8-22. Adding information to the gallery

Gallery and Data Table Relationship

Multiple lines appear when a label is added to the gallery because each box in the gallery connects to a single row of data. In this example, each box in the gallery connects to a single ticket in the SharePoint list. If you configure the gallery to include the following fields, you can see the result in Figure 8-23.

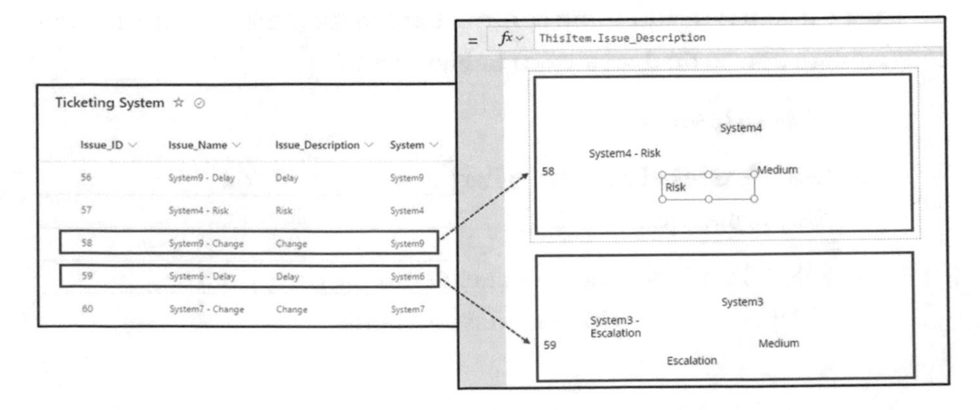

Figure 8-23. Gallery items relate to rows in SharePoint list

As you might have noticed when adding new text labels in the example, a drop-down box appears as soon as you add the dot after ThisItem. Each field in the drop-down refers to a column in the SharePoint List. This is an easy way of navigating the fields without remembering the exact name. You can follow the steps below as shown in Figure 8-24 to check the options.

1. Select a Text Label Control.
2. Choose Text property.
3. Type in ThisItem followed by a dot ".".
4. Check the available options.

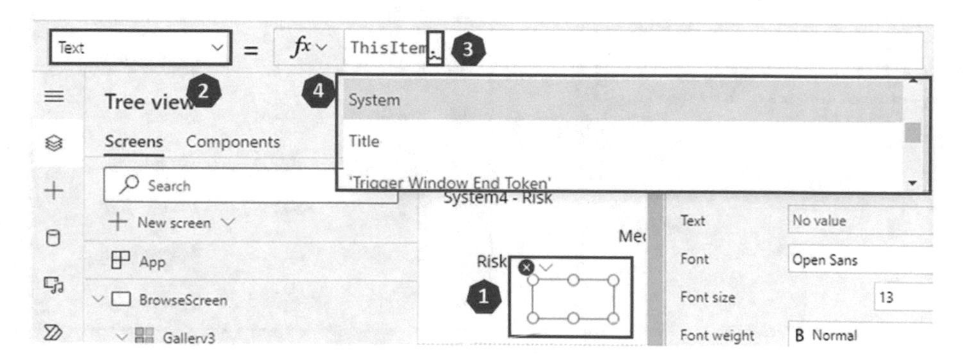

Figure 8-24. *Adding a dot after ThisItem to see available options*

As this case indicated, **ThisItem** refers to a particular row of data. ThisItem is only available in a few controls, the gallery being one of the most common ones that utilize ThisItem.

Note As you can see, one of the benefits of the gallery box is that it allows users to organize the data displays in a flexible way to suit its purpose. If in doubt about the design, the safe option is to limit it to a single row of data.

Organize a Gallery

The gallery box can be configured. The following steps explain how to do it, as shown in Figure 8-25.

1. Change the size and position of the text labels.
2. Adjust the size of the gallery box by reshaping the height of the box.

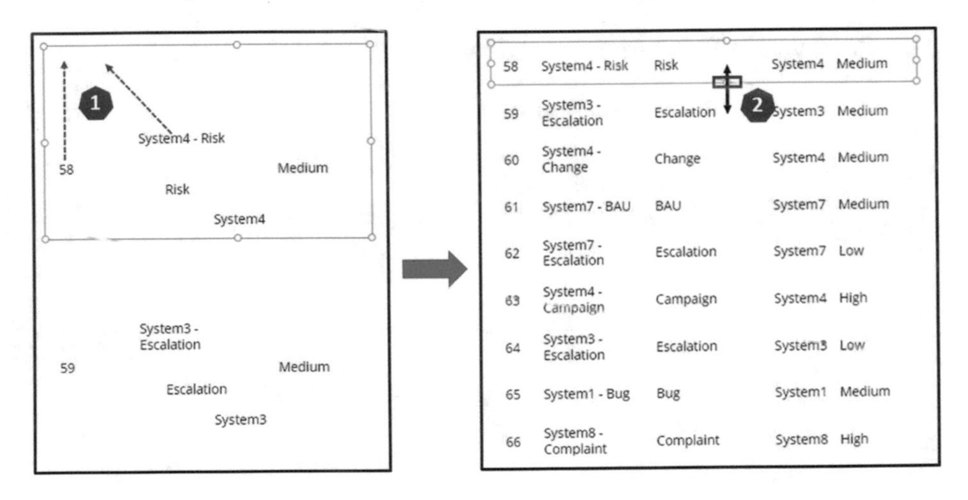

Figure 8-25. *Reorganize the gallery*

As you can see, the gallery repeats the control items and arrangements of the gallery box and repeats it for every row of data from the source.

To separate the different boxes, you can add a separator—a rectangle with a height of 1. You can add the separator by completing the following steps, as shown in Figure 8-26.

1. Drag and drop a Rectangle from Insert View into the *top* gallery box.
2. Change the Rectangle Height property value to 1.

3. Change the Rectangle X position to 0 to start the separator from the left.

4. Adjust the Rectangle Width property to an appropriate value according to the layout.

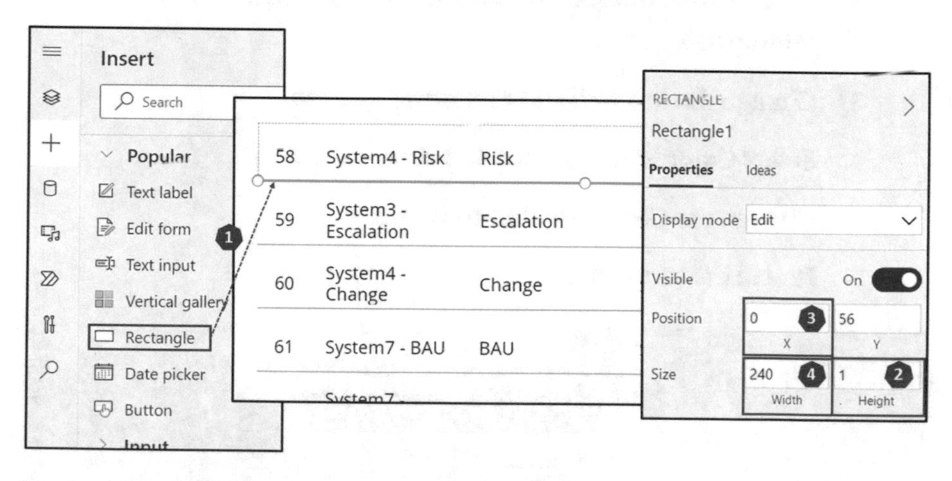

Figure 8-26. Adding a separator to gallery

With the gallery in place, you can add a title control box to the top. The following steps explain how, as shown in Figure 8-27.

1. Drag the Rectangle from Insert View to the top of the screen.

2. Reshape the Rectangle to make it fit the screen.

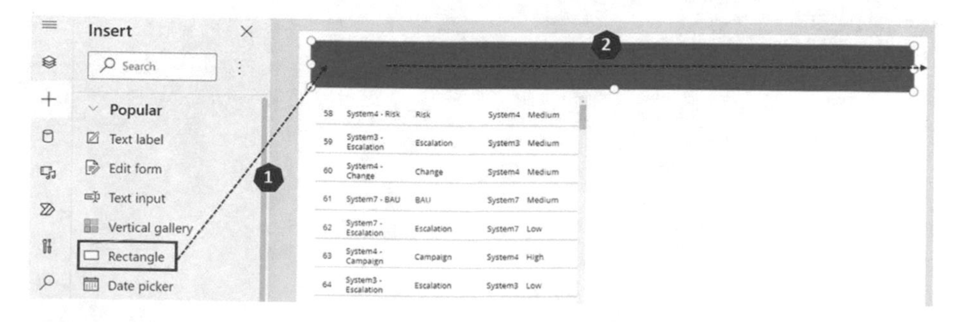

Figure 8-27. Add a title box to the screen

The following steps add another text label to the screen, as shown in Figure 8-28.

1. Drag and drop the text label from Insert View.
2. Change the Font size to 40 (or what you think is appropriate).
3. Change the text to **Issue Ticketing System**.
4. Select Color from the Property drop-down.
5. Change the Color function to White.
6. Resize the text label.

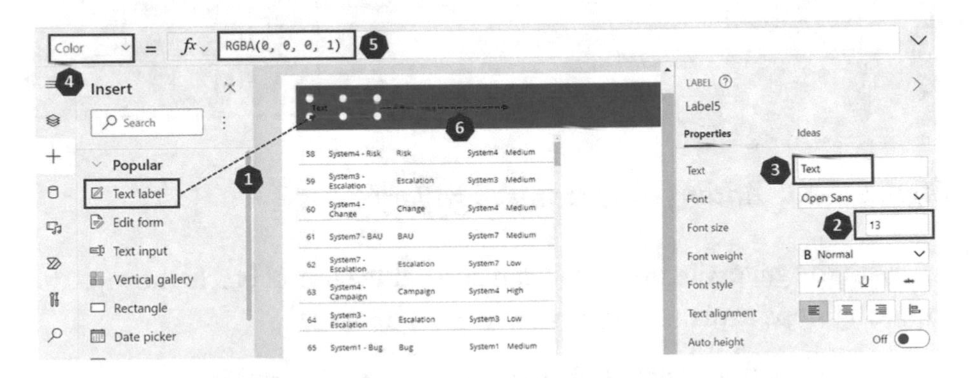

***Figure 8-28.** Add and configure title text*

At this point, you can see the screen that looks like what's shown in Figure 8-29.

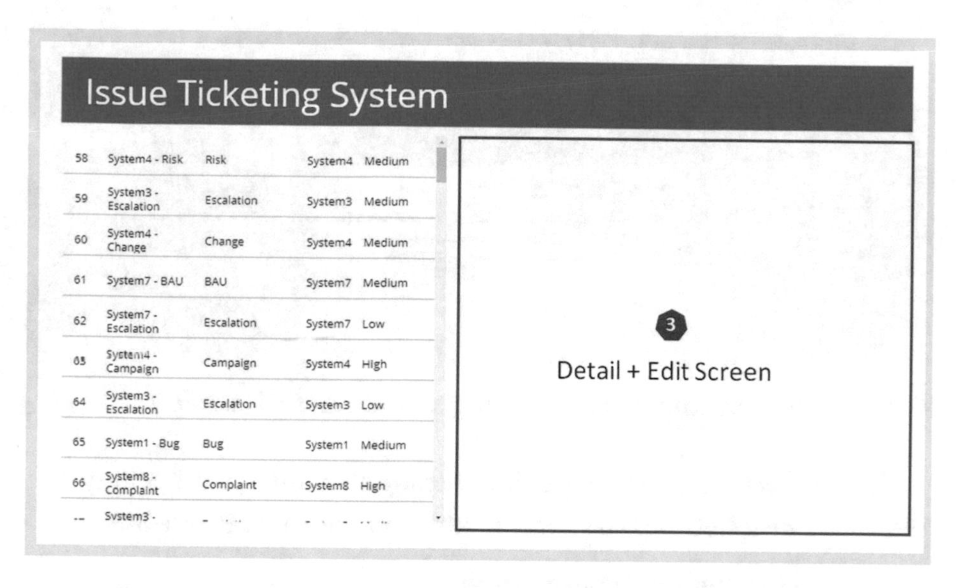

Figure 8-29. *Current main screen WIP*

Edit Screen

The last session is for the Edit and Display controls. Start with the following steps, as shown in Figure 8-30.

1. Drag and drop **Text input** from Insert view.
2. Resize the input control to the appropriate size.
3. Drag and drop **Text label** from Insert View.
4. Change the default text to **Issue ID**.

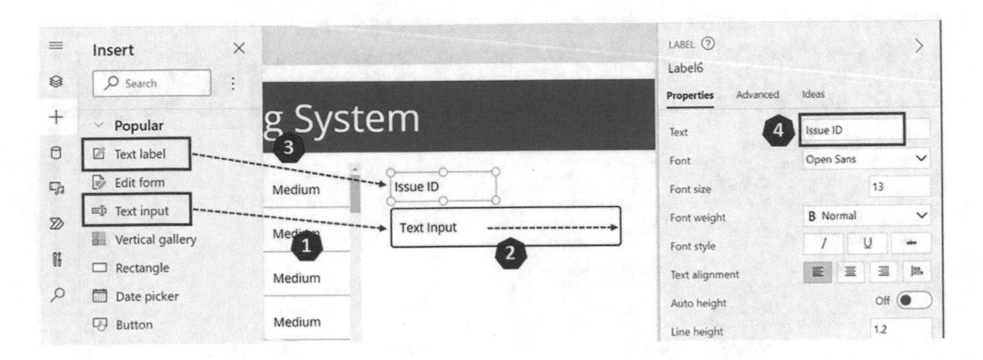

Figure 8-30. *Adding first control items into the Edit session*

The idea is that the Text input box display according to the gallery selection. The following is a simple method, as shown in Figure 8-31.

1. Select the Text Input control.
2. Select Default in the property drop-down.
3. Set the function to *Gallery1.Selected.ID*.

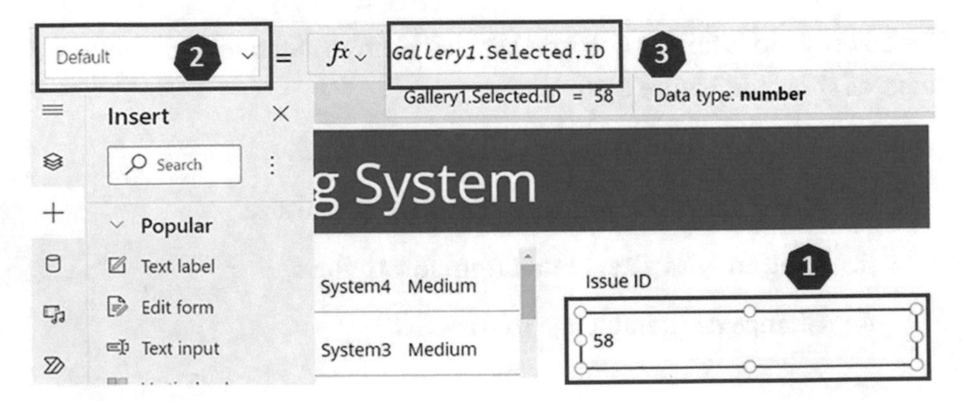

Figure 8-31. *Set up gallery selection control*

You can test the app by using the Play button at the top-right corner of the browser or clicking F5 on the keyboard. Alternatively, you can hold the ALT key on the keyboard and use the mouse to test the interaction.

If the text input control is configured correctly, the Issue ID should change when you click different boxes in the gallery, as shown in Figure 8-32.

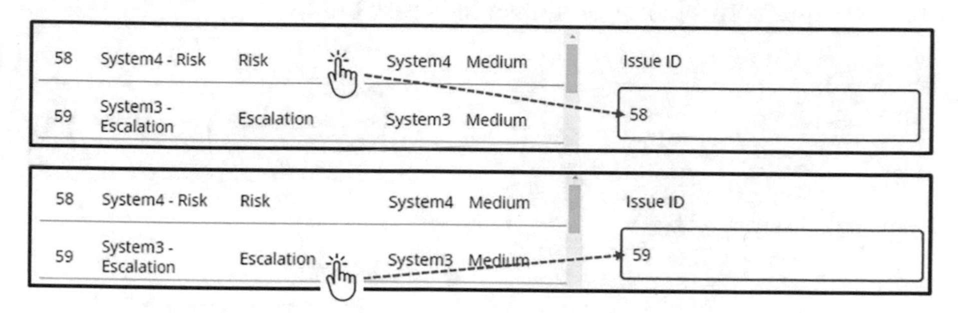

***Figure 8-32.** Testing gallery interaction*

Once it is working, you can copy and paste the text label and text input seven times to create eight input fields, as shown in Figure 8-33. Then change the name to the appropriate fields.

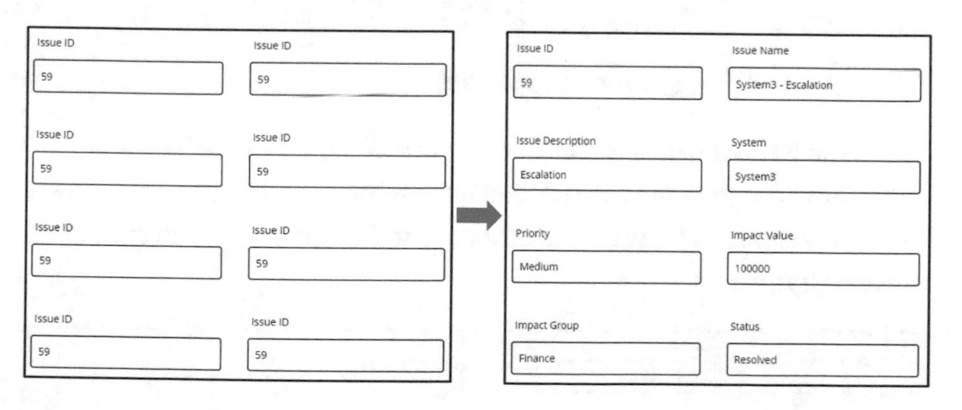

***Figure 8-33.** Copy and configure the control items in the Edit screen*

The current layout in Figure 8-33 looks similar to what you need. But there are two main issues.

- All fields are manual input.
- It does not allow writing back to the SharePoint list.

Edit to Display Switch

You can address the first problem through the **Display mode** property of the Text Input control item, as shown in Figure 8-34.

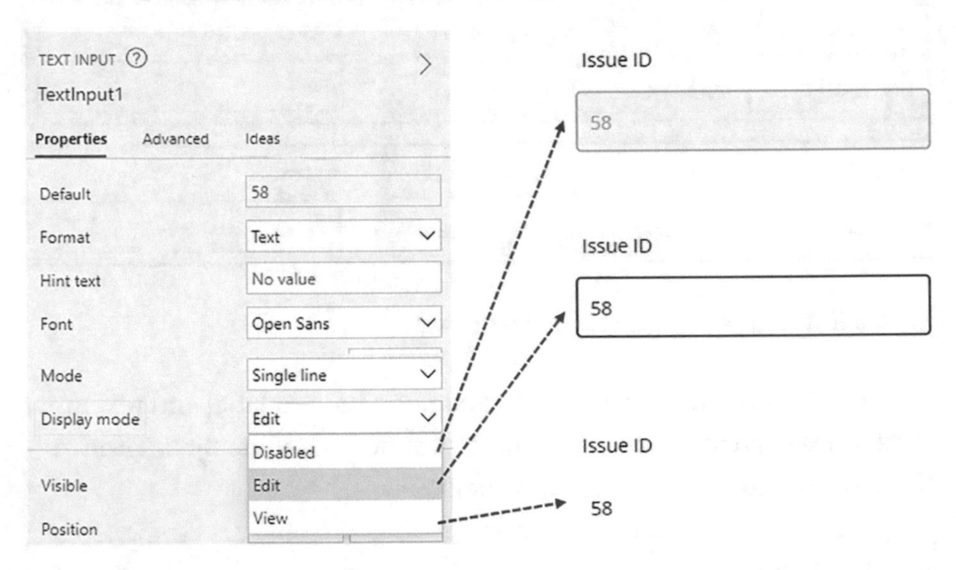

Figure 8-34. *Display mode properties*

For the Issue ID display, you can use the Disabled mode because SharePoint autogenerates the ID. You do not have any control. You can set the Display mode to View for the other control items. The result should look like Figure 8-35.

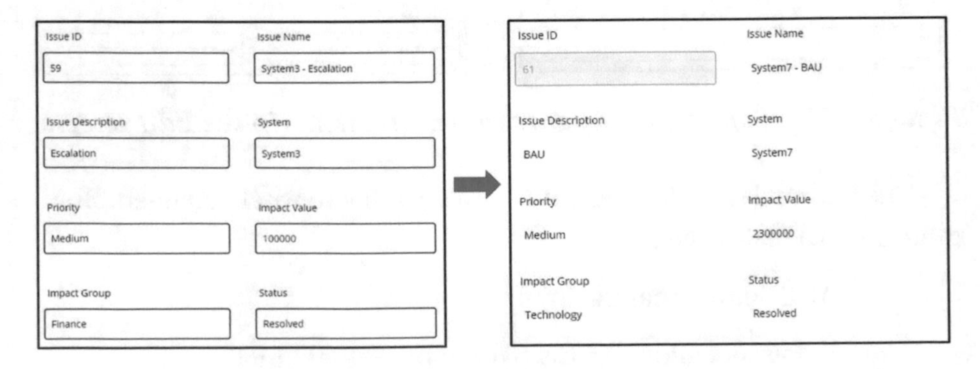

Figure 8-35. *Convert Display mode for the Detail screen*

You can now convert this view into Edit mode by introducing a button and a *global variable*. You can start by adding an Edit button by completing the following steps, as shown in Figure 8-36.

1. Drag and drop Button to the appropriate place.
2. Rename the button **Edit** in the default text.
3. Choose the OnSelect property.
4. Enter thc following function.
    - Set(varEdit, true)
5. Create a Save button using the same steps but using the following function.
    - Set(varEdit, false)

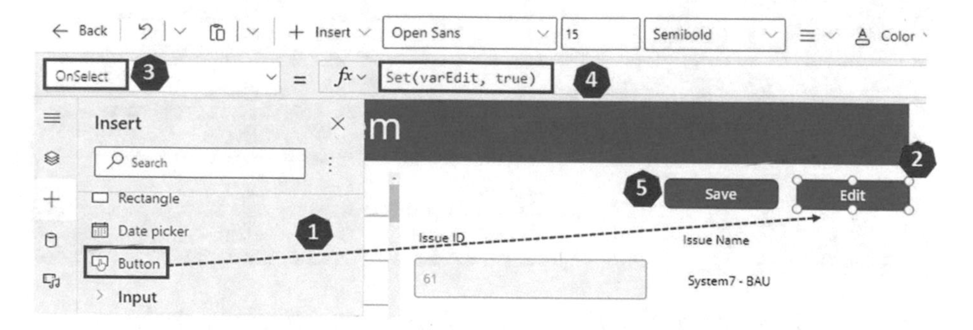

***Figure 8-36.** Create an Edit button to set value for varEdit*

What the Edit button is doing is that when it is selected, it sets the value of varEdit (a global variable) to true. The Save button does the opposite by setting the value of varEdit to false.

Next, you need to change the DisplayMode for the Text inputs based on the value of varEdit. The following steps explain how to do this, as illustrated in Figure 8-37.

1. Select all seven text input control items (except for Issue ID). You can do this by holding down the Ctrl key and left-clicking each.
2. Select DisplayMode from the drop-down property list.
3. Change the function to the following.
   - If(varEdit = true, Edit, View)

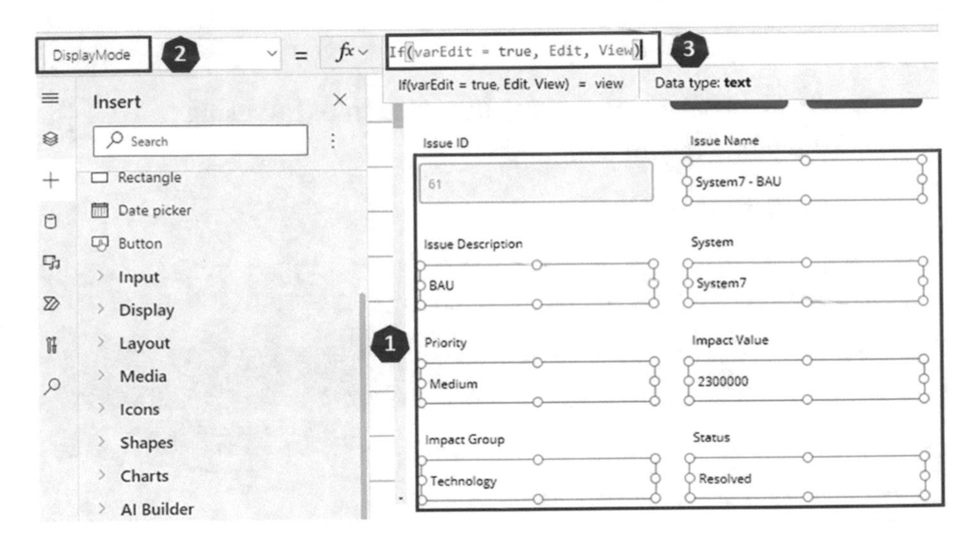

Figure 8-37. *Change the DisplayMode for text input controls*

Now if you click the Edit button and then the Save button, you should see the result shown in Figure 8-38.

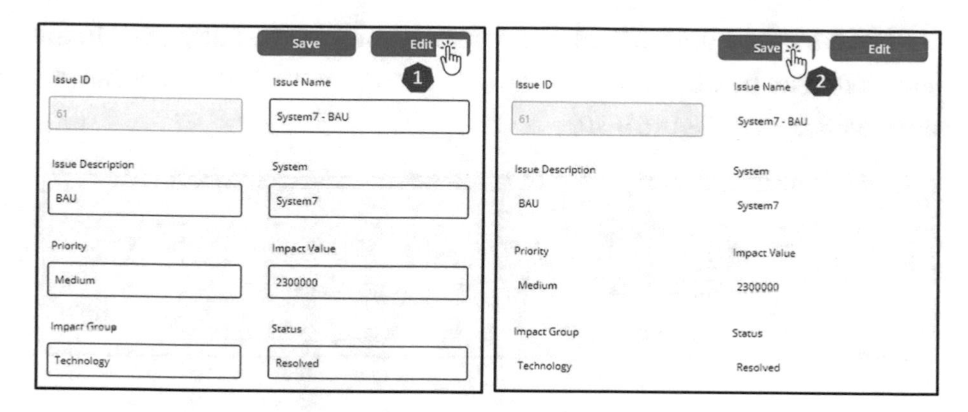

Figure 8-38. *Test the Edit and Save buttons*

Add a New Record Button

In addition to viewing and editing existing tickets, the tools also need to allow the user to add new tickets. You can start by adding a New button. Follow the steps shown in Figure 8-39. Similar to the Edit button, the New button creates a new global variable called *varNewTicket*. The value is set to true on select. Because the New ticket button requires edit fields, the *varEdit* variable must also be set to true.

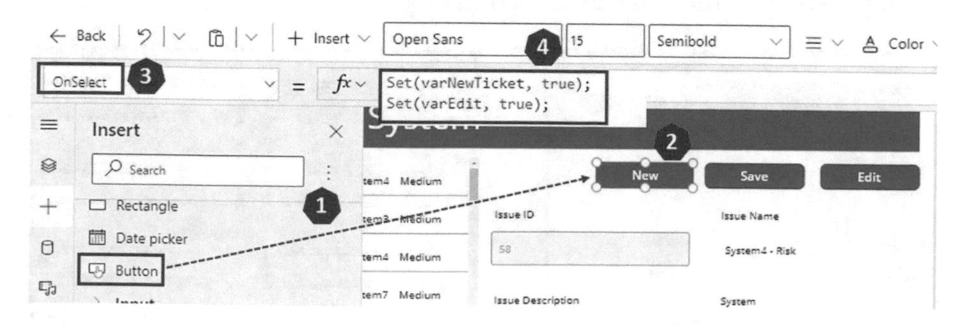

Figure 8-39. *Adding a new ticket button*

When the new button is clicked, it removes all the text displayed in the fields. This can be done by embedding an IF logic with the *varNewTicket* value, as shown in Figure 8-40.

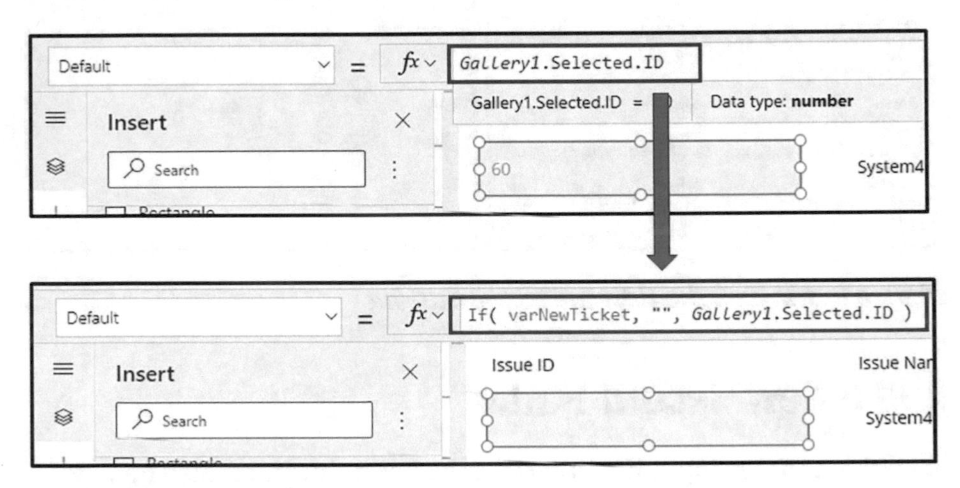

***Figure 8-40.** Change the input default text to display nothing for a new ticket*

Repeat the preceding steps for all text input control items. When the New button is clicked, it should turn the page to switch to Edit mode while displaying no default value, as shown in Figure 8-41.

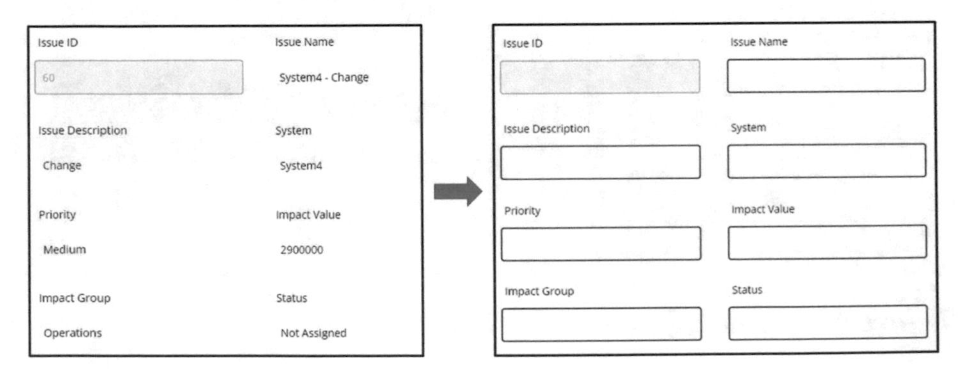

***Figure 8-41.** Effect of new button*

It is always important to find a way to reset the *varNewTicket* and *varEdit* to false. Otherwise, according to the If logic, nothing ever be displayed again. One logical place is to do the reset in the gallery by following these next steps, as shown in Figure 8-42.

1. Click anywhere in the gallery (except for the top box) to select the entire gallery.
2. Choose OnSelect from the drop-down property.
3. Enter the following formulas.
    - Set(varNewTicket, false);
    - Set(varEdit, false)

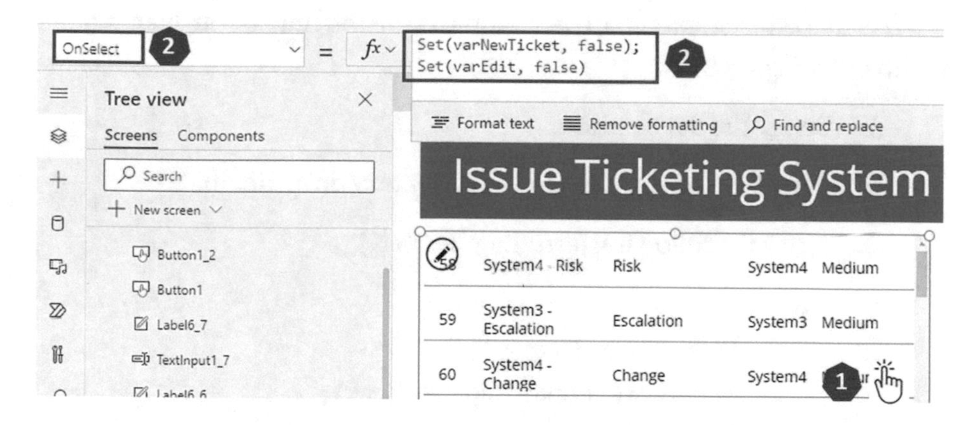

Figure 8-42. *Use gallery OnSelect to reset varNewTicket*

At this point, the custom ticket app is pretty much all done. The only thing missing is the configuration to save it to a SharePoint list.

Save to a SharePoint List

In the prior example, the Save button only switches the Display mode from Edit back to View. You need to configure the Save button to save the value to the SharePoint list. To do this, you need to use the *Patch*() function.

The Patch() function can be used in two ways: to add a new record to the data source or to change an existing data source.

Use Patch to Add New Data

The Patch() function takes three parameters as follows.

- Data source (in this case, ticketing system connection to the SharePoint list)
- Default/Record (uses Default for a new item or specifies the record for edit)
- New records data

The following steps set up the Patch to save the new Issue_Name field, as shown in Figure 8-43.

1. Select the Save button.
2. Select the oneOnSelect in the property drop-down.
3. Enter the following formula after Set().

```
Patch(
      'Ticketing System',
      Defaults('Ticketing System'),
      {
            Title: "X"
            Issue_Name: TextInput1_1.Text
      }
)
```

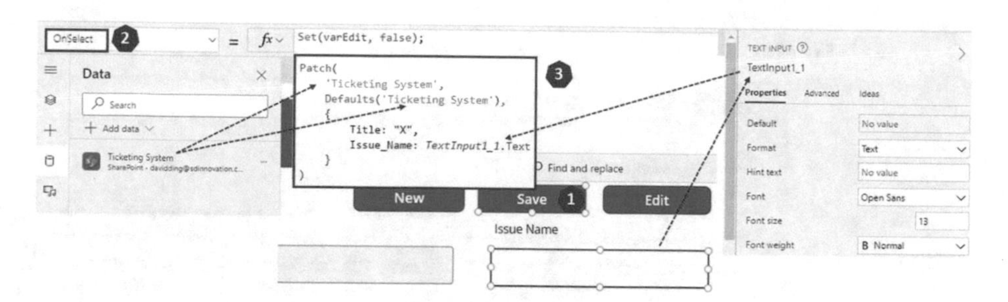

Figure 8-43. *Saving a new ticket with the Issue Name field*

The semicolon is a separator between different functions. This allows the same property to contain multiple actions.

Caution The Title record is required because it is a mandatory field in a SharePoint list. You receive an error when clicking the Save button without the Title field.

You can test this code by following these next steps.

1. Click the New button.

2. Enter a value under Issue Name.

3. Go to the SharePoint list and check the new row of data (try Sort descending by ID).

If you follow these steps, you should see a new line of record in the SharePoint list, as shown in Figure 8-44.

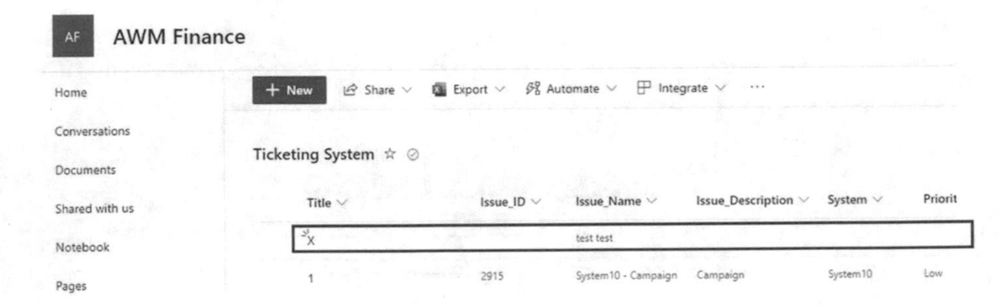

***Figure 8-44.** New record created*

You can add the other Text Input fields into the Patch function. Once you have done that, assuming the SharePoint list data type for Impact_Value is set to Number, you should see an error message, as shown in Figure 8-45.

```
Patch(
    'Ticketing System',
    Defaults('Ticketing System'),
    {
        Title: "X",
        Issue_Name: TextInput1_1.Text,
        Issue_Description: TextInput1_2.Text,
        System: TextInput1_3.Text,
        Priority: TextInput1_4.Text,
        Impact_Value: TextInput1_5.Text,
        Impact_Group: TextInput1_6.Text,
        Status: TextInput1_7.Text
    }
)
```

The type of this argument 'Impact_Value' does not match the expected type 'Number'. Found type 'Text'.

The function 'Patch' has some invalid arguments.

***Figure 8-45.** Adding all input fields into Patch*

The cause of the error is indicated in the error message. The Impact_Value field in the SharePoint list is a number format, but the TextInput control items produce a text format. To fix this, you can simply use the Value function to convert the Text into a Number.

```
Impact_Value: Value(TextInput1_5.Text)
```

The various data types in a SharePoint list can lead to many issues and frustration when transmitting data. That's why you want to limit the data types in the SharePoint list to a single line of text, multiple lines of text, and numbers. Avoid using other types, such as date and person.

Tip It is always a good idea to start troubleshooting by reading the error message. It is useful most of the time. Another common method is copying and pasting the error message into Google Search.

Use Patch to Modify Data

In this ticketing system design, the Save function performs two functions.

- It creates a new row of data if the user selects the New button.
- It updates an existing row of data if the user selects the gallery.

In an earlier example, you implemented the code for the new row. You can change it to update existing rows, as shown in Figure 8-46.

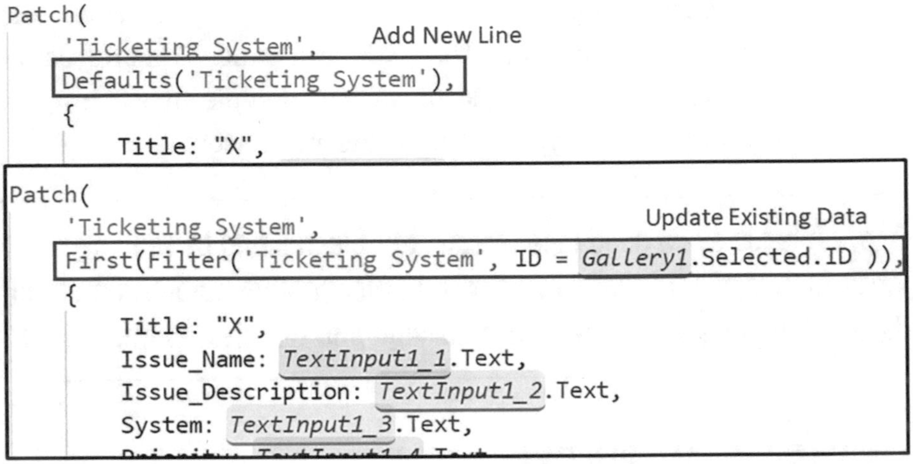

```
Patch(
    'Ticketing System',          Add New Line
    Defaults('Ticketing System'),
    {
        Title: "X",
```

```
Patch(
    'Ticketing System',                          Update Existing Data
    First(Filter('Ticketing System', ID = Gallery1.Selected.ID )),
    {
        Title: "X",
        Issue_Name: TextInput1_1.Text,
        Issue_Description: TextInput1_2.Text,
        System: TextInput1_3.Text,
```

Figure 8-46. *Change Patch from new line to update existing line*

The First and Filter function combination ensures that a single row of the selected record is selected for update. You need two parameters within the Filter function: the data source and the filter logic. It is generally a good idea to implement a unique key for this. A SharePoint list automatically generates a unique ID column for all records. This ID column is perfect for record update purposes.

Now you have the Patch function working for new and updated records. The final challenge is to determine when to use which. Remember that you can get to the edit page in two ways: selecting the gallery or hitting the new button. This means you need a global variable to track which method was selected. Luckily, you have already created a global *varNewTicket* variable, for that purpose. Now you can use it in an IF function, as shown in Figure 8-47.

```
If(
    varNewTicket,
    Patch(
        'Ticketing System',
        Defaults('Ticketing System'),
        {
            Title: "X",
            Issue_Name: TextInput1_1.Text
            ...
            ...
            Impact_Group: TextInput1_6.Text,
            Status: TextInput1_7.Text
        }
    ),
    Patch(
        'Ticketing System',
        First(Filter('Ticketing System', ID = Gallery1.Selected.ID)),
        {
            Title: "X",
            Issue_Name: TextInput1_1.Text
            ...
            ...
            Impact_Group: TextInput1_6.Text,
            Status: TextInput1_7.Text
        }
    )
)
```

Add New Line

Update Existing Data

***Figure 8-47.** Add and Update code in IF function*

Note The varNewTicket variable is already a boolean, meaning it can either be true or false. That's why you don't need to run further checks in the IF logic, such as varNewTicket = true.

Kim now completed the initial migration of the ticketing system from a SharePoint list to Power Apps. The initial app is shown in Figure 8-48.

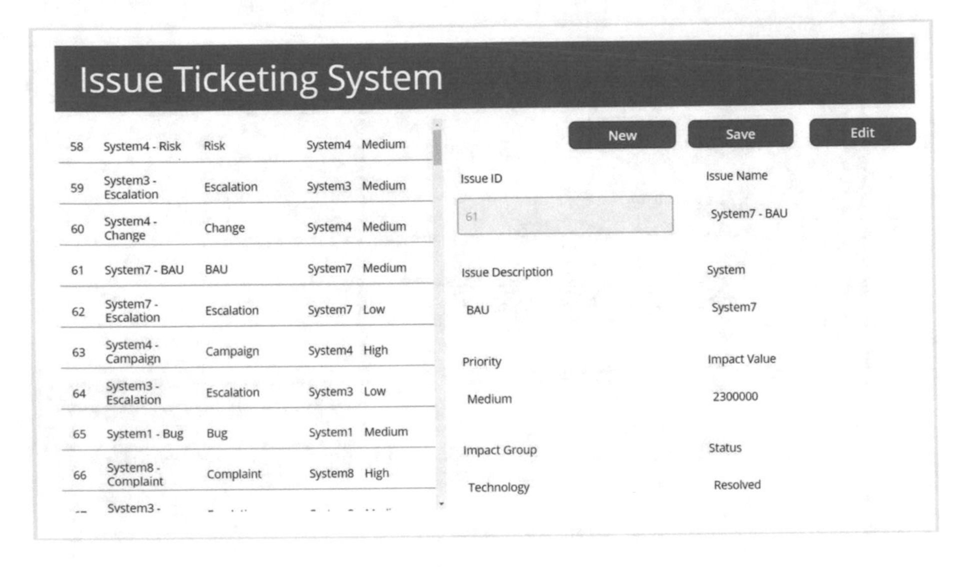

Figure 8-48. *Ticketing System Power Apps*

Conclusion

Congratulations on creating the first Power Apps. Kim is proud too. Power Apps seem to be easier to develop than she originally anticipated. In this chapter, you have explored the following areas of Power Apps.

- Licensing and pricing
- Creating and understanding a default app
- Creating a custom app

You also explored many key components in Power Apps, including data source, screen, gallery, buttons, and various control and connection between the different elements.

It is also important to note how Agile methodology is applied in the creation process. In the next chapter, you make a few more improvements to this original app.

CHAPTER 9

Improve Power Apps

In Chapter 8, Kim created a custom ticketing system app using Power Apps. Although it is working, many aspects of this app can be improved. These improvements can help to turn a barely functioning application into a proper solution.

In this chapter, you go through a few improvements. More importantly, you dive into the details and figure out how to fine-tune various aspects of the app.

The Business Scenario

Kim has decided to share the custom ticketing system (see Chapter 8) with Timothy, a colleague in the technology team. Tim is impressed with her ability to turn around a functioning application in three days.

When pressed for feedback, Tim suggested she investigate the following areas.

- Adopting standards
- Adding some assistance features such as filters, drop-down boxes, and maybe a search bar
- Adding dynamic calculation
- The user experience

D. Ding, *Transitioning to Microsoft Power Platform*,
https://doi.org/10.1007/978-1-4842-9239-6_9

In addition to Tim's feedback, Kim has read more online material and training courses. Two of the concepts that appeared in many trainings are *variables* and *delegation*.

Variables and Delegation

It is natural for beginner Power Apps developers to focus on the application's formulas, interactions, and design aspects. Two other fundamental pieces often get overlooked: variables and delegation.

Variables in Power Apps

Before jumping into the improvements, it is important to understand the three types of variables in Power Apps: local variable, global variable, and collection variable. Variables are a way to store temporary values or tables in the application. In the ticketing system application, you've used *varEdit* and *varNewTicket* to store values that allow other controls to react to value change.

Local Variable

A local variable (or context variable) can only be applied on a single screen. You can create and update the local variable using the UpdateContext() function. The following are examples.

```
UpdateContext({myLocalValue: 123});

UpdateContext({myStr1: "Str1", myStr2: "Str2"});

UpdateContext({myLocalDate: Today()});
```

The development best practice recommends using local variables for more complex applications where possible. But note that there are two main restrictions with context variables, as follows.

- A local variable cannot be accessed from other screens.
- Local variables are not designed to store data tables.

Global Variable

A global variable can be applied from all screens. You can create and update the global variable using the Set() function. For most applications, you may prefer to use the global variables as it is more flexible and can be accessed from other screens. The following are examples.

```
Set(myGlobalValue, 123);

Set(myGlobalString, "String");

Set(myGlobalDate, Today());

// store a filtered table as a variable
Set(myGlobalTable,
     Filter('Ticketing System', Status <> "Closed"));
```

The global variable contains fewer restrictions than local variables, but it does not allow users to edit data tables such as add or remove rows of data. Global variables are commonly used to store single values, (e.g., varEdit and varNewTicket in the ticketing app).

Collection

Collection is the most flexible table structure in Power Apps. You can use ClearCollect() to create a collection and use the Collect() and RemoveIf() functions to add/remove data. The following are examples.

```
ClearCollect(Table1,
      Filter('Ticketing System', Status <> "Closed"));

ClearCollect(Table2,
      Filter('Ticketing System', Status == "Closed"));

// adding table2 to table1
Collect(Table1, Table2);

// remove all rows of data where Status is "Closed"
RemoveIf(Table1, Status=="Closed")
```

Although a collection is best for storing data tables, it is inefficient for storing simple values or parameters. When developing apps, you need to use a combination of these variables to create the desired application.

Create Variables

It is important to understand that a trigger is required for the initial creation of the variables. This could be an action trigger or an event trigger.

An *action trigger* occurs when a button is pressed or a change is made in a control. This includes the OnSelect property in button controls and the OnChange property in various input controls (e.g., Text Input or Drop-down box).

An *event trigger* occurs when an event occurs. For example, when the app is started or when a screen is displayed.

Most new variables are created using event triggers. Let's continue with the completed ticketing system application from Chapter 8. Figure 9-1 shows that the following steps create a new collection variable called *colOpenTickets*. The collection contains all tickets that have not been assigned.

1. Select App under Tree view.

2. Choose the **OnStart** property.

3. Use the following ClearCollect() function to create a collection.

```
ClearCollect(
        colOpenTickets,
        Filter('Ticketing System', Status <> "Not
        Assigned")
);
```

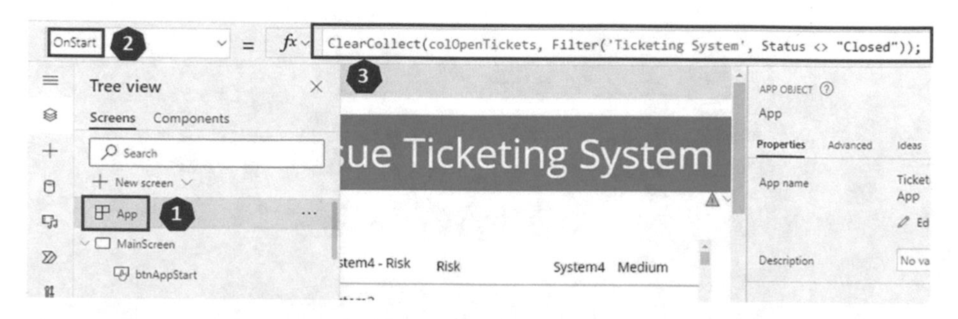

Figure 9-1. *Create a new collection during app start*

One issue with creating the collection this way is that the code is executed initially during app start. It is very inconvenient to close and re-open the app to test the functions in the App OnStart property every time. You can create a button (Action trigger) for testing purposes, as shown in Figure 9-2.

1. From the Insert view, drag and drop **Button** onto the screen.
2. Change the Text property to **App Start**.
3. Select OnSelect property from the drop-down.

4. Copy and paste the same ClearCollect() function into the function area.

```
ClearCollect(
    colOpenTickets,
    Filter('Ticketing System', Status <> "Not
    Assigned")
)
```

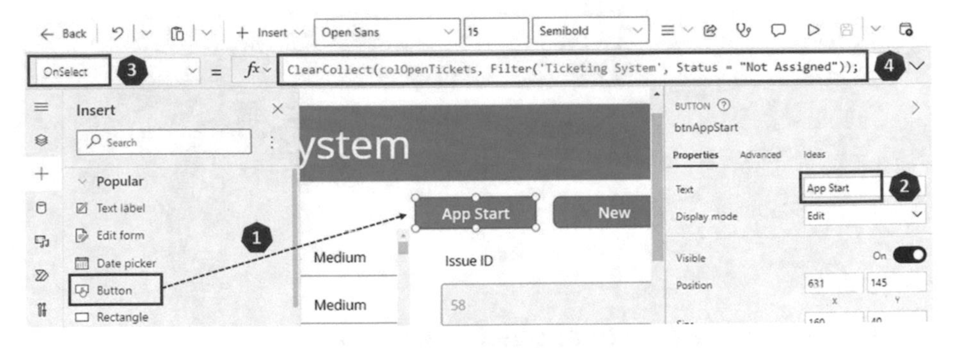

Figure 9-2. *Create a button to simulate app start behavior*

Now you can click the App Start button to create the collection. When the codes are working, you can copy them in the property and paste them onto the App OnStart property formula bar. You can check the collection by completing the following steps, as shown in Figure 9-3.

1. Choose the more options icon.
2. Select Collections.
3. Choose the collection you just created.
4. Review the first five items in the collection.

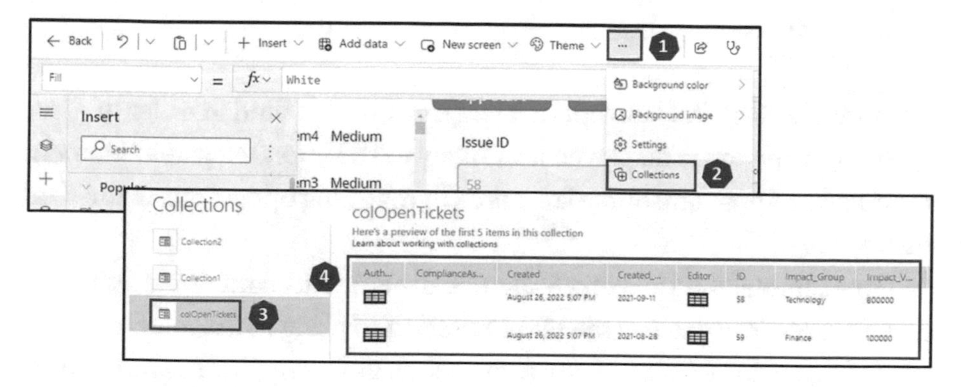

Figure 9-3. *Check the collection*

The other common event trigger is when the screen become visible. You can also create the same collection in the OnVisible property. This brings us to the execution order of Power Apps from a developer's perspective, as shown in Figure 9-4. The key difference between App OnStart and Screen OnVisible is that the app only starts once, but the screen may be loaded multiple times. This means it is more efficient for many initialization tasks to go into App OnStart.

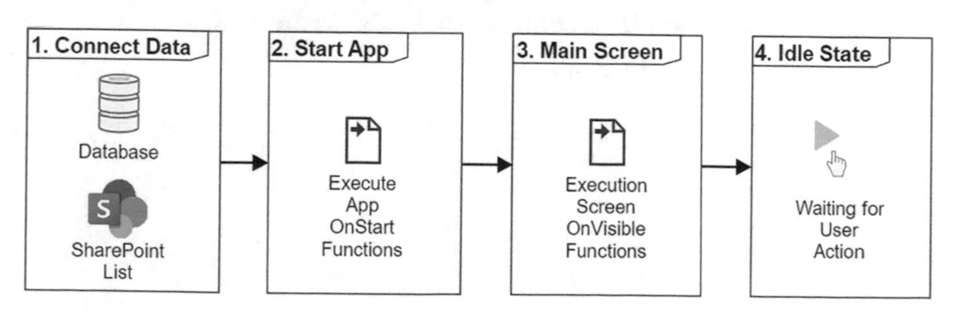

Figure 9-4. *Power Apps execution order*

Delegation

Delegation is a critical concept that you must understand in order to ensure information is displayed accurately in the app. Delegation happens when Power Apps "outsource" resource demanding operations to the data source server.

For example, the Ticketing System SharePoint list contains around 3,000 records. You can use the Filter function to bring in the relevant tickets (e.g., ticket created by the same user). Power Apps *delegate* the filter operation to be performed in SharePoint server rather than inside the app.

What if the data source does not support delegation (e.g., Excel file in OneDrive)? In such situations, Power Apps only take the first 500 rows of data by default, as shown in Figure 9-5. This means *only* the first 500 rows of data are loaded into Power Apps. All other data is dropped.

The following steps (see Figure 9-5) change this limit. The maximum limit of data load is only 2,000 rows.

1. Go to Settings.
2. In General settings, scroll down to **Data row limit** and change it to 2000.

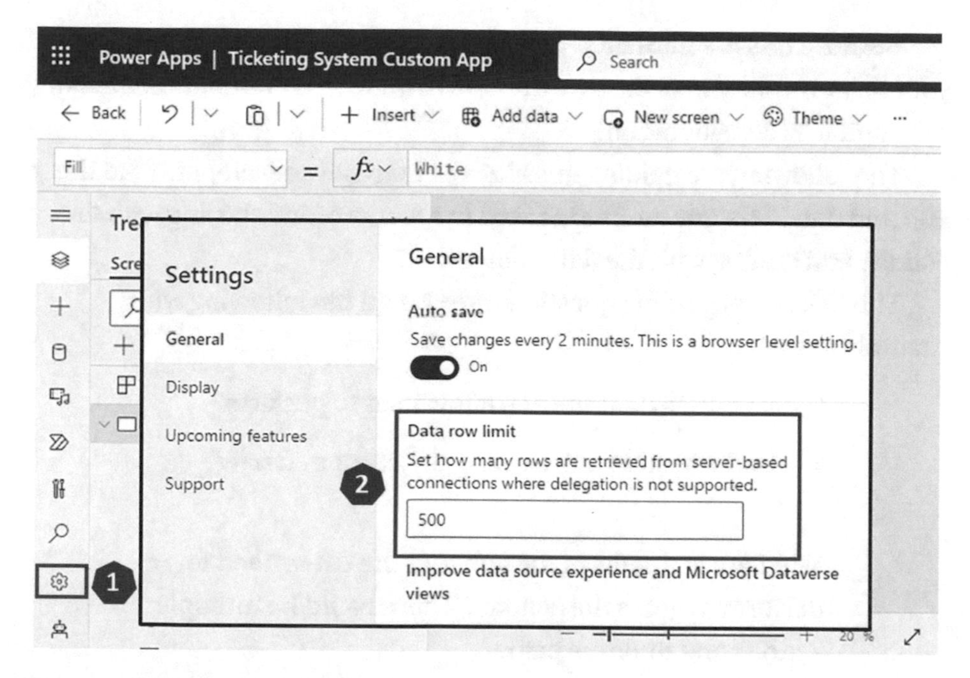

Figure 9-5. *Data row limit*

It is good to increase this limit with caution, as it may slow down the application and impact on overall user experience.

Tip To speed up development, you can also set the limit smaller to make the overall application more responsive.

Reason for Row Limit

While this row limit may seem strange from a business analysis perspective, it is actually standard practice to reduce the amount of data points in the front end application. Self-service BI solutions such as Power BI did an excellent job in hiding this complexity in resource optimization from BI developers, so that the developers do not need to worry about this level of detail.

Because this is a hard limit of Power Apps, as an app developer, it is your responsibility to make sure the data displayed is complete after taking row limit into consideration.

The following are guiding questions to help new developers plan the app and data. The questions are asked in reverse order, starting with the actions and ending with the data sources.

The high-level guiding questions are shown the following with example answers.

- What actions does the app allow user to perform?
    - Users need to edit existing tickets or add new tickets.
- What are the paths in the app that the users need to perform to get to the actions? There could be multiple actions and multiple paths.
    - Choose action type ➤ find ticket ➤ see details ➤ perform the action.
- Which step requires the most rows of data to be displayed?
    - Find ticket is where most data need to be displayed.
- Is there a way to restrict the app to only display relevant data (i.e., any data display that is not actionable should not be displayed)?
    - Only display open tickets that's created by the user.
- Does the data storage allow for delegation? If not, is it possible to migrate?
    - Data is stored in SharePoint list which already support delegation.

If you follow the prior example, it should lead to the decision to apply restrictions at the OnVisible setting of the screen where a *find ticket* action is happening.

Warning and Error Messages

When delegation is not handled properly, Power Apps sends a warning message. The following steps let you see the warning details, as shown in Figure 9-6.

1. Select the warning sign drop-down list and choose **Get help for this warning**.

2. Note that this is a *delegation warning*.

3. Go to the highlighted formula.

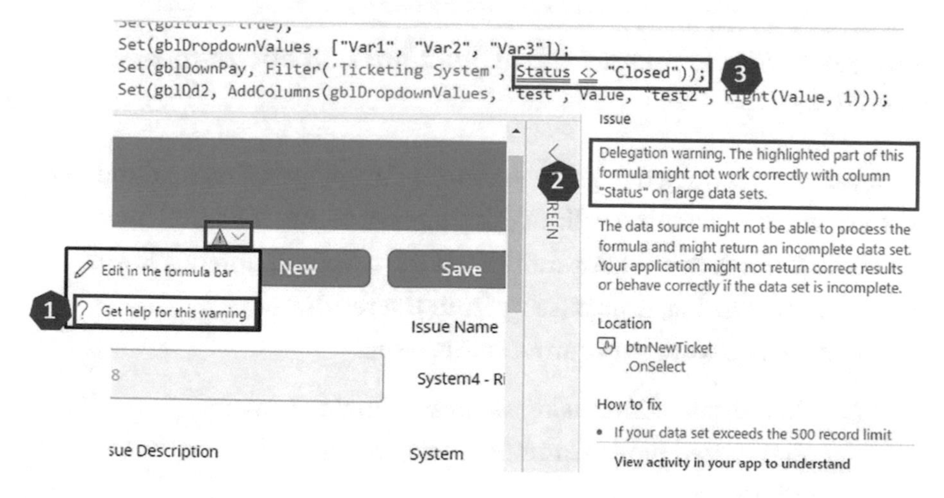

Figure 9-6. *Delegation warning*

The problem is that the <> (not equal) sign in the Filter function is not an acceptable delegation operation. If you Google the term: *Understand delegation in a canvas app*, you find the Microsoft document with link to delegation in SharePoint. At the time of writing in Dec 2022, the list of delegable functions and operations for SharePoint are shown in Figure 9-7.

Operation/Function	Number	Text	Boolean	DateTime	Complex [2]
`Filter`	Yes	Yes	Yes	Yes	Yes
`Sort`	Yes	Yes	Yes	Yes	No
`SortByColumns`	Yes	Yes	Yes	Yes	No
`Lookup`	Yes	Yes	Yes	Yes	Yes
`=`	Yes	Yes	Yes	Yes	Yes
`<, <=, <>, >, >=`	Yes [3]	No	No	Yes	Yes
`StartsWith`	-	Yes	-	-	Yes
`IsBlank`	-	No [4]	-	-	No

Figure 9-7. *SharePoint delegable functions and operations (Dec-2022)*

Because <> is not a delegable function, Power Apps first load the first 500 rows of data (default row limit), then perform the not equal filter operation. If the data contains more than the row limit, the rest is *not* loaded before filtering, thus missing from the result.

There are two ways to get around this issue.

1. Create collections using separate = logic to cover all required types. Combine them into a final collection.

```
ClearCollect(
    colOpenTickets,
    // dataset 1 - not assigned
    Filter('Ticketing System', Status = "Not
    Assigned"),
```

```
        // dataset 2 - in progress
        Filter('Ticketing System', Status = "In
        Progress"),
        // dataset 3 - pending
        Filter('Ticketing System', Status = "Pending")
    )
```

2. Assign a numeric identifier (StatusID) for the different types of texts so that you can use <> on numbers. The following are examples.
   - 1 = Not Assigned, 2 = In Progress, 3 = Pending, 4 = Closed
   - Use StatusID <> 4 to exclude closed tickets

Even when it does not give you the warning, you still need to make sure the filter result is less than the row limit; for example, 10,000 rows of data in the source, filtered down to 3,000 rows. Even when row limit is set to max of 2,000, you still have 1,000 rows of records missing. This means further filtering is required. The additional filter can be added to the Filter() function as follows.

```
ClearCollect(
    colHighClosed,
    Filter(
        'Ticketing System',
        Status = "Closed",
        Priority = "High"
    )
);
```

User Level Filter

With a good understanding on the reason and limitation of row limit in Power Apps, you learn to implement the *user-level filter* (ULF) as a common solution. The ULF does the following.

- Restricts the number of rows of data following into the app (resolve the row limit issue)
- Provides a better user experience by showing only the relevant information.

The ULF's concept is similar to the row-level security, you can also use this as a security measure. The User() function is used to identify the person who is using the app.

The user function has three attributes, as shown in Figure 9-8. They are the full name, email, and image. You can use the full name or email for the ULF, depending on the comparison field. The Ticketing System SharePoint list contains a Created_by column containing the user name, so it is the user name used in a later example.

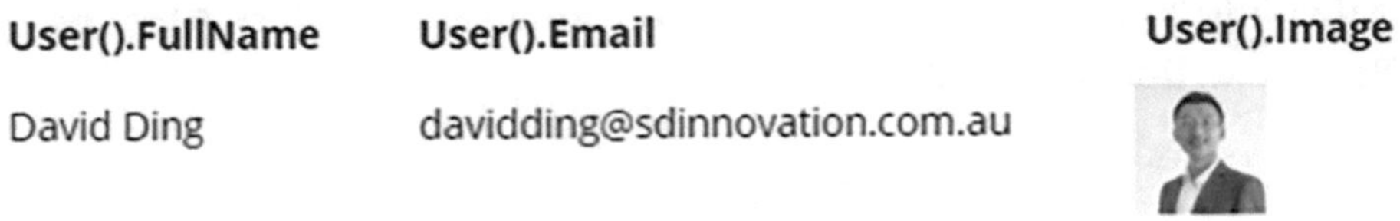

User().FullName	User().Email	User().Image
David Ding	davidding@sdinnovation.com.au	

Figure 9-8. *Attributes of User() function*

The following steps initialize a collection test, as shown in Figure 9-9.

1. Select the App Start button.
2. Choose OnSelect.

3. Change the ClearCollect() function to the following.

```
ClearCollect(
  colUserData,
  Filter(
    'Ticketing System',
    Created_By = User().FullName
  )
);
```

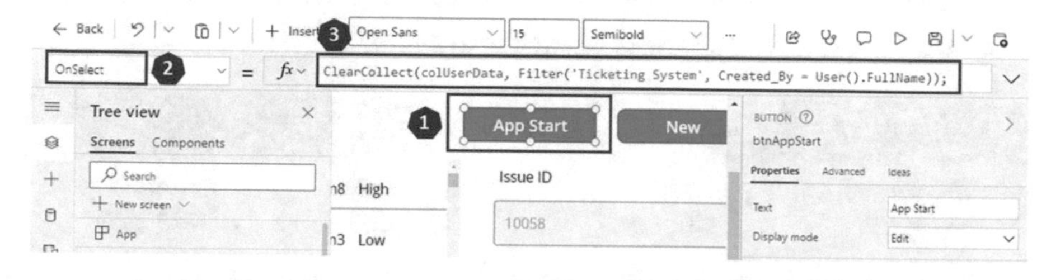

Figure 9-9. Implement user-level filter

Once the new collection has been created, you can change the items property in the gallery to use the colUserData as shown in Figure 9-10.

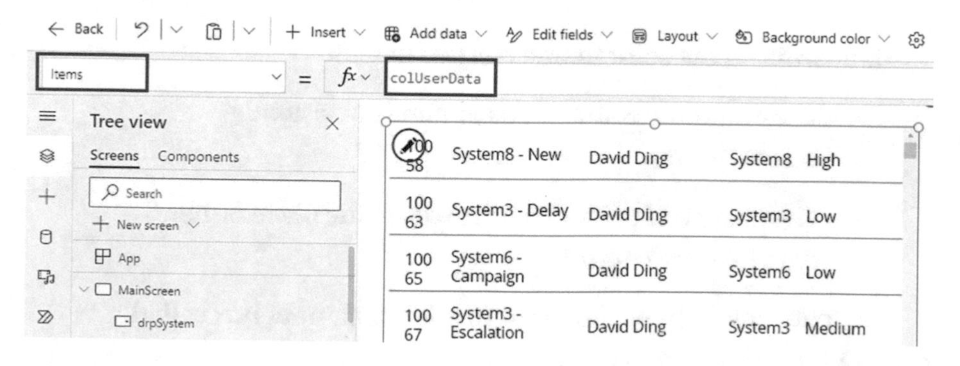

Figure 9-10. Update gallery to display user data

Once this is successful, you can now update the App OnStart property to the same function as in the App Start button.

Color Palette

Most organizations have their own marketing material which typically involve some standard colors. In Power BI, it is easy for a user to create a color theme, save it, and adopt the same theme for all reports. At the time of writing, there no similar functionality in Power Apps. This means the user need to customize the color for each control. One way to simplify this process is to use *color theme items*. These are normal control items that can be used to hold colors.

AWM's corporate color palette is shown in Figure 9-11. It contains both color and the preferred shades of black and silver.

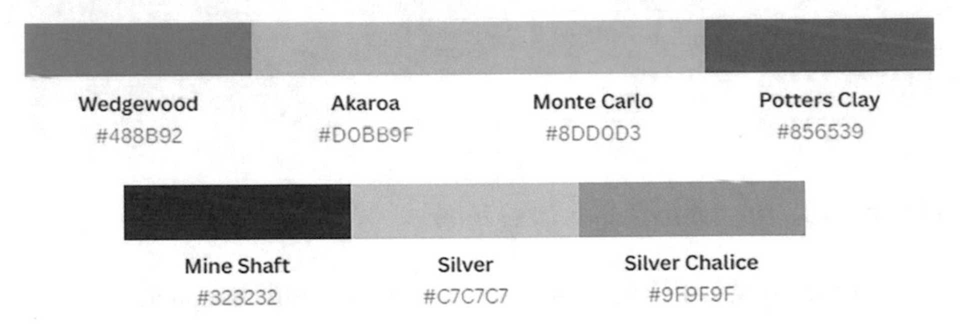

Figure 9-11. *Example corporate color palette*

Standardizing the color theme has two benefits.

- The corporate color can be applied consistently across the app
- The change of color can be made in one place rather than in every control

The following steps explain how to add a text input box called themeColor, as illustrated in Figure 9-12.

1. Select the Insert view.
2. Drag and drop **Text input** to the bottom of the Browse screen.

3. Click the color setting.

4. Click Custom.

5. Enter Hex color **488B92** for Wedgewood.

6. Change the control item name to **themeColor**.

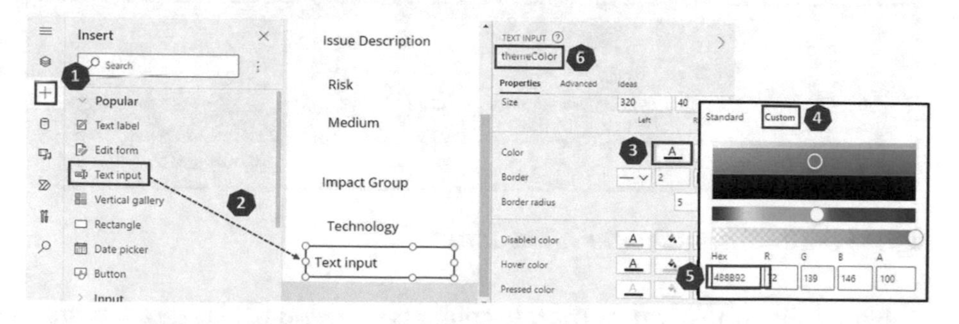

***Figure 9-12.** Add ThemeColor Text input control*

You have now stored the Wedgewood color in themeColor.Color. You can now use this color by completing the following steps, as shown in Figure 9-13.

1. Hold the Ctrl key and select the control items that require the corporate color.

2. Choose **Fill** from the property drop-down.

3. Change the formula to **themeColor.Color**.

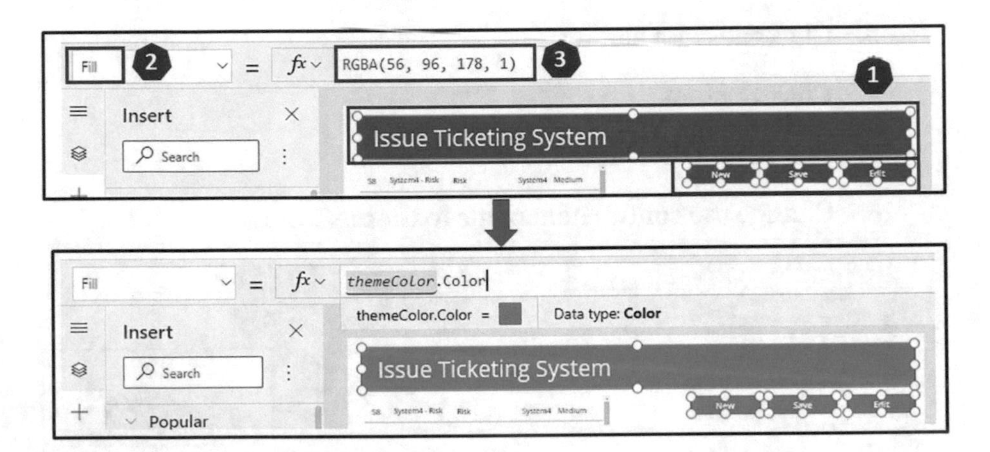

***Figure 9-13.** Apply themeColor.Color*

Notice that one of the corporate colors (Wedgewood) is stored in the Color property of the themeColor text input control. This means you can store other colors in other properties, as shown in Figure 9-14.

1. Select the themeColor text input control.

2. Choose the Color property.

3. Observe the RGBA() color function and use // to add a color comment.

4. Use Fill and ColorValue() function to store the Akaroa color.

5. Use BorderColor to store the Monte Carlo color.

6. Use HoverColor to store the Potters Clay color.

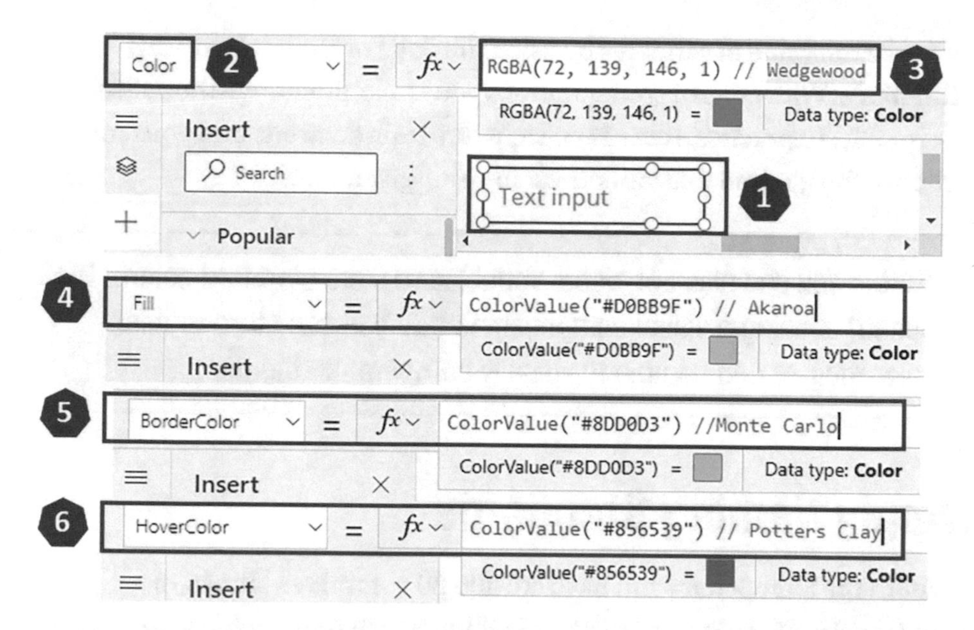

***Figure 9-14.** Store corporate colors in themeColor properties*

You can create a themeBW (black and gray) by following the previous steps. Once you have themeColor and themeBW, you can make them invisible by setting the Visible property to false, as shown in Figure 9-15.

1. Select the two theme controls.
2. Turn off Visible.

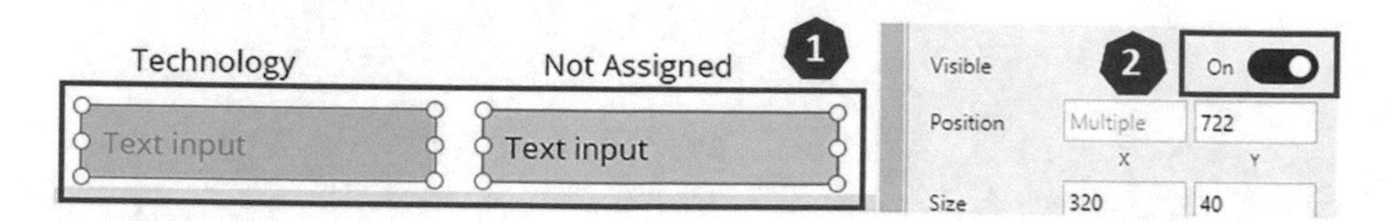

***Figure 9-15.** Hide themes*

The method described to store color palettes can be applied more broadly to other configurations. It is important to note that this is different to storing values in global and local variables. With variables, you can change the value of variables by using the Set() or UpdateContext() functions. With control configurations you can only change the values manually. In practice, there is no right or wrong answers, and you can choose the method that works best for your given scenario.

Note You can also set global variables to store standard colors. The benefit of using a visual component is that you can store multiple color themes and be able to visually compare the themes.

Apply Naming Standards

Within the MainScreen you have created 30 controls as shown in Figure 9-16. There is no easy way of telling which one is which. You can select the control on the screen and it give you the name of it. This can be problematic when you try to refer to the controls. For example, when saving the values to the SharePoint list, you need to remember the name of the control that contains the input values.

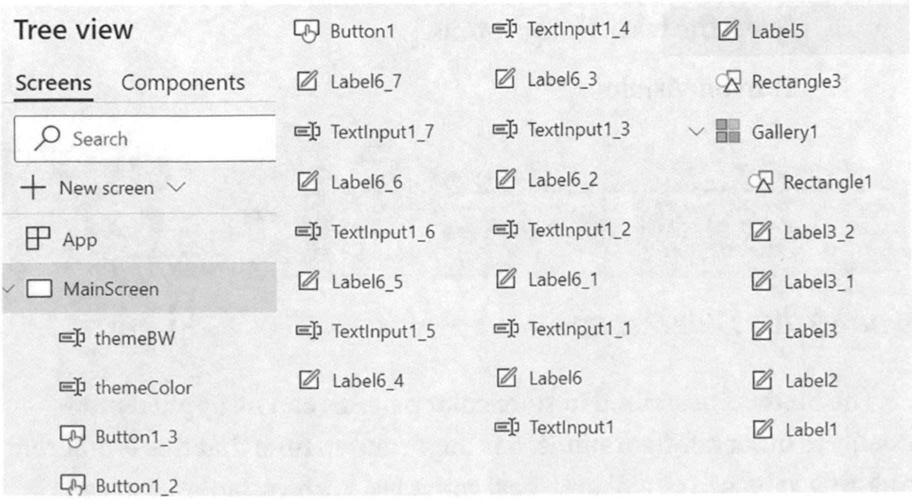

Figure 9-16. *Control names in MainScreen*

It is strongly recommended that you adopt some standard way of naming these controls and variables.

Luckily Microsoft has already developed the coding standards. It is best to follow the official standards than create your own because it is better for everyone if we adopt the same standard. If you Google "Power Apps coding standards and guidelines", you should find this PDF document. In the session on control names in the guide, you should see the following abbreviation table as shown in Figure 9-17.

Control name	Abbreviation
button	btn
camera control	cam
canvas	can
card	crd
collection	col
combo box	cmb
dates	dte
drop down	drp
form	frm
gallery	gal
group	grp
header page shape	hdr
html text	htm
icon	ico
image	img
label	lbl
page section shape	sec
shapes (rectangle, circle, and so on)	shp
table data	tbl
text input	txt
timer	tim

Figure 9-17. *Abbreviation table for control names*

You can take the NEW ticket button as an example as shown in the table in Figure 9-17. The control is a button; the abbreviation for a button is *btn*. You can then add a description to the button such as *NewTicket*. This makes *btnNewTicket* the final name for the control, as shown in Figure 9-18.

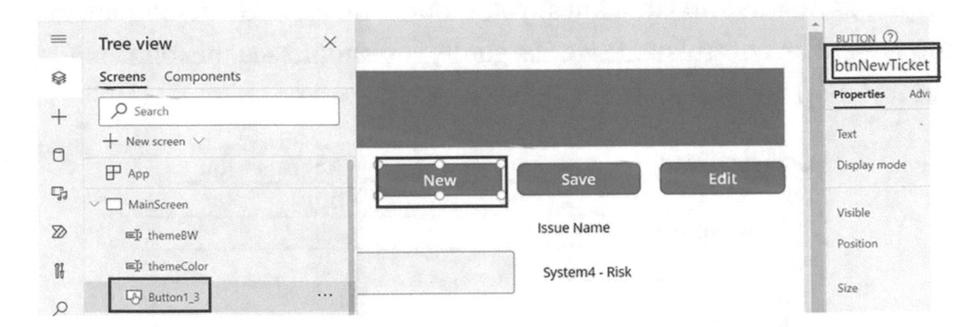

Figure 9-18. *New ticket control button*

The naming style is called Camel casing. Starting with the lower case abbreviation and follow by its purpose. You can have multiple words for the purpose. The sequence looks like lowerProperProper. You need to apply this to all names.

The variable abbreviations are *glb* for global variables and *loc* for local variables. You can now change the names of the variables.

Additional UI Features

While the ticketing system app is ready to be consumed, it is not easy for users to navigate to the required tickets. Three additional User Interface (UI) Features could help the search process.

- Drop-down filter
- Sorting the gallery
- Search bar

Drop-down Filter

Drop-down filter is a common method to help users to narrow down the focus area. There are three types of common drop-down filter scenarios.

- Single drop-down filter
- Multiple drop-down filters
- Create a drop-down list from the data

Single Drop-down Filter

The following steps create a simple drop-down, as shown in Figure 9-19.

1. From the Insert view, drag and drop the **Drop down** control into the screen.
2. Choose **Items** from the drop-down control's property.
3. Enter the list of options as follows.

```
["Not Assigned", "In Progress", "In Review",
"Pending", "Resolved", "Closed"]
```

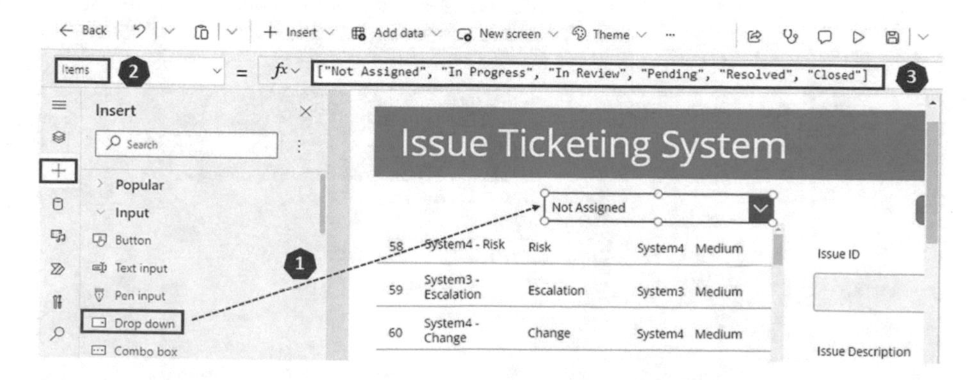

Figure 9-19. *Add drop-down control into screen*

These steps create the list of drop down selections. The first item is the default selection for the drop-down box.

Once the drop-down box has been created, the next step is to create the interaction with the gallery. When a drop-down item has been selected, the gallery only displays matching tickets. You can identify the selected item in the drop-down box using the following code.

```
drpStatus.SelectedText.Value
```

You can use this value inside a Filter() function for the gallery display by completing the following steps, as shown in Figure 9-20.

1. Select the gallery.
2. Choose the **Items** property for the gallery.
3. Apply the following Filter() function.

```
Filter(
  'Ticketing System',
  Status = drpStatus.SelectedText.Value
)
```

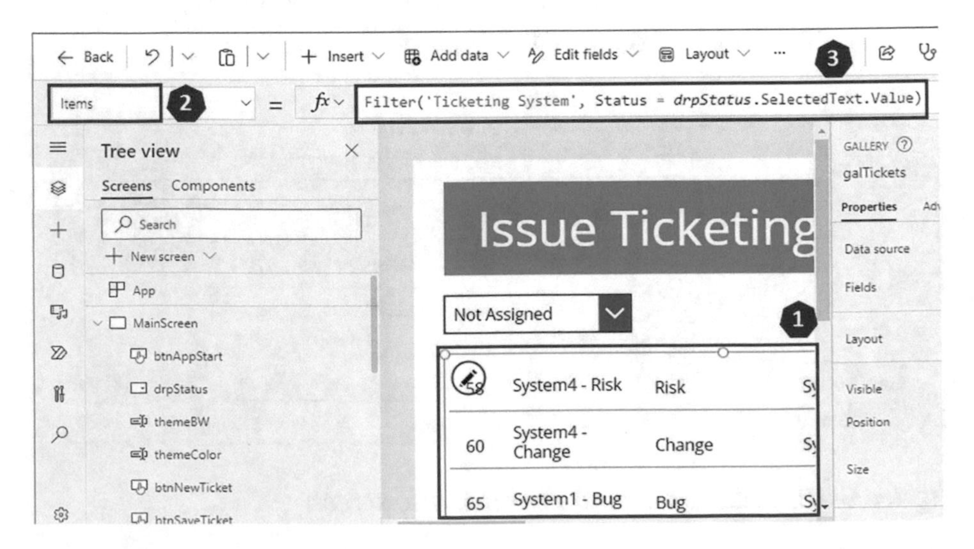

***Figure 9-20.** Drop-down controlled gallery*

Now when you change the selection in the status drop-down, the gallery changes according to the selection.

Multiple Drop-down Filters

Once you understand how a single drop-down filter works, it is easy to add another one for systems. Use the following steps, as shown in Figure 9-21.

1. Copy the Status drop-down, rename it **drpSystems**, and change the Item property to the following.

```
["System1", "System2", "System3", "System4", "System5",
"System6", "System7"]
```

2. Select the gallery.

3. Go to the **Items** property of the gallery.

4. Change the Filter() function to the following.

```
Filter(
    'Ticketing System',
    Status = drpStatus.SelectedText.Value,
    System = drpSystem.SelectedText.Value
)
```

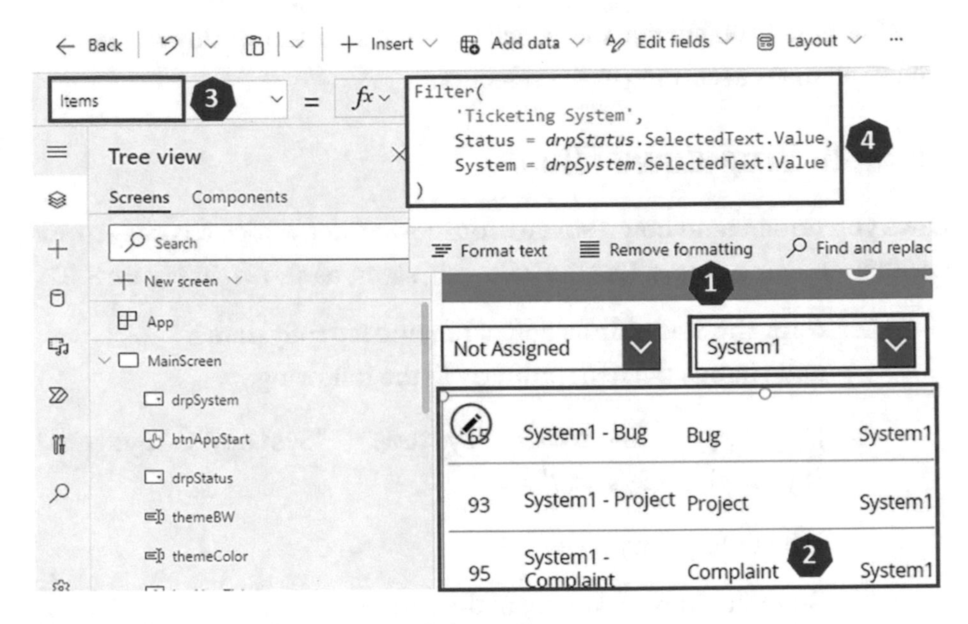

***Figure 9-21.** Adding a second drop-down*

Now you have a gallery that filtered by two drop-downs controls. Everything should work well, except this approach has a small problem. This implementation prevents the user from seeing all issues in system 1 or all systems with tickets Not Assigned.

You can add an ALL selection option in both filters using the following code.

```
// Items in Status Drop-down:
["All Status", "Not Assigned", "In Progress", "In Review",
"Pending", "Resolved", "Closed"]

// Items in System Drop-down:
["All Systems", "System1", "System2", "System3", "System4",
"System5", "System6", "System7"]
```

The resulting drop-down filters should look like Figure 9-22.

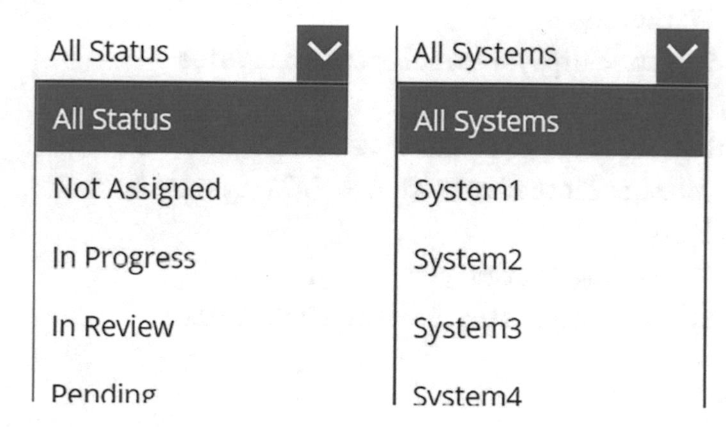

Figure 9-22. *All Selection options in drop-down filters*

Since the **All selection** option cannot match any values, the gallery no longer displays any data. There are two options for solving this problem.

- Use nested If() and Filter() functions in the gallery items
- Implement filter into the collection when triggered by a change in the drop-down

Option 1: You can change the gallery item formula to the following. While it looks complicated, it is just checking three conditions: both Status drop-down and System drop-down with All selections, only Status drop-down has All selections, only System drop-down has All selection; otherwise, apply both filters.

```
If(
  drpStatus.SelectedText.Value = "All Status" &&
    drpSystem.SelectedText.Value = "All Systems",
      'Ticketing System',
  drpStatus.SelectedText.Value = "All Status" &&
    drpSystem.SelectedText.Value <> "All Systems",
```

```
      Filter(
        'Ticketing System',
        System = drpSystem.SelectedText.Value
      ),
  drpStatus.SelectedText.Value <> "All Status" &&
    drpSystem.SelectedText.Value = "All Systems",
      Filter(
        'Ticketing System',
        Status = drpStatus.SelectedText.Value
      ),
  Filter(
    'Ticketing System',
    Status = drpStatus.SelectedText.Value,
    System = drpSystem.SelectedText.Value
  )
)
```

Even though the formula may look complicated, this approach does not require additional collections to be created. When there are two drop-down filters, you need to run three condition checks, but if there are three drop-down filters, you need to do eight condition checks, which is not recommended. As a result, this option should only be considered for up to two drop-down filters.

Option 2: You can add an OnChange property to each drop-down filter to adjust the collection that the gallery is linked to. You create a colDisplay collection for this. This solution can be implemented using the following steps, as shown in Figure 9-23.

1. Select the Status drop-down filter.
2. Go to the OnChange property.

3. Implement the following two filter functions. Note that the first step is based on the source data, but the second step is based on the result from step 1.

```
// Status Filter
If(
  drpStatus.SelectedText.Value = "All Status",
  ClearCollect(colDisplay, 'Ticketing System'),
  ClearCollect(
    colDisplay,
    Filter(
      'Ticketing System',
      Status = drpStatus.SelectedText.Value
    )
  )
);

// System Filter
If(
  drpSystem.SelectedText.Value = "All Systems",
  colDisplay,
  ClearCollect(
    colDisplay,
    Filter(colDisplay, System = drpSystem.
    SelectedText.Value)
  )
);
```

4. Select the gallery and change the Items property to colDisplay.

5. Copy and paste the formula in Step 3 to the System drop-down filter.

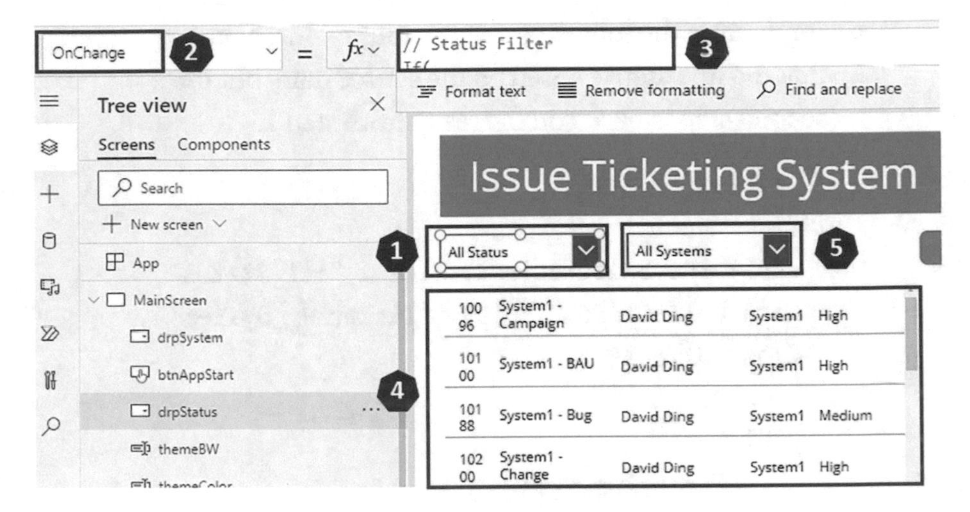

Figure 9-23. *Use OnChange property in drop-down filters*

The code being translated into plain English is performing the following tasks.

1. Check whether the selection in the status filter is All Status.
    - If yes, create colDisplay based on source data.
    - If no, create colDisplay based on filtered source data.
2. Check whether the selection in the system filter is All Systems.
    - If yes, do not change the colDisplay.
    - If no, update colDisplay based on filtered colDisplay.

This formula must also be copied into the OnChange property in the system filter, because the change can be triggered from all filters. If you have three drop-down controls, the same code is in the OnChange property of all drop-down controls.

The only extra step required is to initialize the colDisplay collection at the beginning. In the current setup, the collection is first created when the user selects the drop-down filter. This means nothing is displayed in the gallery until the trigger. Add the following code to the App OnStart property to initialize the colDisplay.

```
ClearCollect(colDisplay, 'Ticketing System')
```

If you adopt this method, you can introduce additional filters without worrying about exponential growth in complexity.

Note As you can see from the long code, code formatting (with proper indentation) is extremely important in writing readable code. This not only helps other developers to understand your code, but it also helps you to write more reliable code as well as the troubleshooting process.

Create a Drop-down List from Data

In the previous examples, you created the drop-down list by manually keying in the selection options. In some cases, you may want to get the unique values from a column from a data source for the drop-down items. You can achieve this by using the Distinct() function and storing the data in a collection by completing the following steps, as shown in Figure 9-24.

1. Go to the App Start button and add the following code to create colDistPriorities.

    ```
    ClearCollect(
        colDistPriorities,
        {Result: "All Priorities"},
        Distinct(colUserData, Priority)
    )
    ```

2. Add new drop-down control and set Items to colDistPriorities.

3. Test the drop-down values.

4. Migrate the App Start button code into App OnStart.

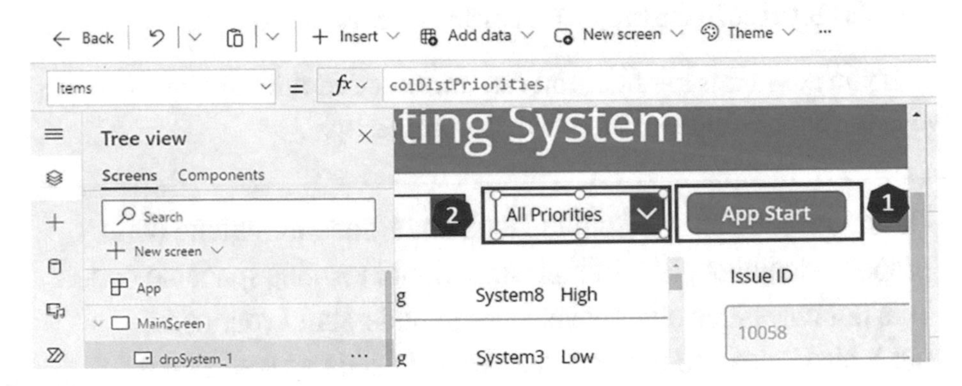

Figure 9-24. *Use values from data in drop-down*

This summarizes a few variations of the drop-down filter. The drop-down is one of the most common UI features in many apps. A simple drop-down control can greatly enhance a user's ability to narrow the focus to the relevant data.

Sorting the Gallery

Another common feature to assist with the navigation process is sorting. Before implementing sorting, add the column names and ID sorting button. The following steps explain how to do this, as shown in Figure 9-25. Note that these controls are added *outside* the gallery.

1. Drag and drop **Text label** from Insert View to the position above the ID field, and change the Text property to **ID**.
2. Repeat step 1 four times (once for Name, Creator, System, and Priority).
3. Change the control name to a standard naming format.
4. Search for the Sort icon, drag and drop it next to the ID field, and name it **icnSortID**.

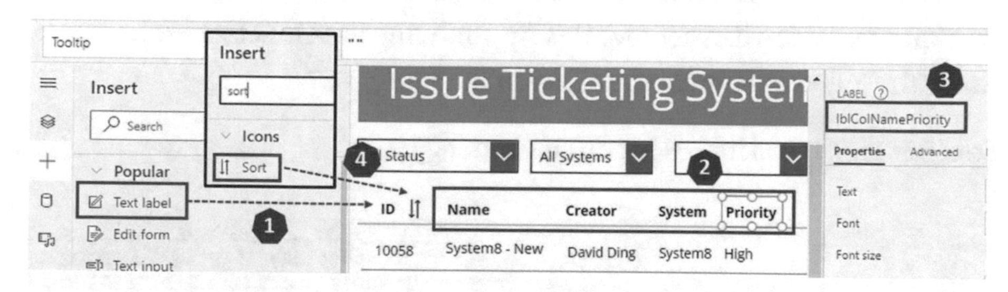

***Figure 9-25.** Adding title and sort icon*

Now you can try to configure the ID sort button. You need to first create a global variable SortIDAscending in the App Start button using the following code, as shown in Figure 9-26. You must also click the App Start button to create the variable first.

```
Set(SortIDAscending, true);
```

Figure 9-26. Add sort order to OnSelect property

The icnSortID icon switches the value of SortIDAscending between true and false. You can achieve this by changing the OnSelect property using the following function, as shown in Figure 9-27.

```
Set(SortIDAscending, !SortIDAscending);
```

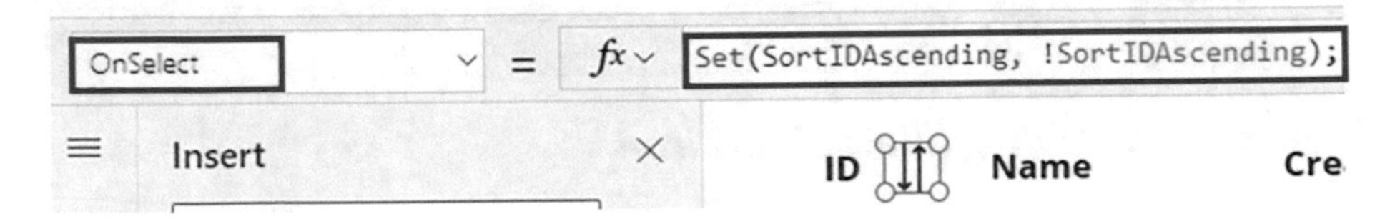

Figure 9-27. Switch the value of SortIDAscending

With the changing variable, you can set the gallery Items property using the following code, as shown in Figure 9-28.

```
SortByColumns(
    colDisplay,
    "ID", If(SortIDAscending, Ascending, Descending )
)
```

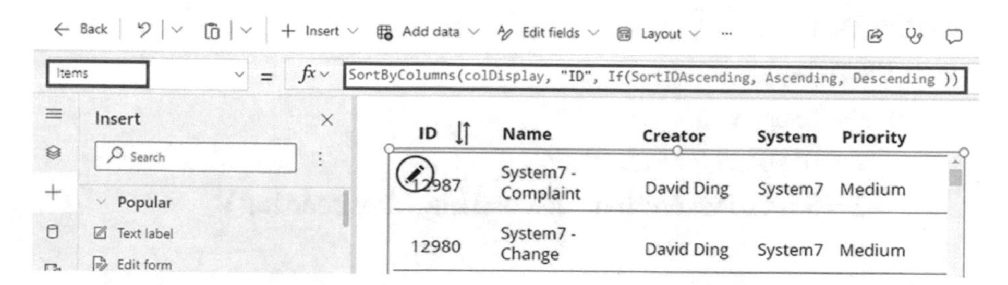

Figure 9-28. *Sorting gallery according to Sort button*

SortByColumns() function can be used to rearrange table orders. The If() function sets the ascending or descending order based on the column. Now when you click the Sort button, the gallery items display in a different order by the ID column.

While this method works well with a single Sort button, it does not work well when additional Sort buttons are added, as shown in Figure 9-29. This is because the gallery item function becomes much more complicated if it needs to check the sorting status for each button.

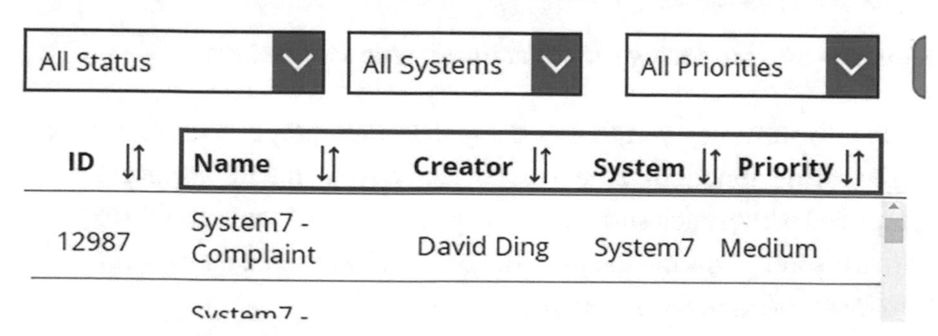

Figure 9-29. *Additional sort buttons*

You can set the sorting order independently by updating the colDisplay collection using the following code inside the OnSelect property of each sort button, as shown in Figure 9-30. Then the gallery can simply display the colDisplay collection.

```
ClearCollect(
    colDisplay,
    SortByColumns(
        colDisplay, "Priority",
        If(SortIDAscending, Ascending, Descending)
    )
);

Set(SortIDAscending, !SortIDAscending);
```

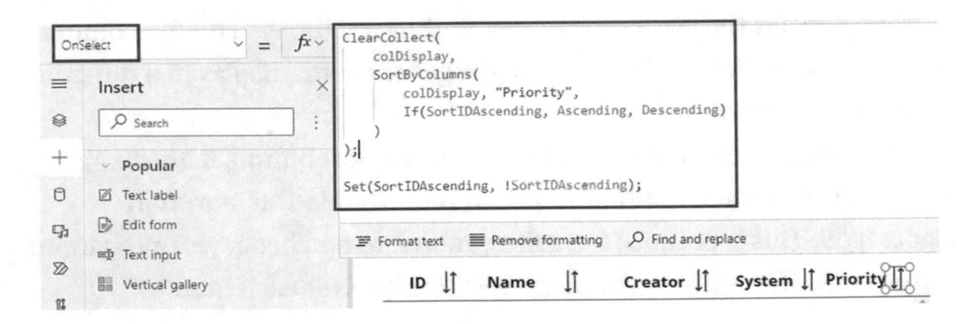

***Figure 9-30.** Updating the collection in the sort button*

Please make sure you change the gallery Item property to colDisplay for the sorting functions to operate correctly. Note that the sorting is completed within each sort button, so you only need one variable to store the status. This means you can change the variable name from *SortIDAscending* to *SortAscending*.

Tip As you have seen from the drop-down and sorting examples, the same logic can be implemented differently to achieve the same result. When logic becomes overly complicated, you may want to *ask yourself if there is a simpler way to solve the same problem.*

Search Bar

The last common improvement feature is the free text search capability. This allows the user to type in a free text input box, and the gallery is updated dynamically by comparing the input text with the Name field.

You can follow the steps to add the search bar, as shown in Figure 9-31.

1. Put the **Text input** box onto the screen, clean the default text to "" and name it **txtSearchText**.

2. Place **Text label** next to the search bar and name the text **Search**.

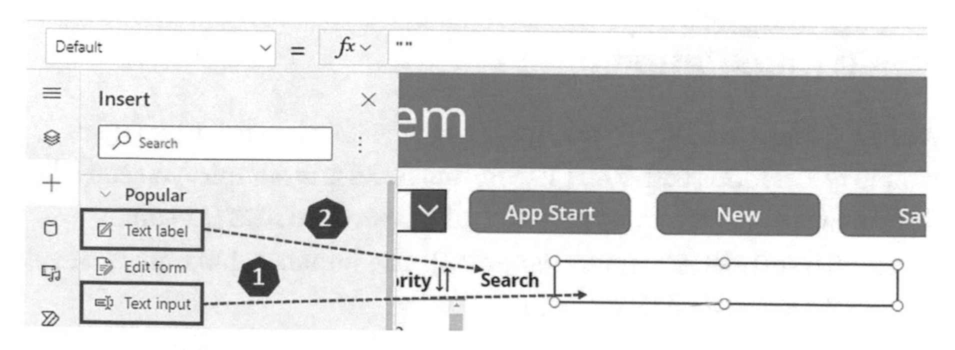

Figure 9-31. *Adding search bar to app*

Once completed, you can update the gallery item property using the Search() function with the following code.

```
Search(colDisplay, txtSearchText.Text, "Issue_Name")
```

The search function searches the Issue_Name column in the colDisplay collection and displays all records that contain the search text.

You can also easily modify the Search() function to search for multiple fields by adding column names at the end. For example, the following code searches the name in both Issue_Name and Priority Fields.

```
Search(colDisplay, txtSearchText.Text, "Issue_Name",
"Priority")
```

Dynamic Calculation

Another common requirement is to run dynamic calculations in the app. If you are familiar with Excel formulas, this should be easy to understand. There are three main types of calculations.

- Basic calculations
- Calculation in a gallery
- Aggregation
- Calculation in a collection

Basic Calculations

A basic calculation directly manipulates the values in a cell by applying simple calculations with Excel-like formulas. As an example, you can take the impact value (assuming in US$) and convert it to AU$ (Australia, Forex 1.5) and SG$ (Singapore, Forex 1.3). The impact value is placed in a text label, as shown in Figure 9-32.

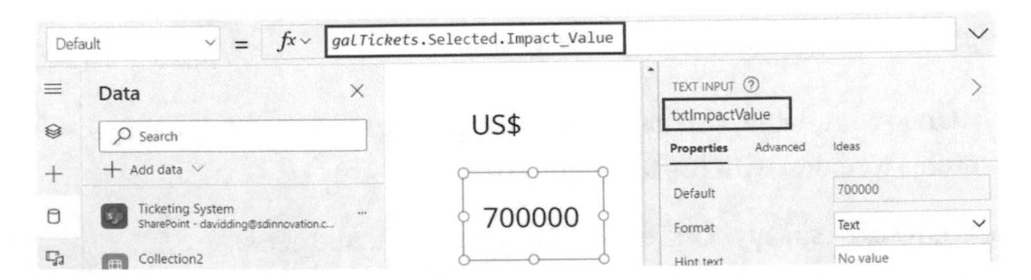

Figure 9-32. *Displaying the basic impact value*

Note This value is based on the gallery selection, so it may differ in your case.

You can calculate an AU$ by applying the exchange rate to the selected Impact_Value, as shown in Figure 9-33.

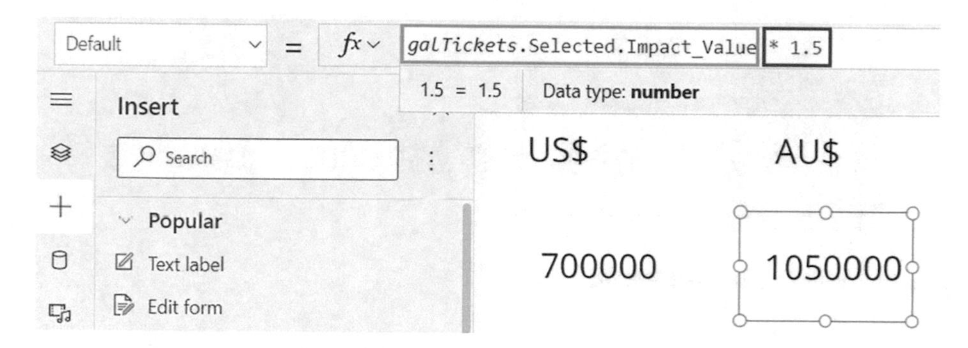

Figure 9-33. *Convert to AU$*

For the SG$ calculation, instead of using the Impact_Value, you can refer to the control text, as shown in Figure 9-34.

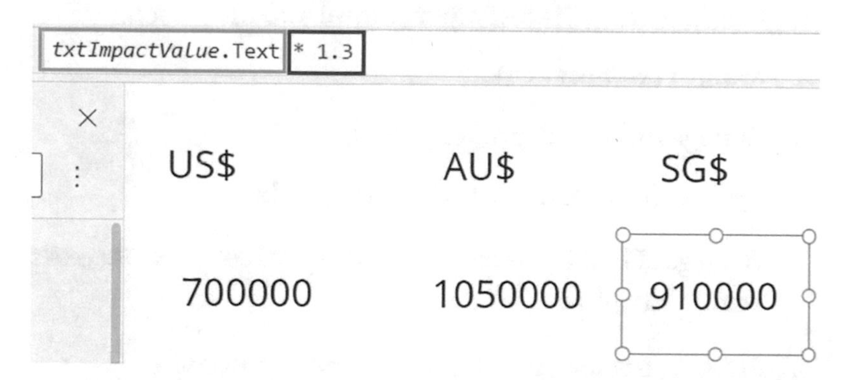

Figure 9-34. *Convert to SG$*

You can also format the value display by using the Text() function, as shown in Figure 9-35.

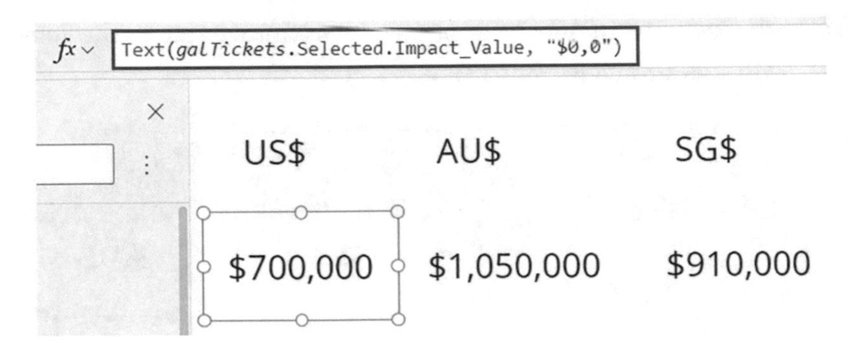

Figure 9-35. *Formatted values*

The basic calculation is the most common type of calculation you use in Power Apps. Similar to the Excel formula, you can also use a control variable, such as the foreign exchange rate, to update the value, as shown in Figure 9-36.

1. Drag and drop the text input control into the screen.
2. Name it **txtForexAUD**.
3. Change the default property to 1.5.
4. Update the formula to the following code.

```
Text(galTickets.Selected.Impact_Value * txtForexAUD.
Text, "$0,0")
```

5. Adjust the value in the input box and observe the change in the $ value.

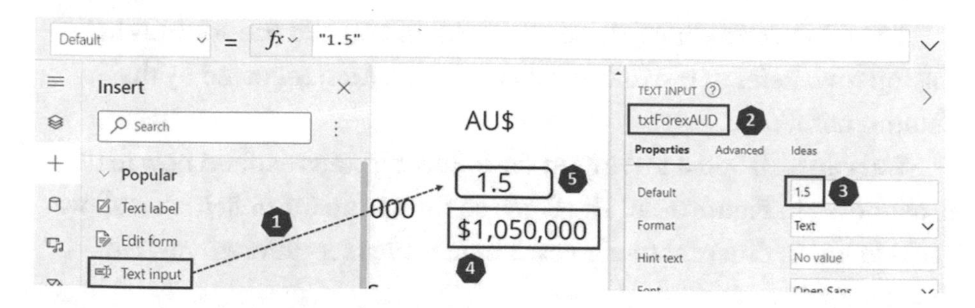

Figure 9-36. Use text input as part of the formula

Calculation in a Gallery

The same basic calculation can be applied in the gallery and by completing the following steps, as shown in Figure 9-37.

1. Drag and drop **Text label** to create the column names.

2. Drag and drop **Text label** to the top row of the gallery.

3. Configure the Text property to use the formula with the appropriate foreign exchange rate.

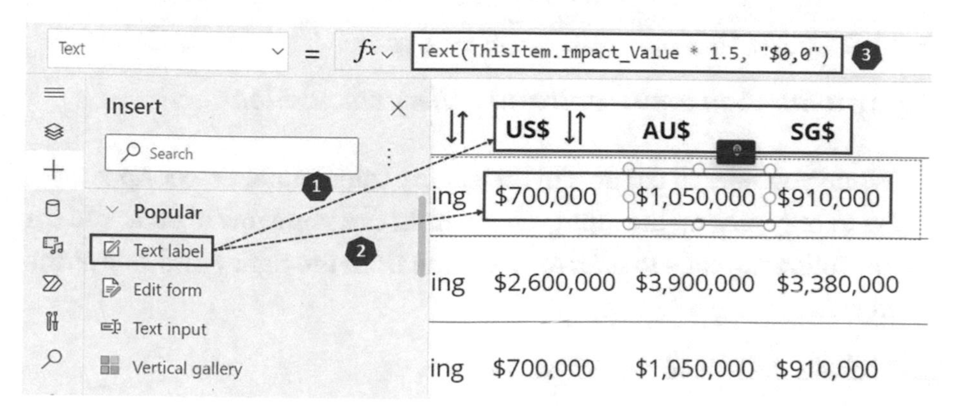

Figure 9-37. Adding calculation in gallery

The other rows change accordingly when you add the controls in the top row. Refer to the row of data using ThisItem followed by the column name.

You can also add a text input; the following steps explain how to do it, as shown in Figure 9-38. The Forex column is an input field that allows users to put in different foreign exchange values for different rows in the data.

1. Add the text label to create new column names.
2. Add Text input under the Forex label, name it **txtForex** and set the default value to 1.5.
3. Change the Text property for the **Currency label** to use the txtForex input value.

```
Text(ThisItem.Impact_Value * txtForex.Text, "$0,0")
```

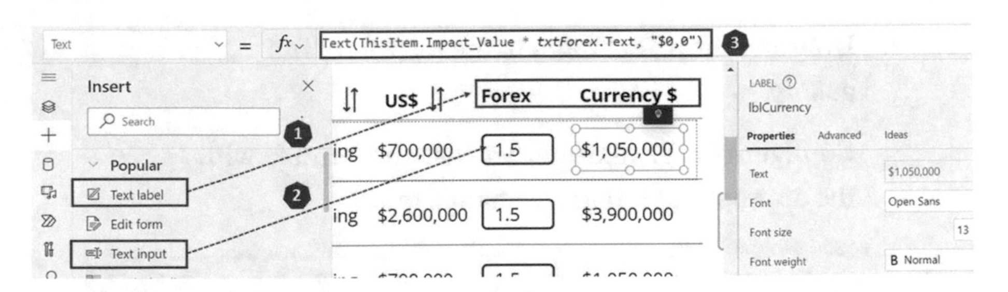

Figure 9-38. *Add input control to gallery calculation*

When you refer to txtForex in the gallery calculation, Power Apps knows to only pick up the input value within the same row of data. You can use the following code to refer to this value from the right-hand side of the detail view.

```
galTickets.Selected.txtForex.Text
```

Aggregation Calculations

In the prior code example, you can see that Power Apps is treating the gallery as a collection-like object. It is a common requirement to display a *total* at the bottom of the gallery. The following steps explain how to do this, as shown in Figure 9-39.

1. Drag and drop **Text label** to the bottom of the gallery and change the Text property to the following code.

```
"Total (" & CountRows(galTickets.AllItems) & " rows)"
```

2. Drag and drop **Text label** below the US$ column of the gallery and change the Text property to the following code.

```
Text(Sum(galTickets.AllItems, galLblValue), "$0,0")
```

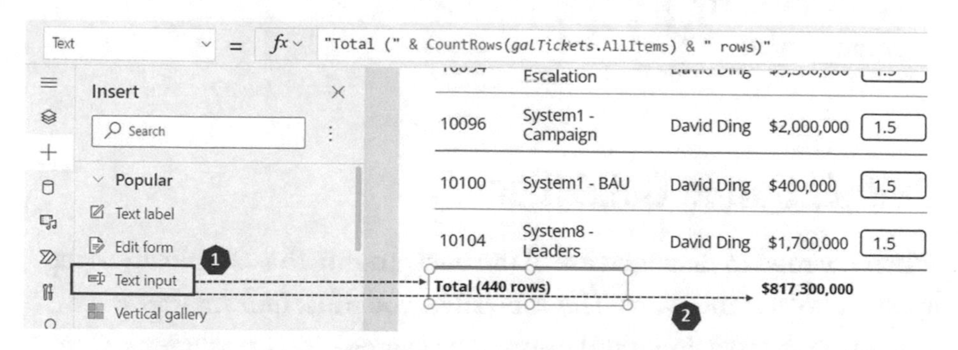

Figure 9-39. *Add total value of gallery*

From the code, you can see that you can refer to all items in the gallery by using the .AllItems extension for the Sum() function. You can then provide the control name.

Here is an interesting but confusing behavior about the gallery. When it loads directly from a SharePoint list (i.e., change gallery Item property to Ticketing System), it only initially displays the top 100 rows of data. This means the total from the method is only for the first 100 rows of data. If you are displaying more than 100 rows of data, the initial total is wrong. This is *not* an issue when you use the collection as the items in the gallery. The collection is only limited by the row limit (default 500) discussed previously.

The other option to produce the same result is to total the collection that feeds the gallery, as shown in the following code. The AllItems method has the advantage of also including any dynamic calculation fields from the gallery.

```
Sum(
  // below is the same as Item property in the gallery
  Search(colDisplay,txtSearchText.Text,"Issue_
  Name","Priority"),
  Impact_Value
)
```

Calculation in Collection

When the rows of data increase in the background, the complexity of the application also increases. This is because you must find smarter ways to manage the row limits and the various aggregation calculations. One robust method is as follows.

1. Set up a custom collection.
2. Use the custom collection in the gallery.
3. Use the gallery input controls to update the collection.

You can use the following code to create a custom collection by adding two more columns of data to colDisplay.

```
ClearCollect(
    colUsrNewColumns,
    AddColumns(
        colUserData,
        "Forex", 1.5,
        "Currency", Impact_Value * 1.5
    )
);
```

You can then update the gallery Items using the following code.

```
Search(
  colUsrNewColumns, txtSearchText.Text,
  "Issue_Name", "Priority"
)
```

Next, you can update the default value of txtForex control to ThisItem.Forex and lblCurrency to ThisItem.Currency, as shown in Figure 9-40.

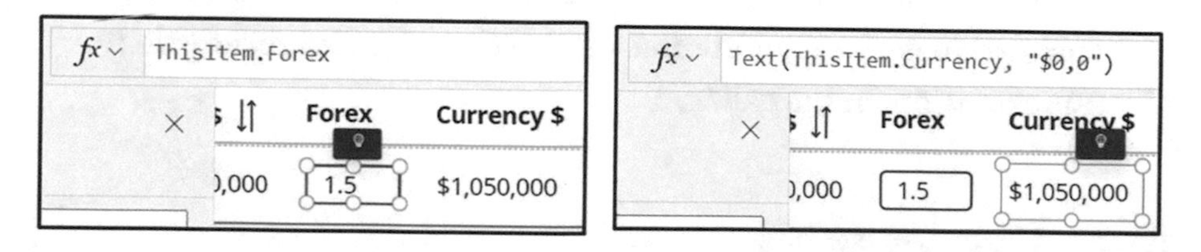

***Figure 9-40.** Update gallery control with new columns*

The final step is to update the OnChange property of the txtForex control to the following, as shown in Figure 9-41.

```
Patch(
    colUsrNewColumns,
    First(Filter(colUsrNewColumns, ID = ThisItem.ID)),
    {
        Forex: Value(Self.Text),
        Currency: Value(Self.Text) * ThisItem.Impact_Value
    }
)
```

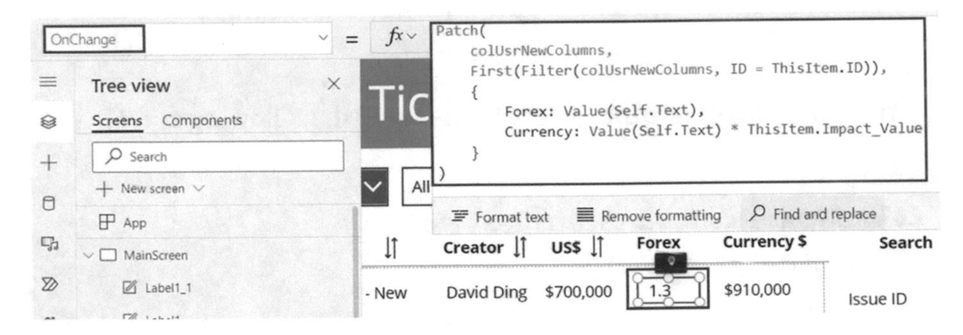

***Figure 9-41.** Updating the OnChange property of txtForex control*

Knowing all the changes are stored in the collection, you can use the new columns to perform additional aggregations without relying on the AllItems extension for the gallery.

User Experience

A good user experience (UX) helps users to easily navigate the app for its designed purpose. This session focuses on the following two practical methods to improve UX.

- Have a clear purpose
- Add interaction feedback
- Input validation

Have a Clear Purpose

Before you can start on UX design, you must have a *clear* purpose for the app. When the developer is unclear about the purpose, it confuses the users.

The following three questions help to clarify the purpose.

- What are the final desired actions for the app? If there are multiple actions, what are the priorities?
- What information is required for the user to take the desired action?
- What prevents the users from taking the desired action?

Once you have the answers to these questions, you can try to reduce the friction by trying to present the right information at the right time and eliminating any areas of confusion.

For the example ticketing system, the answers are as follows.

- Desired actions
    - create a new ticket
    - update an existing ticket
- Information that leads to actions
    - current ticket status and other information
- Roadblocks
    - not able to find the ticket (solution: build navigation)
    - key ticket information missing (solution: build validation)

Add Interaction Feedback

In addition to the border considerations, there are also some techniques to make certain actions more intuitive. Two useful technique is covered in this session.

- Use a rectangle to mimic the hover effect in the gallery.
- Use a pop-up window to acknowledge important actions.

Hover Effect in Gallery

You can add a hover box by completing the following steps, shown in Figure 9-42.

1. Move the rectangle to the first row of the gallery, and resize it to fit the full row.
2. Change the Color property to **RGBA(0, 0, 0, 0)**.
3. Change the Hover color property to **RGBA(252, 151, 104, 0.2)**.
4. Use Play (or hold the ALT key) to test the hover effect.

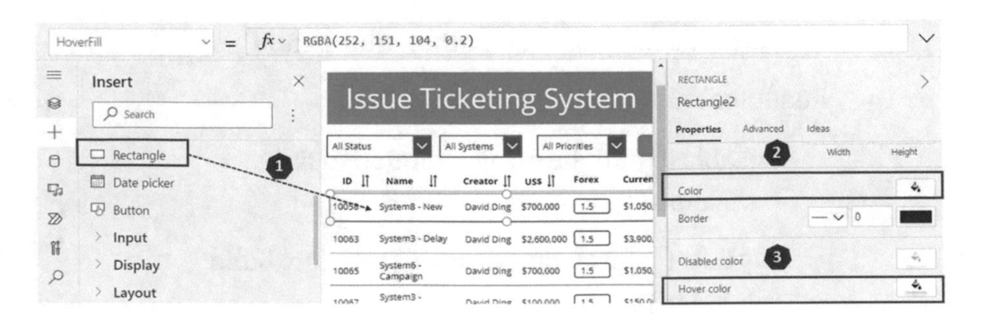

Figure 9-42. *Adding a hover box to the gallery*

The first three numbers in the RGBA() function represent the red, green, and blue color range, while the last digit is for alpha (transparency between 0 and 1).

Because this is a rectangle, if you still want to use the text input from the gallery, you need to change the order of the input box. The following steps explain how to do this, as shown in Figure 9-43.

1. Right-click the input box.
2. Select Reorder.
3. Select **Bring to front**.

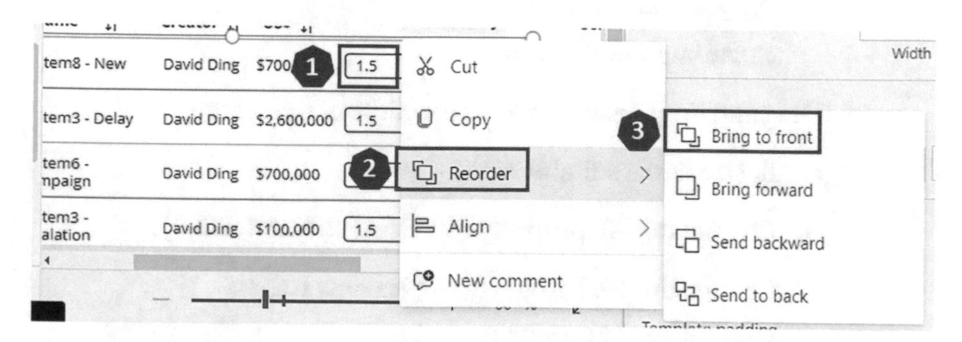

***Figure 9-43.** Changing display order*

Pop-up Window

The second interactive feature is a pop-up window. This helps acknowledge important actions, such as submitting a new ticket or submitting a change. In the ticketing app, the trigger point for both actions is the Save button. When the Save button is pressed, if there is no acknowledgment provided, the user is confused about if the right process has been followed.

You can use the following steps to create a pop-up window, as shown in Figure 9-44.

1. Set a new variable called varAcknowledgement in the OnSelect property of the Save button. *Set(varAcknowledgement, true)*

2. Drag and drop a Rectangle onto the screen.

    a. Size it to the same size as the screen.

    b. Change the Color property to **RGBA(0,0,0,0.2)**.

    c. Change the Visible property to **varAcknowledgement**.

3. Drag and drop **Text label** to the middle of the screen.

    a. Resize it to a suitable size.

    b. Change the Fill property to **RGBA(255,255,255,1)**.

    c. Change the Text property to **The new ticket/ change has been saved**.

    d. Change the Visible property to **varAcknowledgement**.

    e. Add OnSelect property to Set(varAcknowledgement, false).

4. Add a Cancel icon into the top right-hand corner of the text label.

    a. Resize it to a suitable size.

    b. Change the Visible property to **varAcknowledgement**.

    c. Change the OnSelect property to Set(varAcknowledgement, false).

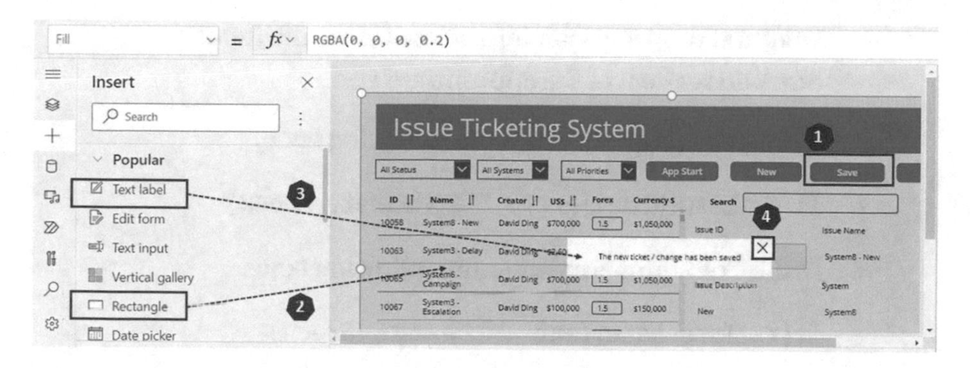

***Figure 9-44.** Add a pop-up window*

When the user clicks the Save button, an acknowledgment pop-up window confirms the action. When the Cancel button is pressed, the window disappears.

Input Validation

To prevent users from entering incorrect data into the system, input validation is often used to guide the data entry process.

Input validation contains two parts.

- Display validation warning
- Deny saving

There are multiple methods for displaying the validation warning. You can do a check vs. cross or change the input box border to red.

You can start with a simple check to see if there is any value in the box. The following steps display the check or cross icons for the Issue Name input, as shown in Figure 9-45.

1. Drag and drop any icon into an area near the input box, and resize it to the appropriate size.
2. In the Icon property, enter the following code.

   ```
   If(txtIssueName.Text = "", Cancel, Check)
   ```

3. In the Color property, enter the following code.

   ```
   If(txtIssueName.Text = "", Red, Green)
   ```

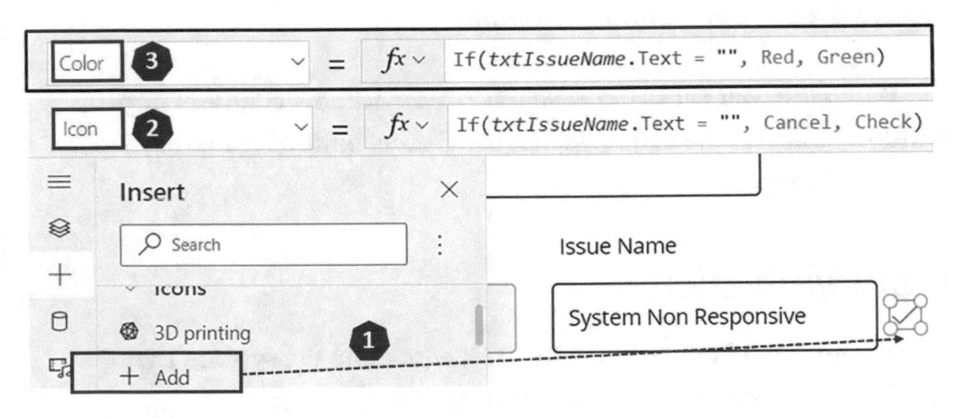

Figure 9-45. *Adding check vs. cross icon*

You can then change the BorderColor property of the input box to the following code, as shown in Figure 9-46.

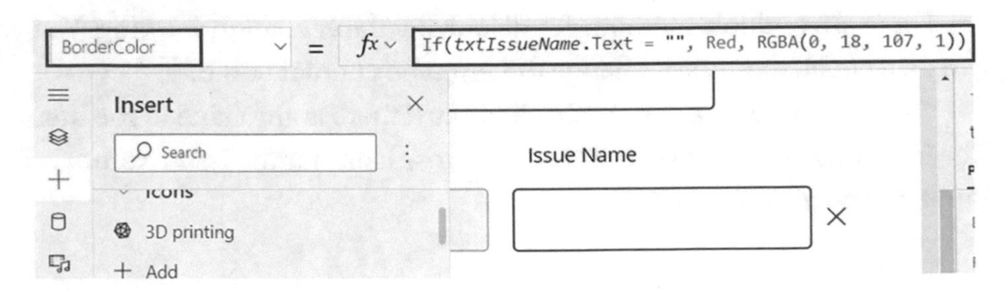

Figure 9-46. *Change BorderColor property of the input box*

You can also run more sophisticated checks, such as email. Even though there are no contact email and phone number fields in the data, you can still create the fields to learn about validation, as shown in Figure 9-47.

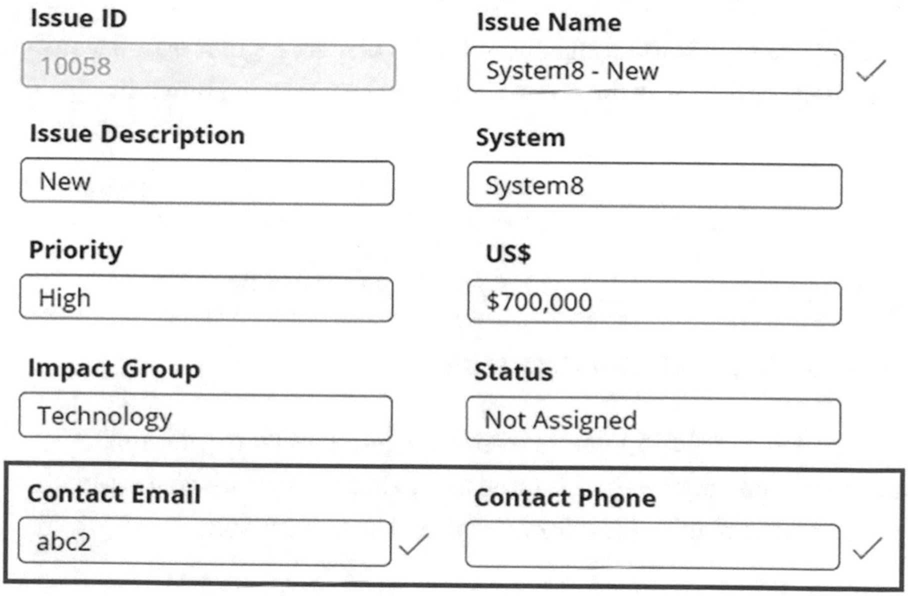

Figure 9-47. *New contact input*

You can then configure the Icon box and the colors using the following code.

The following is for check vs. cross.

```
If(IsMatch(txtContactEmail.Text, Match.Email), Check, Cancel)
```

The following is for color.

```
If(IsMatch(txtContactEmail.Text, Match.Email), Green, Red)
```

As another example, you can check the Contact Phone number is a nine-digit number using the following code.

```
If(IsMatch(txtContactPhone.Text, Digit & Digit & Digit & Digit
& Digit & Digit & Digit & Digit & Digit)
```

If this method is too restrictive, you can also try regular expressions. Moving the cursor near the email shows the hint text, as shown in Figure 9-48.

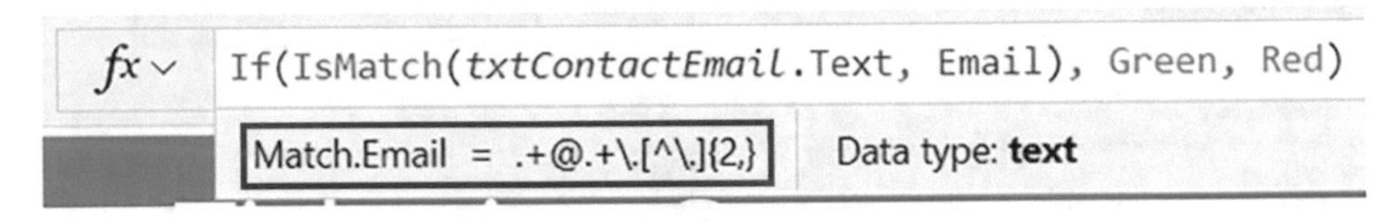

Figure 9-48. *Match.Email expression*

The **.+@.+\.[^\.]{2,}** part is referred to as a regular expression. It is a pattern recognizing method for free text fields. If you replace the Match. Email with the regular (as follows) you get the same result.

```
If(IsMatch(txtContactEmail.Text, ".+@.+\.[^\.]{2,}"),
Green, Red)
```

This adoption of regular expression enables a flexible method for validating data even in a free text field. You can also use regular expressions to shorten the phone number check to as follows.

```
If(IsMatch(txtContactEmail.Text, "\(?[0-9]{3}\)?"), Green, Red)
```

In addition to the validation warning, you also need to prevent the user from saving using a global variable to store the results in the OnSelect property of the Save button as follows.

```
Set(
    varSave,
    // condition 1 and
```

```
    IsMatch(txtContactEmail.Text, Match.Email) &&
        // condition 2
        IsMatch(txtContactPhone.Text, "\(?[0-9]{3}\)?")
);
```

Once the variable has been created, you can wrap another If function outside the existing If() – Patch() logic to prevent the Patch from happening if the mandatory conditions are not met.

```
If( varSave,  //new If() to check if all conditions are met
    If( gblNewTicket,  //old code from here
        Patch(
            'Ticketing System',
            .. .. ..
```

Finally, you must update the message in the acknowledgment text box by changing the Text property to the following code.

```
If(
    varSave,
    "The new ticket / change has been saved",
    "Saving Error. Please check the mandatory fields"
)
```

Conclusion

Congratulations on getting to this part of the chapter. This chapter covered a broad range of practical techniques to help you to improve basic Power Apps, which includes the following.

- How and when to use variables and delegations
- Create color palettes and naming standards

- Add additional UI features
- Dynamic calculation
- User experience

Mini-Hackathon

Once you have a basic understanding of Power Apps, you can build useful applications. In practice, it is often good to classify the requirement into broad categories and learn from established solutions in that category. The following are a few examples.

- Catalog (amazon.com, ebay.com)
- Project management (jira.com, goodday.work)
- Asset management (upkeep.com, pulseway.com)

Objective

Find some online solutions for the ticketing system. Adopt some practices to improve your ticketing system from Chapter 8. Try to rate the identified feature on usefulness and difficulty, and focus on features that are useful but also easy to implement.

Time Limit

Complete this within five hours.

CHAPTER 10

Power Automate

According to Microsoft, Power Automate is an online workflow service that automates actions across the most common apps and services. In the context of this book, Power Automate is used to further enhance your apps in the following ways.

- Send out email notifications.
- Create backup.
- Streamline approval processes.

These three enhancements represent the common applications of Power Automate cloud flows. You learn how to develop cloud flows to suit the needs of your workplace.

One of the interesting components of Power Automate that is not covered in this book is Power Automate Desktop (PAD). PAD is a Robotic Process Automation (RPA) solution that can perform predefined tasks on your local machine.

Business Scenario

Kim has just received her superman badge from Richard, the technology manager for the ticketing system. In the past few weeks, Kim's Power Apps effort paid off. The leadership team approved the proof-of-concept (POC) ticketing app as a high-priority project.

D. Ding, *Transitioning to Microsoft Power Platform*,
https://doi.org/10.1007/978-1-4842-9239-6_10

Kim is working closely with Richard to "productionize" the app in four weeks by adding several final improvement features. Kim is comfortable with most of the features except for the following ones.

- When a new ticket is created or changed, an acknowledgment email is sent to the creator and the support team.
- Create a daily system backup for all tickets.
- When a new high priority ticket is created, an approval request is created for the line manager.

Just when Kim was about to give up on these features and exclude them from the project scope, Richard found some similar workflow templates in Power Automate. They have decided to investigate the three issues separately to test its feasibility.

Power Automate Licensing

Power Automate shares many licensing rules with Power Apps. Instead of repeating the licensing details covered in Chapter 8, you only need to understand the following.

- Power Automate workflows triggered in Power Apps are covered under the Power Apps license.
- Power Automate for Office365 covers the Standard connectors (connector to SharePoint list is classified as Standard connector).

Premium connectors may be required; the pricing is shown in Figure 10-1.

Subscription plans		
Licence by user	Licence by user	License by flow
Per user plan	Per user plan with attended RPA	Per flow plan
$20.60	$54.90	$137.30
per user/month	per user/month	per flow/month; minimum 5 flows[1,2]
Pay-as-you-go plans		
Licence by flow run	Licence by flow run	Licence by flow run
$0.80	$0.80	$4.10
per flow run[4]	per flow run[4]	per flow run[4]

Figure 10-1. *Power Automate licensing plans in USD*

The information is captured from the Microsoft website in Dec 2022. For the most up-to-date licensing information on Power Apps, please google and download the *Power Platform Licensing Guide.*

You can check your access to Power Automate by logging into make.powerautomate.com, as shown in Figure 10-2.

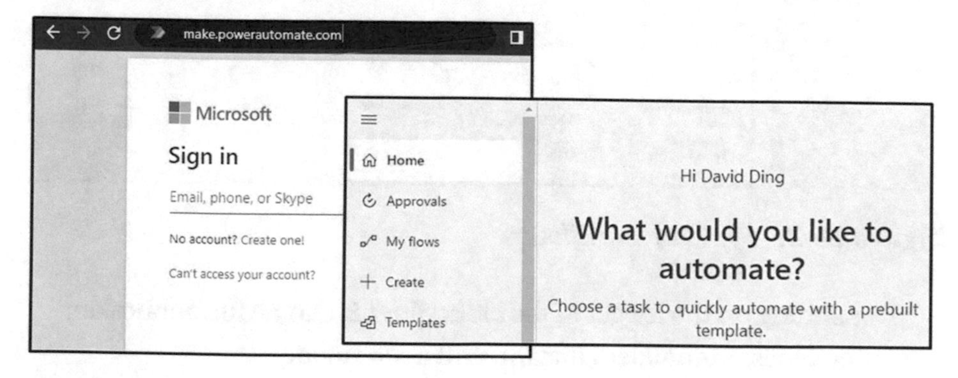

Figure 10-2. *Log in to Power Automate*

Power Automate Flows

Workflows or Flows are the end products in Power Automate. There are two types of components in flow: triggers and actions. Every flow starts from a *trigger*. Once the flow is triggered, it can go through one or multiple *actions*. Both the triggers and the actions require connection to other systems outside of Power Automate. For example, an automated cloud flow may be triggered by an email (in Microsoft Outlook) with a specific title, and the action is to save the attachments of the email into a SharePoint folder (SharePoint Server).

After you've logged in, you can see the three types of Power Automate flows, as shown in the following steps and Figure 10-3.

1. Select Create.

2. Review the blank workflow options.

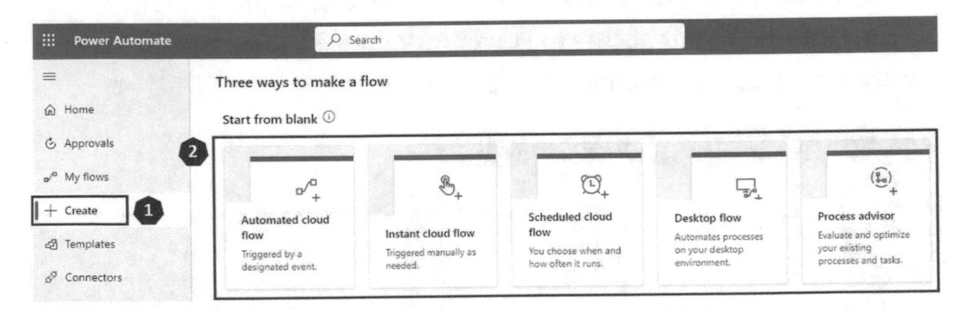

Figure 10-3. *Types of workflows*

Depending on the trigger of the cloud flow, it can be further broken into three types: Automated, Instant, and Scheduled.

- **Automated cloud flow** is triggered by specific events (e.g., receiving an email).
- **Instant clod flow** is triggered manually by the user (e.g., pressing a button in Power Apps).
- **Scheduled cloud flow** is triggered regularly at time specified (e.g., 8 AM every Monday).

The flow can also connect to your local machine by installing the Power Automate Desktop. This can create interesting **Desktop flow** to automate standard human interactions. For example, an admin needs to perform data entry tasks based on information in a standard website. Note that Desktop flow can only apply to highly repetitive tasks.

There is also a type of flow called a *process advisor*, which can help record one or more people performing the same tasks on their own devices. The process advisor analyzes all individual steps, compare them to each other, and finally visualizes the differences or similarities between users.

To fulfill the three requirements, the following flows are used.

- Instant cloud flow for sending out email notifications
- Scheduled cloud flow for creating backup
- Instant cloud flow for streamlining approval processes

Cloud Flow Overview

You can find a simple cloud flow, as shown in Figure 10-4. This flow is triggered by Power Apps and sends out a standard email to the user. The view contains the following elements.

1. Main design screen
2. Flow Checker brings up the errors and warnings
3. Test brings up the testing options

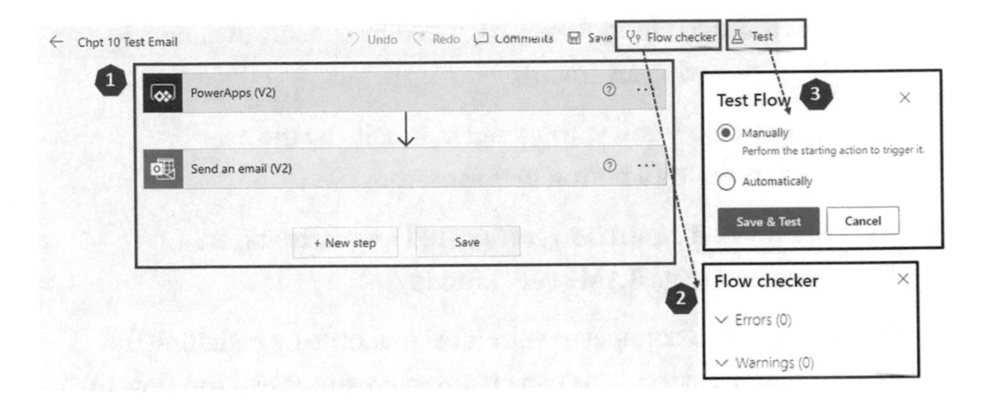

***Figure 10-4.** Simple instant cloud flow*

For flow development, the design screen is the main screen of focus. The item at the top of the design screen is always a trigger. In the prior example, the trigger is from Power Apps (e.g., when a button is pressed). You can add new step(s) after the event, each step is an action. Flow development follows a simple user interface to add and customize events and actions. The idea of flow automation is to develop a chain of actions following an event. In the example, there is only one action in the chain, which is to send out an email.

You can now save the flow and test it in Power Apps, as shown in the following steps and Figure 10-5. Note that it may take a few minutes for the new flow to appear in Power Apps.

1. After opening Power Apps, go to the Power Automate pane.
2. Select **Add flow**.
3. Choose the flow you have just created.
4. Create a new button and name it **Email**.

5. Go to the **OnSelect** property of the new button.

6. Enter the following code.

```
NameOfYourFlow.Run(
    "someone@somewhere.com.au",
    "Here is the email subject",
    "Here is the email body"
)
```

7. Click the button and wait for the email to arrive.

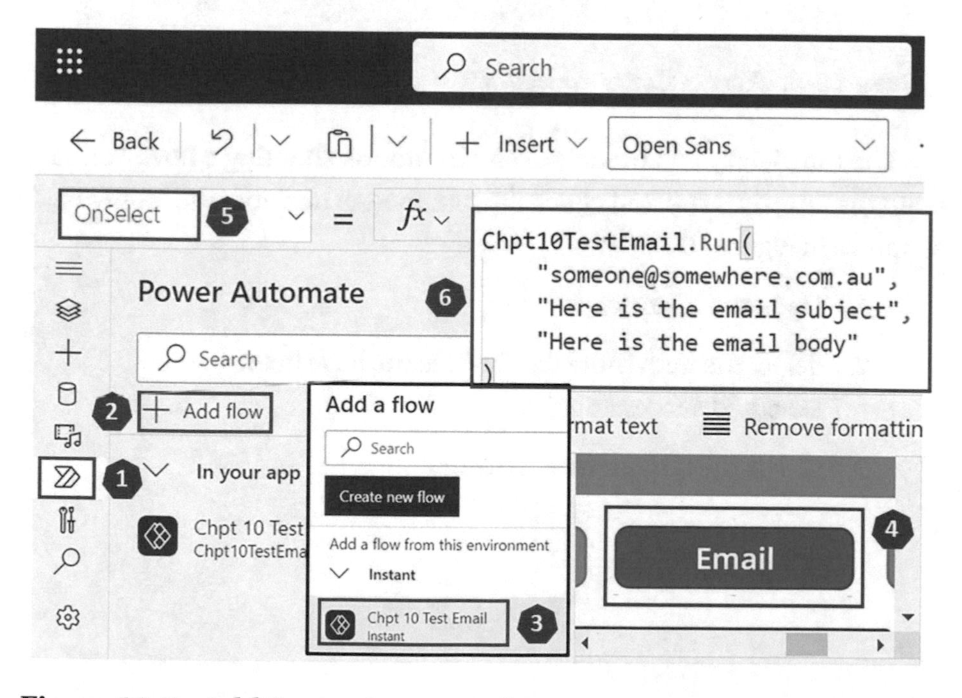

Figure 10-5. Add Power Automate flow

If you have followed the steps closely, an email should arrive at the nominated email address.

To troubleshoot this step, return to the flow page and do the following to see the run history, as shown in Figure 10-6.

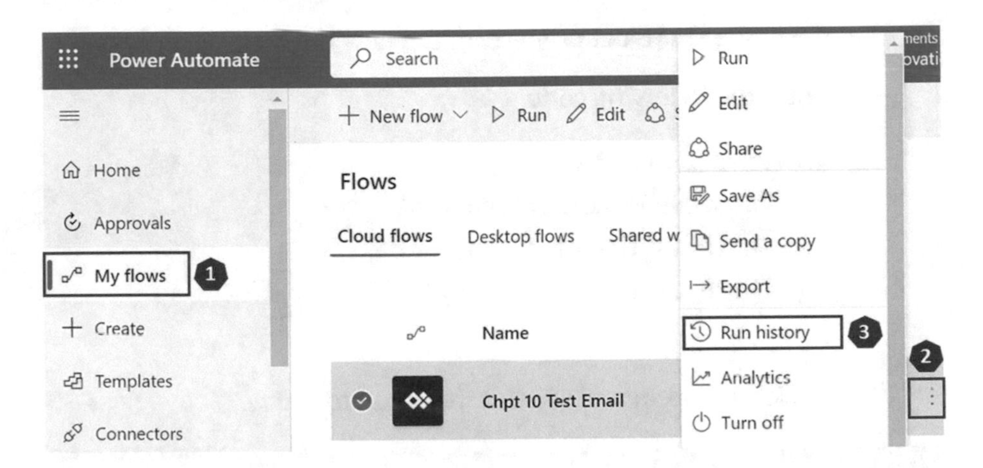

Figure 10-6. *Access flow run history*

The run history is extremely useful for troubleshooting a flow. You can open the historical run and check the parameters by doing the following, as shown in Figure 10-7.

1. Open the historical run.
2. Select the item (note that both items have been executed successfully).
3. Check the parameters of the item.

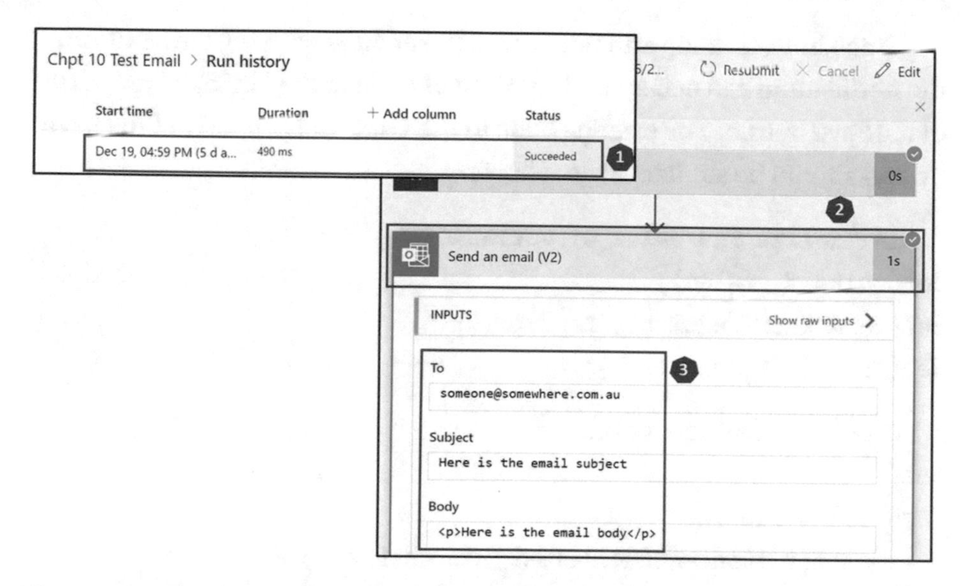

Figure 10-7. *Review flow historical run parameters*

Power Automate has already managed all the complexity with API integration with various systems. This allows the developer to focus on the flow logic and automation.

Email Notification

With a basic understanding of flow development and Power Apps integration, you can now use it to enhance the ticketing system app. The first enhancement is to add an email notification to be issued to the user after he/she has created a new ticket or updated an existing ticket.

Since both creating and updating a ticket ends with the Save button, the notification can occur as a last piece of code in the OnSelect property of the Save button. The existing code in the OnSelect property of the Save button should be similar to the following.

```
// initialize the value of variables
Set(gblEdit, false);
Set(varAcknowledgement, true);
Set(gblValidation, true);

// check validation value
Set(
  gblValidation,
  IsMatch(txtContactEmail.Text, Match.Email) &&
    IsMatch(txtContactPhone.Text, "\(?[0-9]{3}\)?")
);

// does input meet validation requirements?
// if so, consult the gblNewTicket value to determine if to
create a new item or update existing one
If( gblValidation,
  If( gblNewTicket,
    Patch(
      'Ticketing System',
      Defaults('Ticketing System'),
      {
        Title: "X",
        Issue_Name: txtIssueName.Text,
          Issue_Description: txtIssueDescription.Text,
          System: txtSystem.Text,
          Priority: txtPriority.Text,
          Impact_Value: Value(txtImpactValue.Text),
          Impact_Group: txtImpactGroup.Text,
```

```
            Status: txtStatus.Text
          }
      ),
      Patch(
        'Ticketing System',
        First(Filter(
          'Ticketing System',ID=galTickets.Selected.ID
        )),
        {
           Title: "X",
           Issue_Name: txtIssueName.Text,
           Issue_Description: txtIssueDescription.Text,
           System: txtSystem.Text,
           Priority: txtPriority.Text,
           Impact_Value: Value(txtImpactValue.Text),
           Impact_Group: txtImpactGroup.Text,
           Status: txtStatus.Text
        }
      )
    )
);
```

You can add the email notification to the end of the code. The flow is the same as what has been set up in the Chapter 10 Test Email flow. The email address, subject, and body are determined using the following logic.

- **Email address** is the user email, which can be determined using the value in User().Email.
- **Email subject** can be determined using the gblNewTicket variable. If it is a new ticket, the subject contains the phrase *New Ticket*; otherwise, it contains the phrase *Change of Ticket*.

- Part of the **email body** is determined using the gblNewTicket variable like email subject. The other part of the email body comes from the various input control fields.

When running the development, you can create a text label for testing purposes to display the user's email address. The following steps explain how to do this, as shown in Figure 10-8.

1. Drag and drop a Text label to an empty space on the main screen.

2. Select the **Text** property.

3. Enter the following code to check the correct user email address.

```
User().Email
```

4. Migrate the code into the flowName.Run() function in the OnSelect property of the Save button.

```
Chpt10TestEmail.Run(
    User().Email,
    "Work In Progress",
    "Work In Progress"
)
```

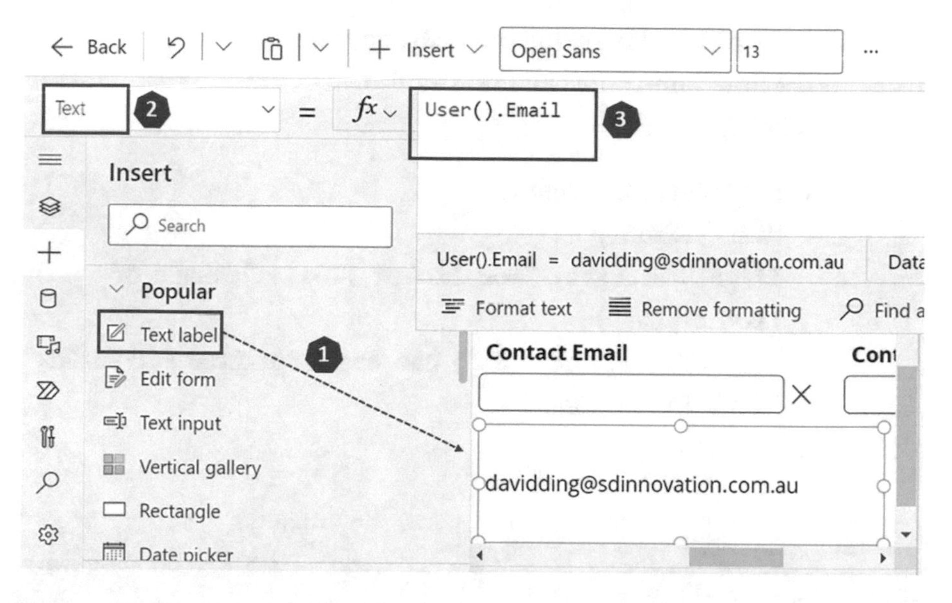

Figure 10-8. *Test email address*

The next item is the email subject. The following steps test the email subject text, as shown in Figure 10-9.

1. Drag and drop a text label to an empty space on the main screen.
2. Select the **Text** property.
3. Enter the following code to determine the wording based on the gblNewTicket variable.

```
If(gblNewTicket, "New Ticket (ID:", "Change of Ticket (ID:")
    & txtIssueID.Text & ") has been submitted for Review"
```

4. Try to click the New button and select an item in the gallery to test if the wording changes according to your expectations.

5. If the display is working correctly, copy it into the flowName.Run() function as shown in the following code.

```
Chpt10TestEmail.Run(
  User().Email,
  If(gblNewTicket, "New Ticket (ID:", "Change of
  Ticket (ID:")
    & txtIssueID.Text & ") has been submitted for Review",
  "Work In Progress"
)
```

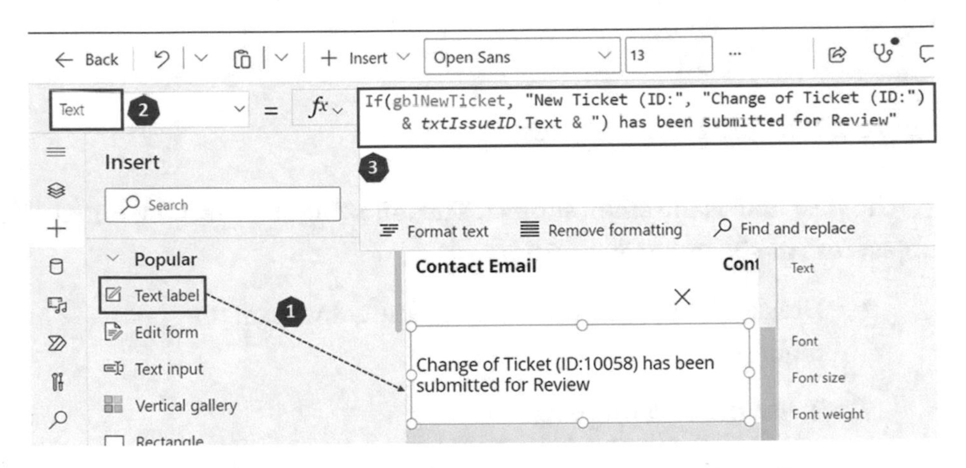

Figure 10-9. *Test email subject text*

The & operator acted as a connector between different text values in the code. This is useful for displaying dynamic text values. You use more of it in the email body text.

If you look at the previous successful flow historical record in the body session of the action item, you can see the <p> and </p> as shown in Figure 10-10.

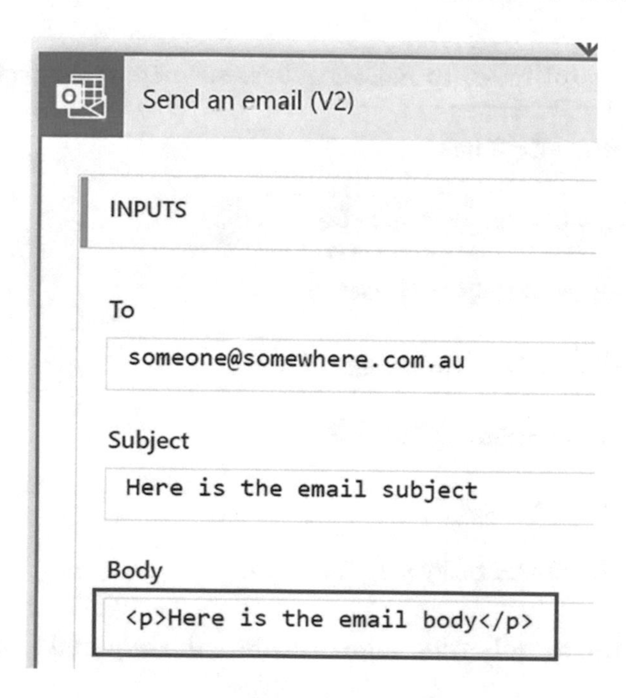

Figure 10-10. *Email body HTML text*

This means the body session accepts HTML text formats. The HyperText Markup Language (HTML) is the standard markup language for documents designed to be displayed in a web browser. While it can be used to display complex text structures, you only need to use the following basic functions to generate the body text.

- Paragraph staring with a new line: <p> ... </p>
- Tab:
- Bold: <b> ... </b>

The target email body format is as shown in Figure 10-11 with the following characteristics.

1. Some of the text is bolded.
2. Use dynamic text to add relevant ticket information.
3. Add a tab in front of every ticket information text.

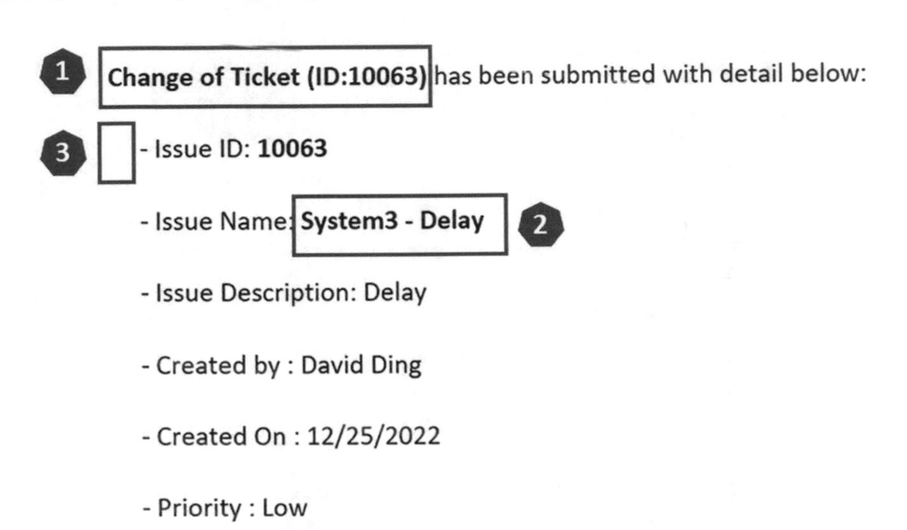

Figure 10-11. Email body target format

You can use the following code to achieve the required format.

```
// part 1: starting sentence of email body
If(gblNewTicket, "<p><b>New Ticket (ID:", "<b>Change of
Ticket (ID:")
    & txtIssueID.Text & ")</b> has been submitted with detail
    below:</p>" & "

// part 2: add ticket information in separate lines with tab
<p>   - Issue ID: <b>" & txtIssueID.Text & "</b> </p>
<p>   - Issue Name: <b> " & txtIssueName & " </b> </p>
<p>   - Issue Description: " & txtIssueDescription & "</p>
<p>   - Created by : " & User().FullName & "</p>
<p>   - Created On : " & Today() & "</p>
<p>   - Priority : " & txtPriority & "</p>"
```

The special HTML format has been bolded in the example. Part 1 of the code is very similar to the email subject, except it has included some HTML formats including new paragraph and bolding. Part 2 of the code adopted dynamic text with a tab at the beginning of each line.

Note Adding HTML format can introduce more complexity into the code. It is better to build this in steps and test each step using the email button created earlier.

The final FlowName.Run() function should look like the following code.

```
Chpt10TestEmail.Run(
  User().Email,
  If(gblNewTicket, "New Ticket (ID:", "Change of Ticket (ID:")
    & txtIssueID.Text & ") has been submitted for Review",
  If(gblNewTicket, "<p><b>New Ticket (ID:", "<b>Change of
  Ticket (ID:")
    & txtIssueID.Text & ")</b> has been submitted with detail
    below:</p>" & "
  <p>   - Issue ID: <b>" & txtIssueID.Text & "</b> </p>
  <p>   - Issue Name: <b> " & txtIssueName & " </b> </p>
  <p>   - Issue Description: " &
  txtIssueDescription & "</p>
  <p>   - Created by : " & User().FullName & "</p>
  <p>   - Created On : " & Today() & "</p>
  <p>   - Priority : " & txtPriority & "</p>"
)
```

This can be added to the end of the Save button to issue an email to the user after the ticket has been submitted.

Create a Backup

A daily or weekly backup can be helpful for the following reasons.

- Restore the system in case of unexpected behavior.
- Track change history for the same issue
- Build KPI reports on the ticketing system and monitor it over time (such as resolution time, systems performance, and teams performance)

When you have data stored in SharePoint list, you can create these backups using the scheduled cloud flow method by setting a backup time (e.g., 2 AM every Monday). The backup can be created in many different types of *destinations*. In this session you use a simple Excel file in a SharePoint folder to store the backup data.

Prepare a Backup Excel File

Power Automate flow can only connect to a table in an Excel file stored in a SharePoint folder or OneDrive. To prepare the file, it is easier to run a simple export from the SharePoint list, as shown in Figure 10-12.

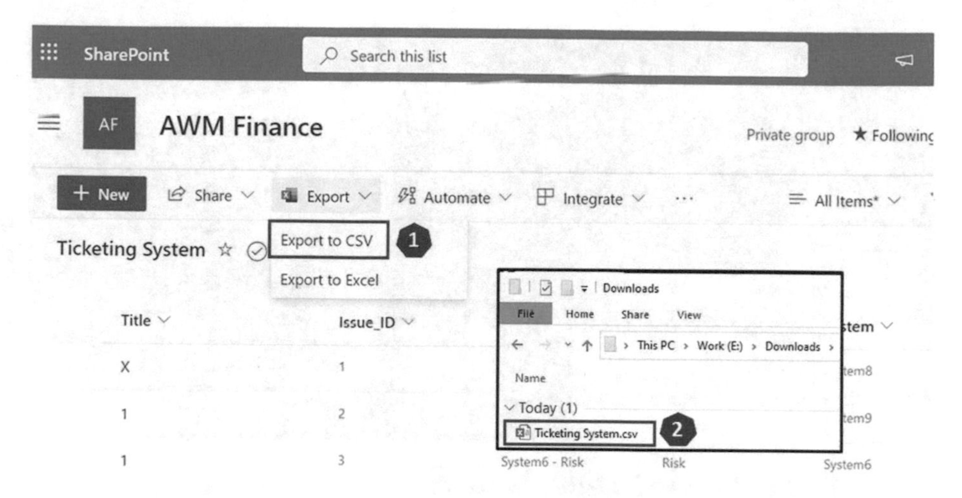

Figure 10-12. Export data from SharePoint list

You are using the Export to CSV (comma-separated values) file instead of Excel because the Excel export creates an Internet Query (.iqy) file in a local folder. This query file does not store any data; instead, it contains a connection method that downloads the required data when you open the file. Many companies block the connection for security reasons. Once you have the CSV file opened, you can convert it into a table by completing the following steps, as shown in Figure 10-13.

1. Make sure a cell with a value is selected (e.g., A1).

2. Under the Insert tab, click Table.

3. Check the data range, A1–M2944, and click OK. You will know the table was created successfully when the display format changes.

4. Change the Table Name to **Backup_Issues**.

5. Insert a new column into the table called Backup_Date and populate it with yesterday's date (e.g., 2022-09-30). Yesterday is used to differentiate from the new backup, which uses today's date.

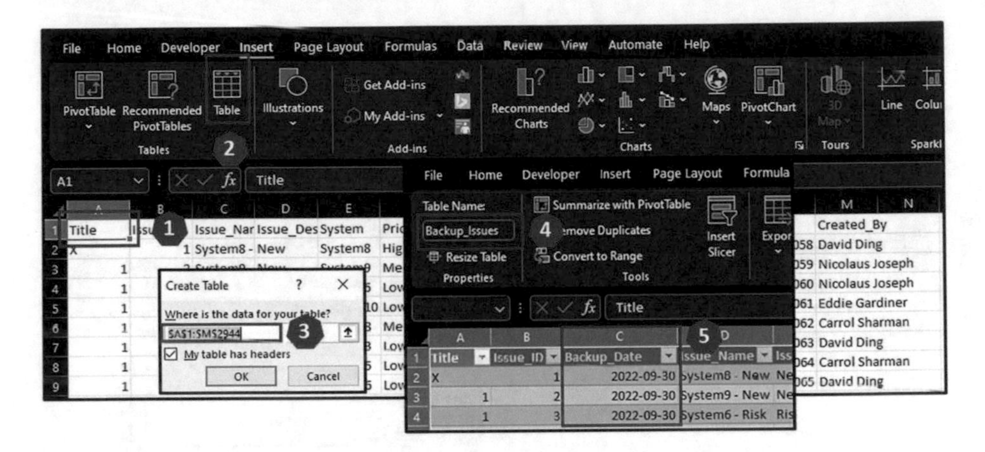

Figure 10-13. *Convert a table in Excel*

You need to create this Backup_Issues table because Power Automate requires a table to connect to. In addition of having a table, the file itself must be loaded to OneDrive or SharePoint folder. Do the following to create a SharePoint folder and load the data into the folder, as shown in Figure 10-14.

1. Go to the AWM SharePoint site previously created.
2. Select Documents.
3. Click New and Select Folder.
4. Name the new folder (e.g., Backup).
5. Navigate to the Backup folder.
6. Drag and drop the Excel file from the previous step into the folder.

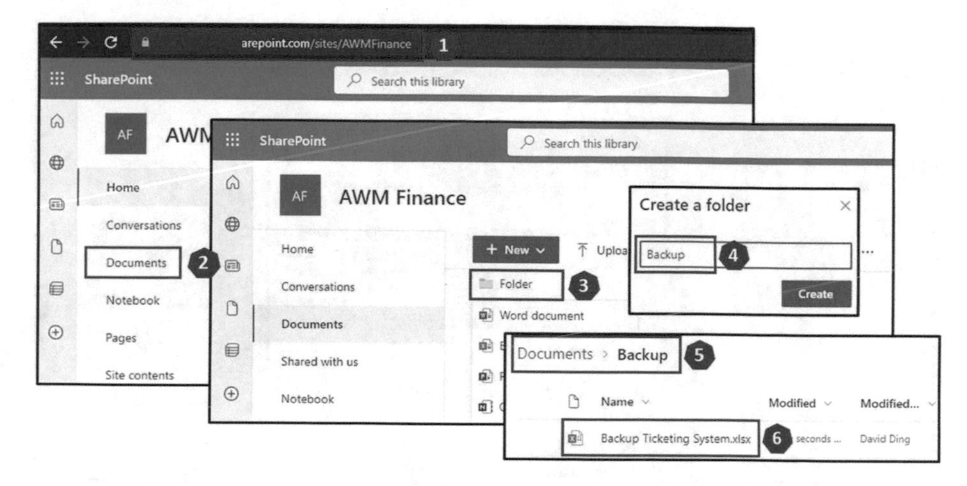

Figure 10-14. *Loading the Excel file into SharePoint folder*

Create the Backup Process

Once the Excel file with table is loaded into the Cloud drive. The data contains in the file can be used as the initial backup. The backup process appends new data into the file. The process also populate the Backup_Date to differentiate the different backups.

The following steps initiate a scheduled cloud flow, as shown in Figure 10-15.

1. Log in to make.powerautomate.com and select Create.
2. Select **Scheduled cloud flow**.
3. Name the flow (e.g., Backup Ticketing System.
4. Create the schedule.
5. Select the day of the week (e.g., Sunday.
6. Check the description.
7. Select Create.

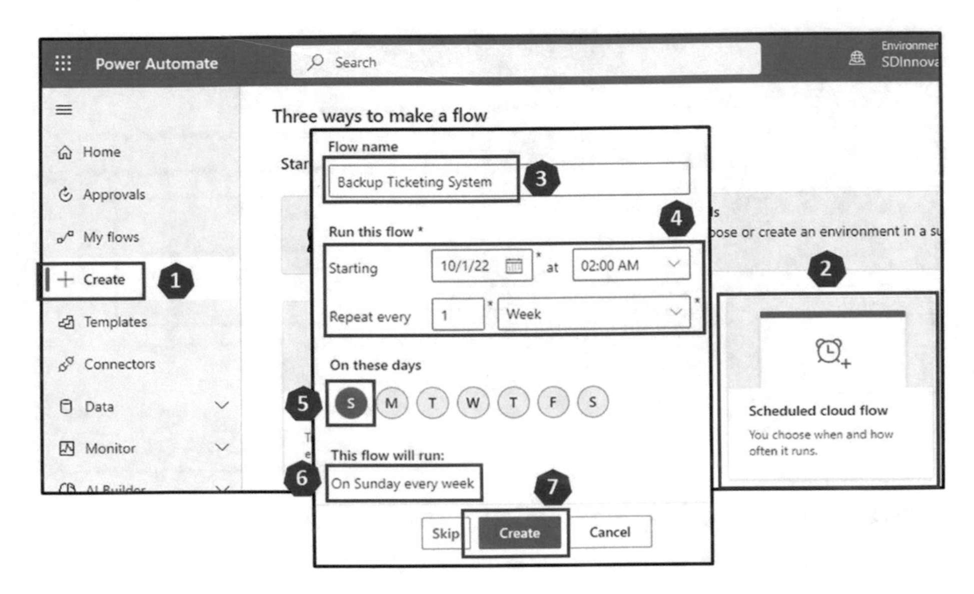

Figure 10-15. *Initiate a new scheduled cloud flow*

Once you click the Create button, you should see a flow screen with a recurrence step in the middle of the screen as shown in Figure 10-16. This is the trigger of the flow.

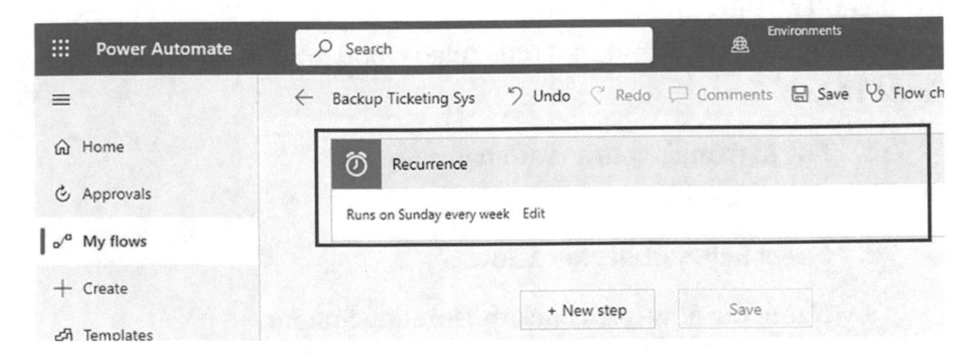

Figure 10-16. *Recurrence trigger in scheduled cloud flow*

Once the trigger is configured, you can add a new step by clicking the **New step** button below the trigger. Once the flow has been triggered, it copies all items from the SharePoint list into the Excel table. In flow, this is broken into two separate actions.

1. Get items from the SharePoint list.
2. Add this item to the Excel table.

The first *Get items* action only gets one item from the SharePoint list at a time but repeats itself by getting the next item from the list until it reaches the end. In programming, this is referred to as a loop.

You can configure the first action by completing the following steps, as shown in Figure 10-17.

1. Enter **SharePoint list** in the search box.
2. Select the **Get items** action.
3. Use the drop-down box to select the SharePoint site where the list is stored.
4. Use the drop-down box to select the SharePoint list.
5. Click **New step** to add the next action.

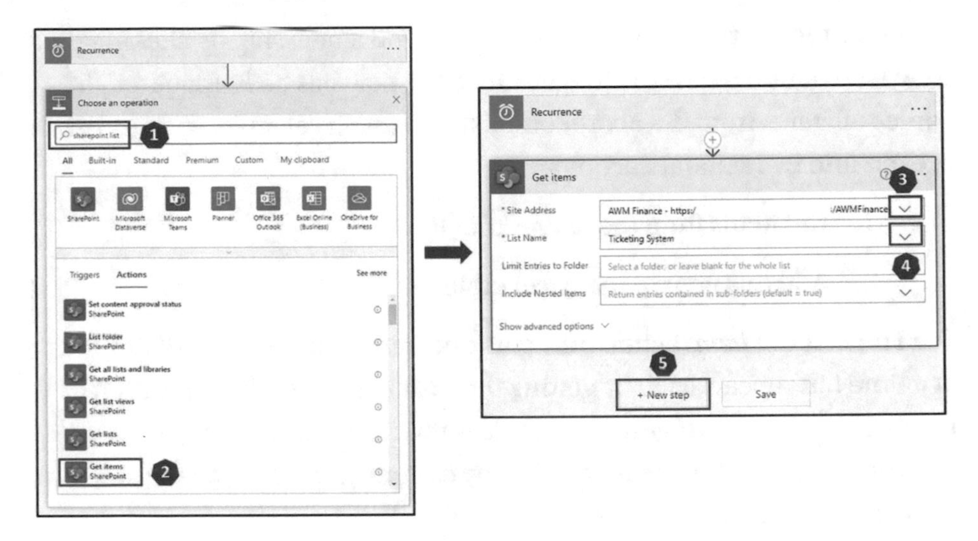

Figure 10-17. *Configure Get items action for SharePoint list*

The prior steps create an action to loop through the SharePoint list. The next step is to copy the item's content into the Excel table. This can be done by completing the following steps, as shown in Figure 10-18.

1. Enter **Excel** in the search box.
2. Select Excel Online (Business).
3. Select **Add a row into a table**.
4. Use the Location drop-down box to select the location.
5. Use the Document Library drop-down box to select the document library.
6. Use the folder box to select the file.
7. Use the Table drop-down to select the table name.

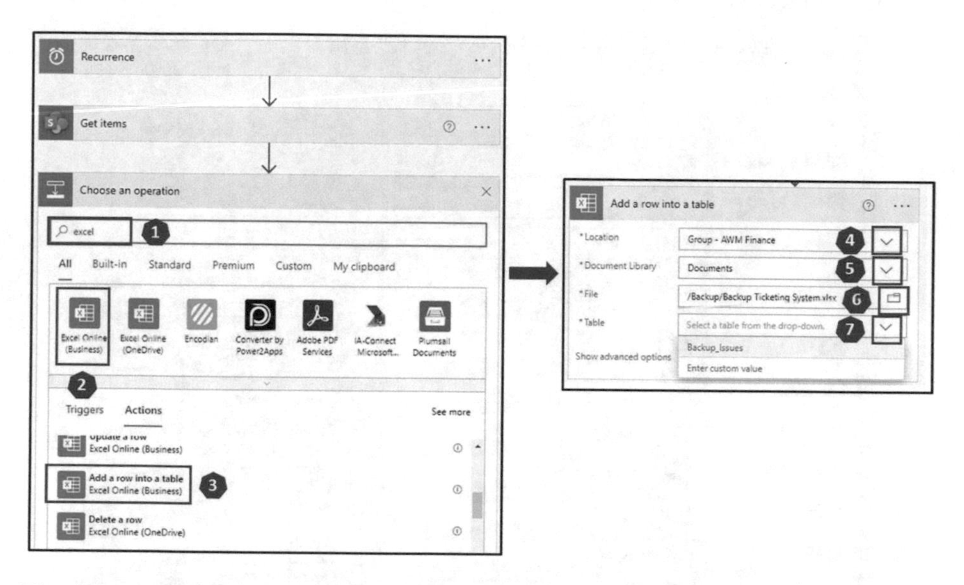

Figure 10-18. Configure Excel add row to a table part 1

Once you select the table, the action box shows more fields. The following steps add the relevant fields from the SharePoint list item. Except for Backup_Date, you can configure these fields. The following steps explain how to do this, as shown in Figure 10-19. Note that the fields in the action box (left) are from the Excel table fields, and those in the dynamic content (right) are from the SharePoint list fields. Your job is to map them.

1. Select each field in the action box.
2. Check **Dynamic content**.
3. Choose the relevant field from the available options.
4. Repeat steps 1–3 for all fields except for Backup_Date.

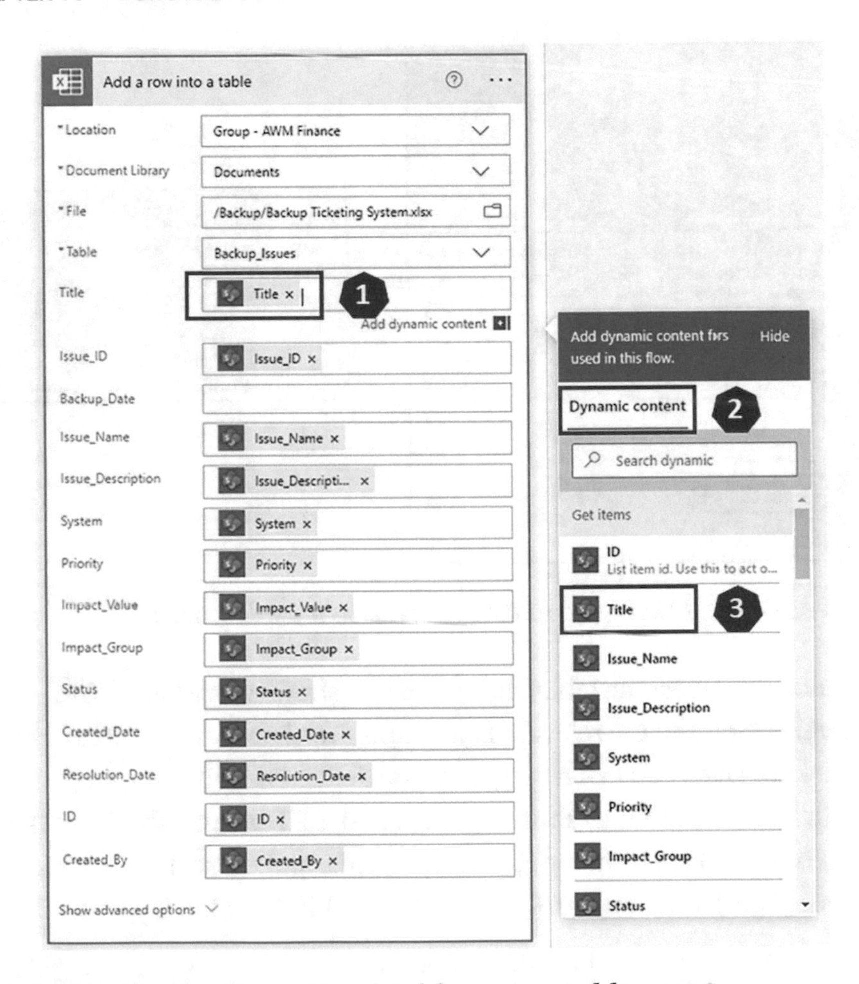

Figure 10-19. Configure Excel Add row to a table part 2

The only field that needs to be populated is the Backup_Date. The value in this field is not from the SharePoint list; instead, it simply takes the date of export. You need to populate this using an *expression* instead of dynamic content. The following steps create the expression, as shown in Figure 10-20.

1. Select the Backup_Date field.
2. Select Expression.

3. Enter **convertFromUtc(utcNow(),)** in the function field.

4. Once you put a comma after the utcNow() function, a hint box should appear. The first paragraph of the hint box tells you the required field (destinationTimeZone), and the second paragraph contains a link. You can copy this link into a browser.

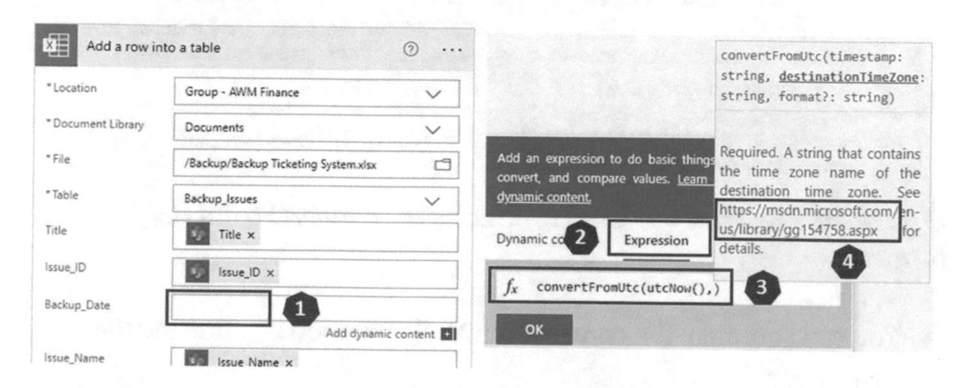

Figure 10-20. *Configure Backup_Date expression*

To obtain the backup date, you need the system to generate a date when the flow was triggered. The function for this is the utcNow() function. UTC is short for Universal Time Coordinated. It is the time in Greenwich, London. You can use different time zone to add or subtract time from the UTC time to obtain your local time.

The link from the previous step takes you to a time zone table. You need to copy the name of the relevant time zone. For example, to find the time zone for Sydney, Australia, do the following, as shown in Figure 10-21.

1. Paste the URL into a browser.

2. Use Ctrl+F to search for the time zone (e.g., Sydney).

3. Copy the Time Zone name, AUS Eastern Standard Time.

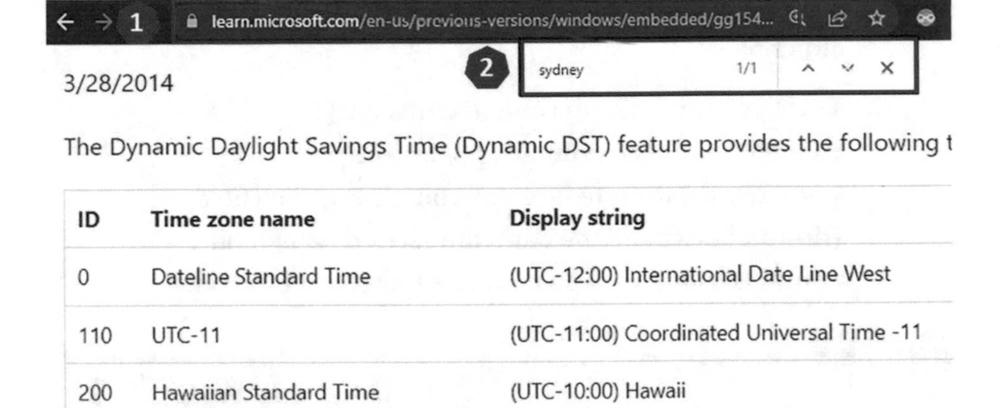

***Figure 10-21.** Find the local time zone for convertFromUtc() function*

You can complete the convertFromUtc() function by entering the following code.

```
convertFromUtc(utcNow(),'Aus Eastern Standard
Time','yyyy-MM-dd')
```

The last parameter in the convertFromUtc() function, 'yyyy-MM-dd', is the format for the output (e.g., 2022-12-30.

After completing all the fields in the **Add a row into a table** action. You may notice a change in the actions shown in Figure 10-22. A new action called Apply to each is automatically wrapped outside the **Add a row into a table** action. You can think about this as preventing the flow from reaching the end after only a single item has been added to Excel.

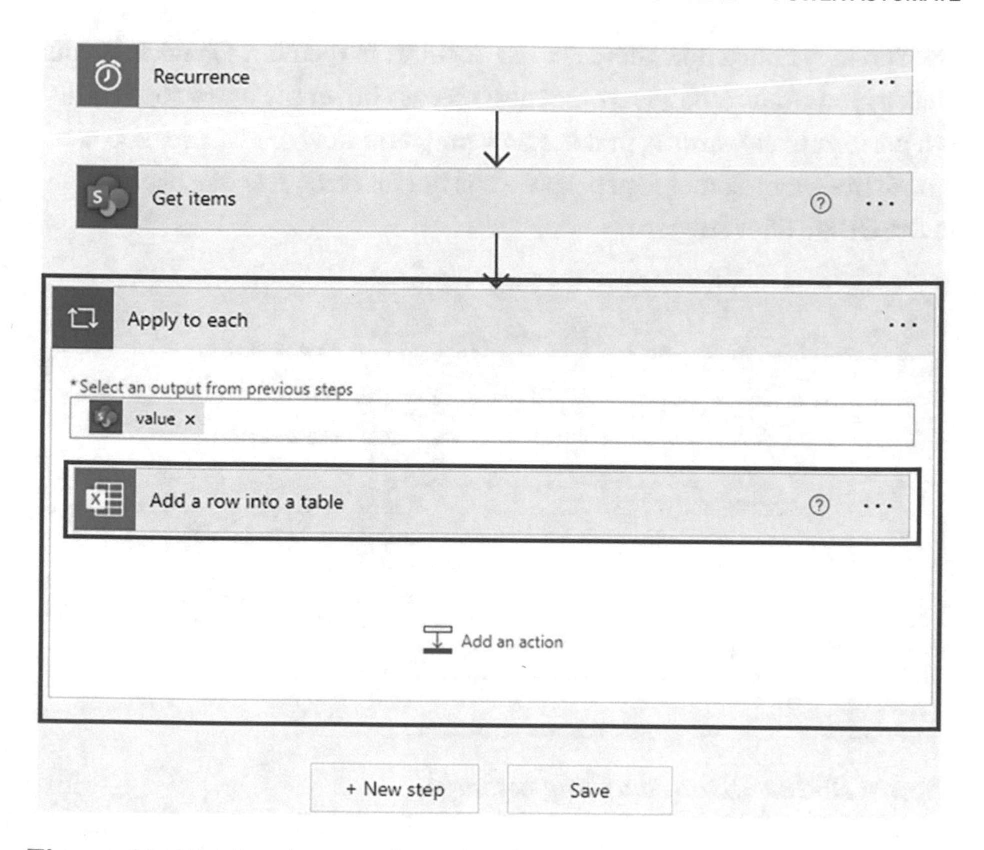

Figure 10-22. *Apply to each action box*

At this point, the flow is complete. If you are to explain this flow in plain English, it probably still makes a lot of sense as follows.

1. Set a recurrence timer trigger for the flow to start (e.g., every Monday at 2 AM).

2. Get all the items from the ticketing system SharePoint list.

3. Copy the various fields into an Excel table for each item and add a backup date field.

You may notice now there is a red dot alert in the Flow Checker. If you click it, you should see a warning message, as shown in Figure 10-23. This is a performance warning that the flow might be slow to execute. You can ignore the warning message because it is not an error. It is discussed later in the OData filter session.

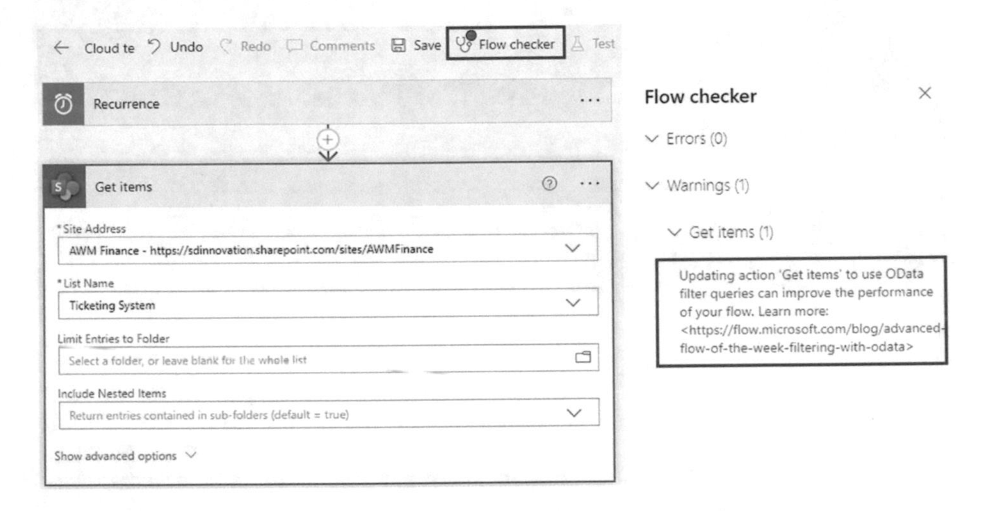

***Figure 10-23.** OData warning message*

Test the Backup Process

You can now test the process by selecting the Test button and choosing **Manually** as the test option. This means you manually trigger the test by doing the following, as shown in Figure 10-24.

1. Click the Test button.
2. Choose Manually.
3. Click Save & Test.
4. Wait for the flow to be ready and click **Run flow**.

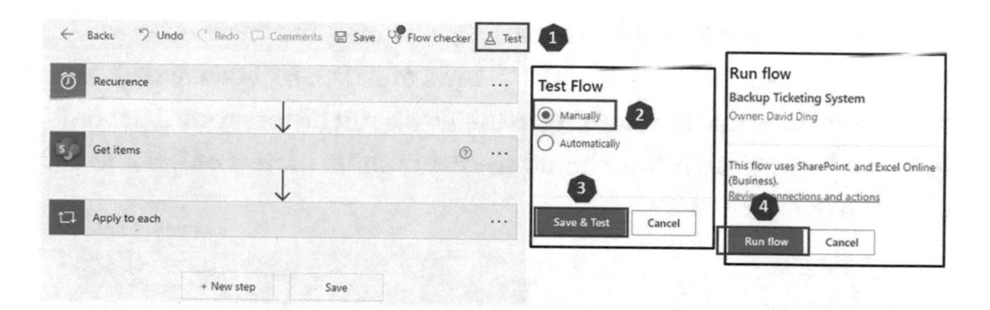

***Figure 10-24.** Testing the flow*

Once you hit the **Run flow** button, you see the flow being executed live, as shown in Figure 10-25. If you now open the Excel Online file in the Backup folder, you can see new rows of data being added to the bottom of the table.

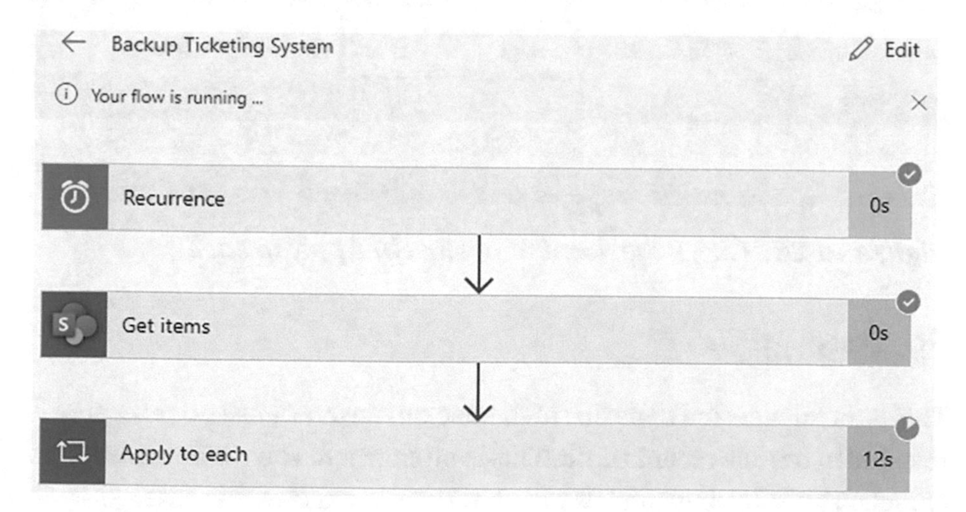

***Figure 10-25.** Live testing view*

This is helpful in understanding how long things take. The trigger and the *Get items* actions are finished within 1 second, but applying to each (adding a row to the Excel table) is much slower. The test should stop around 3 minutes.

The SharePoint list contains nearly 3,000 items, but if you open the Excel Online file, you can see only 100 rows of data have been added. You can also validate this by returning to the flow's run history record. If you open the Apply to each, you should see the number of runs only got to 100, as shown in Figure 10-26.

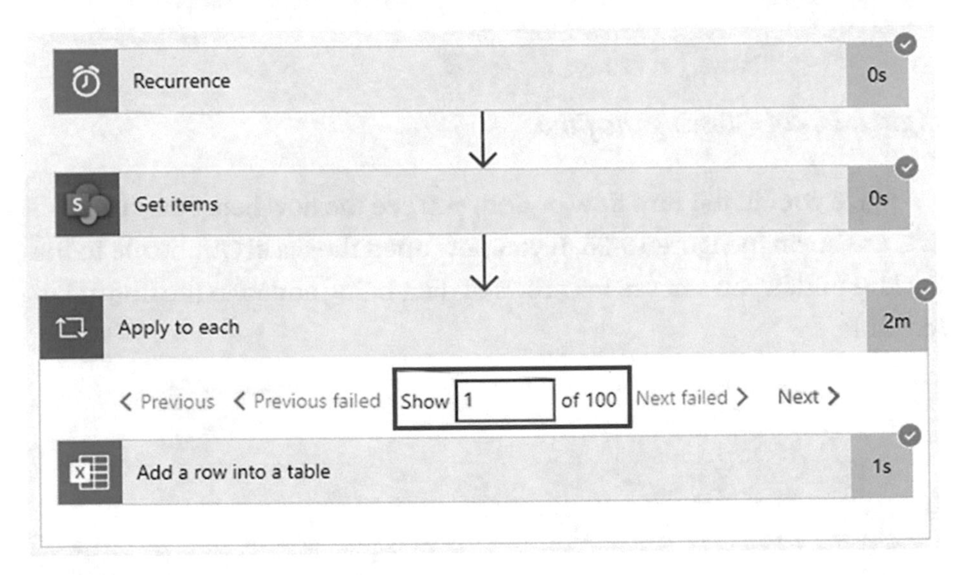

Figure 10-26. *Check the number of runs in Apply to each*

Fix Issues

This is an unexpected behavior of the backup flow because no prior step resulted in the 100 records limit. This is often where you would consult Dr. Google. The following query is one that I would make in this case.

Power Automate only adding 100 rows into Excel from SharePoint list.

You should find two different solutions.

- **Solution 1**: In the **Get items** settings, set the Pagination Threshold to a higher number (up to 5000).
- **Solution 2**: In the **Get items** advanced options, set the Top Count property to a higher number (no limit given).

Both solutions may seem to work, but there is no way of knowing which one works, maybe both. While you may prefer solution 2 because there is no limit, you don't know if it would work. This is where further testing is required. The following steps implement solution 1, as shown in Figure 10-27.

1. Click the More Options button.
2. Select Settings.
3. Turn on Pagination.
4. Change the Threshold to 5000.
5. Run the test again.

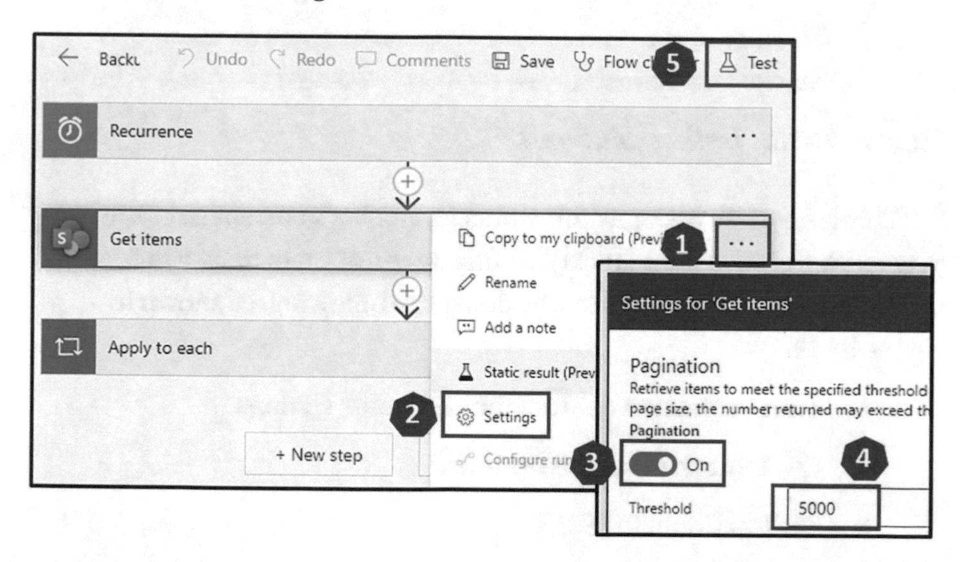

***Figure 10-27.** Change pagination threshold in Get Items action*

From the testing view of solution 1, you notice two things, as shown in Figure 10-28.

1. Get items takes much longer to run.
2. The **Apply to each** step covers the full list.

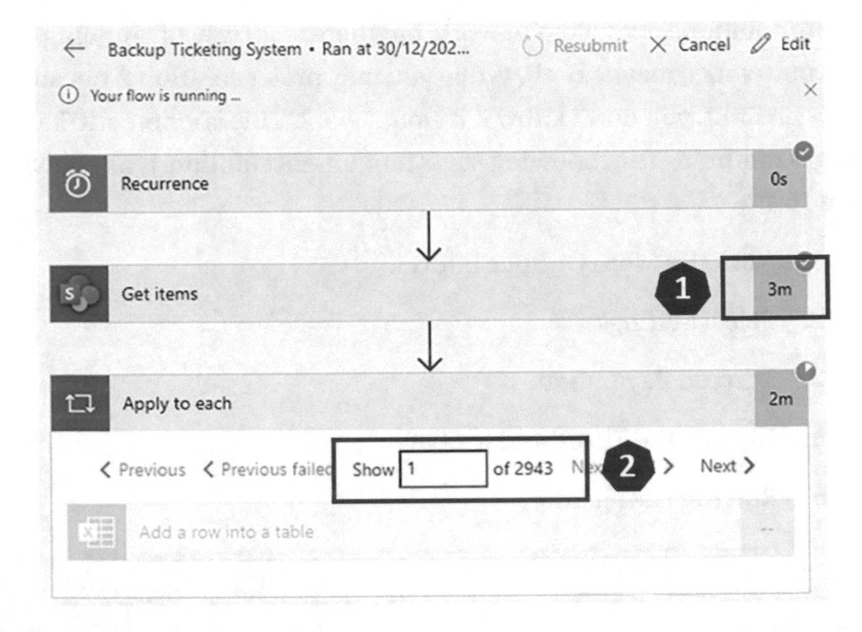

Figure 10-28. *Testing solution 1*

This is good, as we know this works for the list's current size. You may reach the 5000 limit quickly, so this does not seem to be a long-term solution. You can test solution 2 by doing the following, as shown in Figure 10-29.

1. Turn pagination off from the **Get items** settings.
2. Click **Show advanced options**.
3. Set Top Count to 99999.
4. Run the test again.

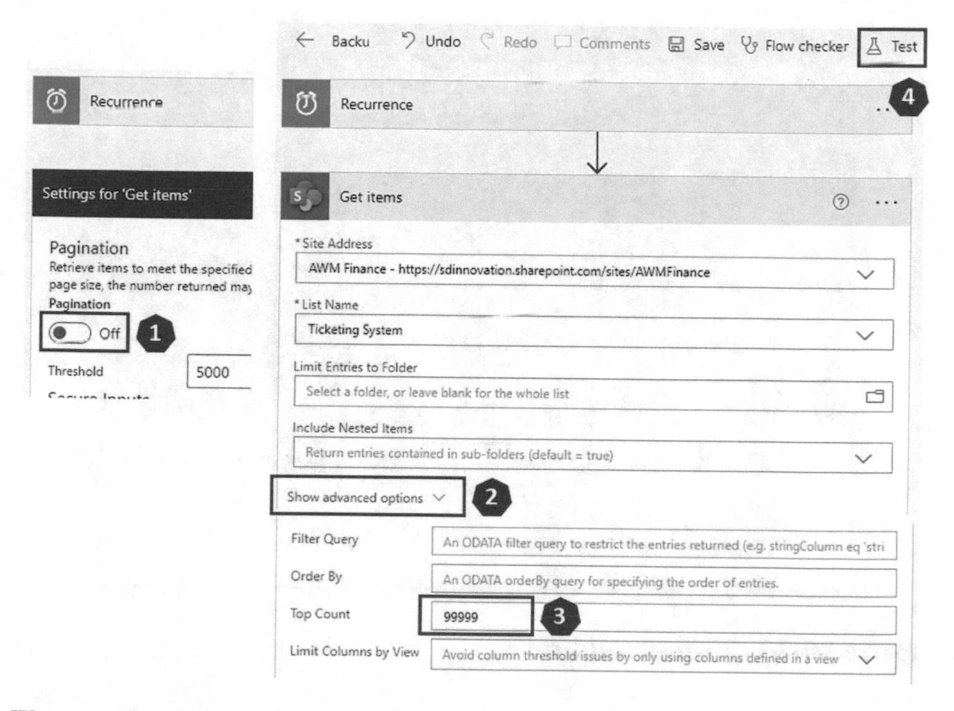

Figure 10-29. *Change top count in Get items action*

From the testing view of solution 2, you notice two things, as shown in Figure 10-30.

1. The **Get items** action finished in 2 seconds, which is much faster than solution 1.

2. The **Apply to each** step also covers the full list.

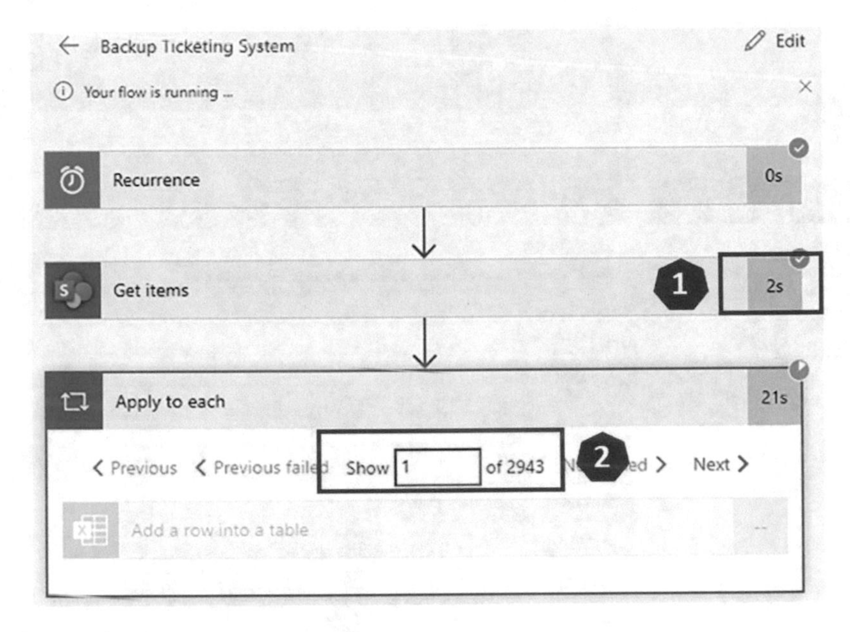

Figure 10-30. *Testing solution 2*

Based on the testing result from solutions 1 and 2, you should use the Top Count method for the following reasons.

- **Get items** takes much less time.
- There is no such limit, such as 5,000 items.

If you keep the test to continue to run, it took 1 hour and 8 minutes to finish loading all 2,943 items into backup.

Note When doing the testing, you don't need to wait for the whole thing to finish to obtain the required information. You can then cancel the test from the flow testing view. Remember that the items added to the backup file do not reverse when the flow is canceled.

You may be wondering why this is so slow. You can finish the job within a minute if you run a manual copy and paste. To answer this question, you must go deeper into how Power Automate works. Power Automate use API (Application Programming Interface) to interact with various system and data sources. In the backup flow, Power Automate uses APIs to interact with SharePoint list and Excel Online. **Get items** is a SharePoint API. This API sends back all the list items in a single API call. In contrast, **Add a row into a table** must be repeated for each row of data. Each **Add a row into a table** API call takes about 1 second to execute, and this is repeated 2,943 times.

While the flow is slow, you do now have a working solution for the backup process. Before you end this session, there is one more important aspect relating to the warning message about the OData filter.

OData Filter

Flow Checker shows a warning message about OData filtering, as shown again in Figure 10-31. OData stands for Open Data Protocol and is a standard to be used for API and other types of data queries. The SharePoint list **Get items** API accepts filters in the OData format.

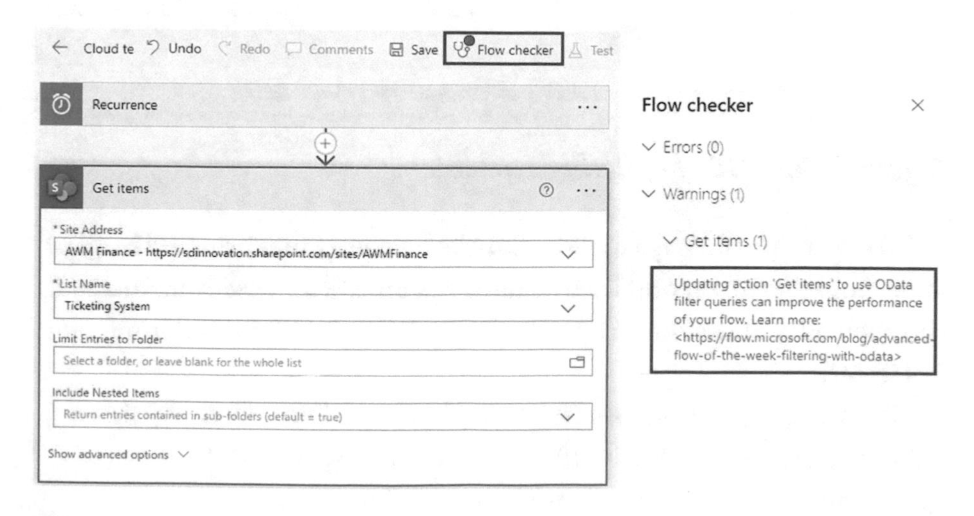

Figure 10-31. *OData warning message*

If you search *OData filter URIs* in Google, you should come across the Microsoft learning page listing the various filter expressions, as shown in Figure 10-32.

Using Filter Expressions in OData URIs

Article • 06/09/2022 • 3 minutes to read • 4 contributors

Feedback

Definition	Example and explanation	Equivalent C/AL expression
Select a range of values	`filter=Entry_No gt 610 and Entry_No lt 615` Query on GLEntry service. Returns entry numbers 611 through 614.	..
And	`filter=Country_Region_Code eq 'ES' and Payment_Terms_Code eq '14 DAYS'` Query on Customer service. Returns customers in Spain where Payment_Terms_Code=**14 DAYS**.	&
Or	`filter= Country_Region_Code eq 'ES' or Country_Region_Code eq 'US'` Query on Customer service. Returns customers in Spain and the United States. **Alert:** You can use OR operators to apply different filters on the same field. However, you cannot use OR operators to apply filters on two different fields.	|
Less than	`filter=Entry_No lt 610` Query on GLEntry service. Returns entry numbers that are less than 610.	<

Figure 10-32. *OData filtering expressions*

The warning message in Flow Checker suggests that bringing in all items from the SharePoint list is usually not an efficient method. Instead, you should try to use the OData filter to limit the data returned from **Get items**.

For example, you can use OData to filter open items by doing the following, as shown in Figure 10-33. The OData expression only keeps *open* items by excluding the Closed and Resolved items.

1. Go to the **Get items** action box and expand to Show advanced options.

2. Enter the OData filter using the following expression.

```
(Status ne 'Closed') and (Status ne 'Resolved')
```

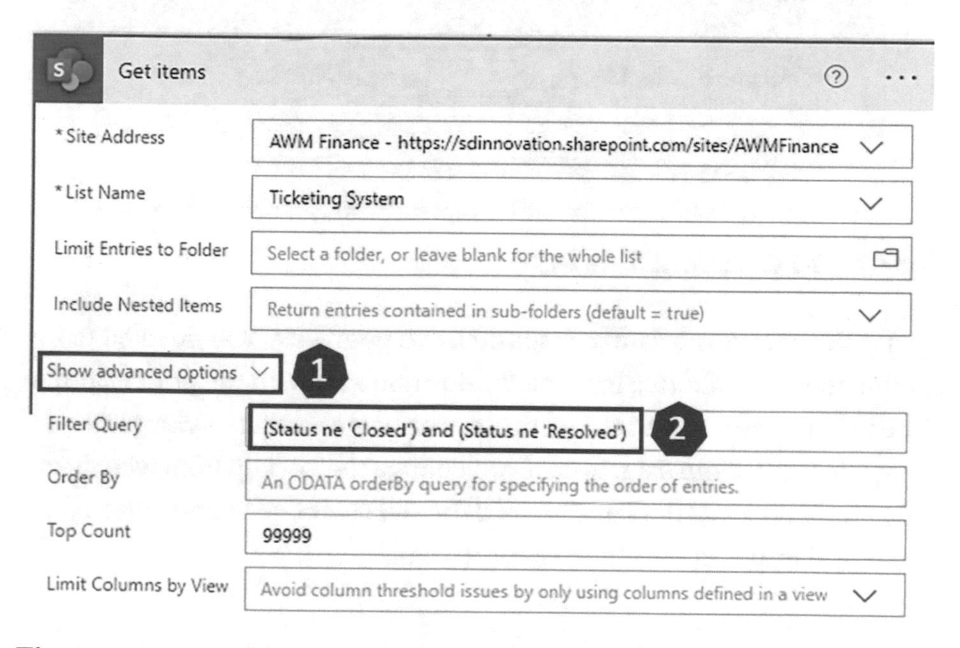

Figure 10-33. *Adding OData filter to Get items*

Once you've added the filter query, the warning message should disappear. If you rerun the test, you can see the number of rows reduced from 2,943 to 1,988, as shown in Figure 10-34. This is because all the closed and resolved items are not included in the output from Get items.

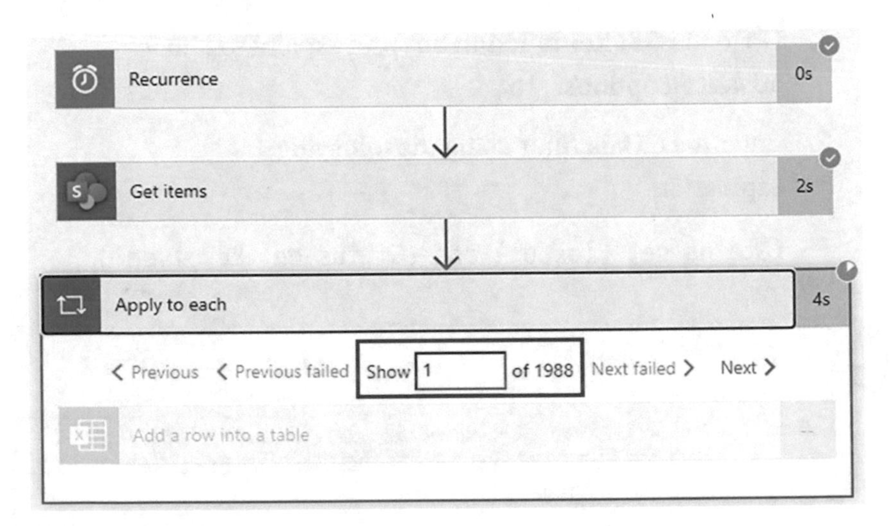

Figure 10-34. *OData filter test*

While limiting the data may sound like a good idea, you do need to be careful about not filtering important information out. In the prior example, if the ticket is created and closed during the same week. It is completely missed in the backup data. Even if you change the backup from weekly to daily, it still presents the same issue: if the ticket is created and closed on the same day, it is missed in the backup data.

Streamline Approval Processes

The last scenario you build is an approval process. Approval is a common requirement in business applications. Power Automate approvals enable approvals in workflows by integrating with two Office Online approval functions.

- Approvals for Teams
- Actionable messages in Outlook

When an approval is triggered, it can generate an alert in Teams and an actionable email in Outlook to the approver. When the approver decides on either Teams or Outlook, the decision can be fed back into Power Automate to drive further action(s) based on the decision.

Let's use the ticketing system as an example. The approval process may be required in the following scenarios.

- When a user submits a ticket, the manager can approve or reject it.
- When a user submits a ticket, the system admin can approve or reject it.
- Before a ticket is resolved, the system admin can approve the resolution.

While the process of implementing them are very similar, the 1st scenario is used for demo purposes.

Prepare the Approver Email

Before implementing the approval process, you need to create an approver email list. In practice, this can be copied from one of the HR data sources, but you manually create the following table in Excel, as shown in Figure 10-35.

1. Enter the data in an Excel spreadsheet.
2. Convert it into a table.
3. Name the table approver_email.
4. Save and load it into the AWM Documents SharePoint folder.

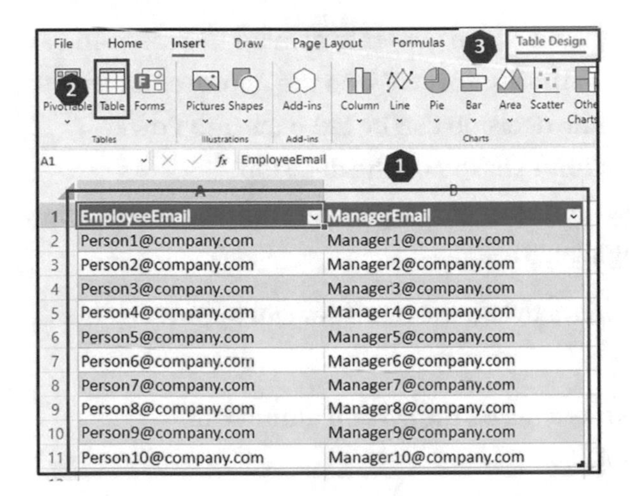

Figure 10-35. *Prepare the approver email file*

Once the data is added to the SharePoint folder, you can connect to it in Power Apps by doing the following, as shown in Figure 10-36.

1. From the data source pane, select **Add data**.

2. Choose Excel Online (Business); a search might help.

3. Find the right site; a search might help.

4. Select the Documents folder.

5. Choose the Excel file.

6. Select the table name.

7. Choose **Employee name** as the unique identifier.

8. Select Connect.

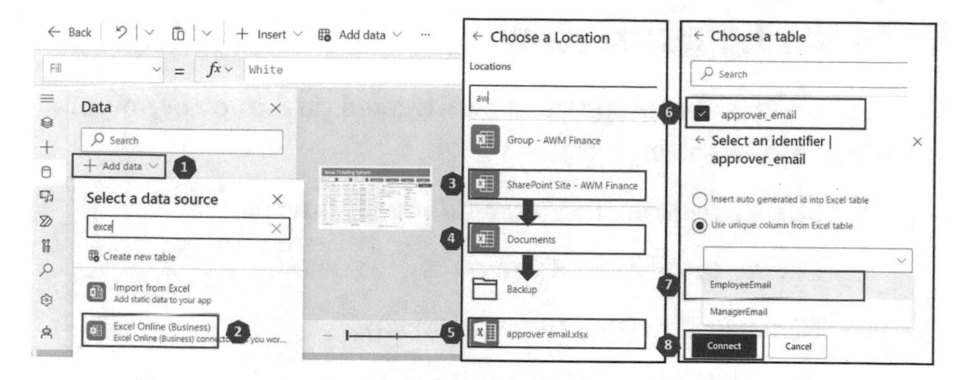

Figure 10-36. *Connect to the Excel data source*

The approver_email table is loaded into the app as a data source. This is important as the approver lookup step is performed inside Power Apps.

You can now look up the manager's email based on the user's email. This is similar to the VLOOKUP() function in Excel but implemented in Power Apps using the following code.

```
Set(
  gblApproverEmail,
  First(Filter(
    approver_email,
    EmployeeEmail = gblUserEmail
  )).ManagerEmail
);
```

The First(Filter(...)) part of the function keeps the first row of data that meets the user's email. You can then get the ManagerEmail column and store it in a global variable using the Set() function. This code can be placed into app start, so the manager ID (gblApproverEmail) is captured at the very beginning and used in later operations.

Build the Approval Flow

You can develop a simple approval flow triggered via the app by doing the following steps, as shown in Figure 10-37.

1. Select **Create** on the Power Automate home page.
2. Choose **Instant cloud flow**.
3. Name the flow (e.g., Ticket Approval).
4. Choose **Power Apps (V2)** as the trigger.
5. Select the Create button.

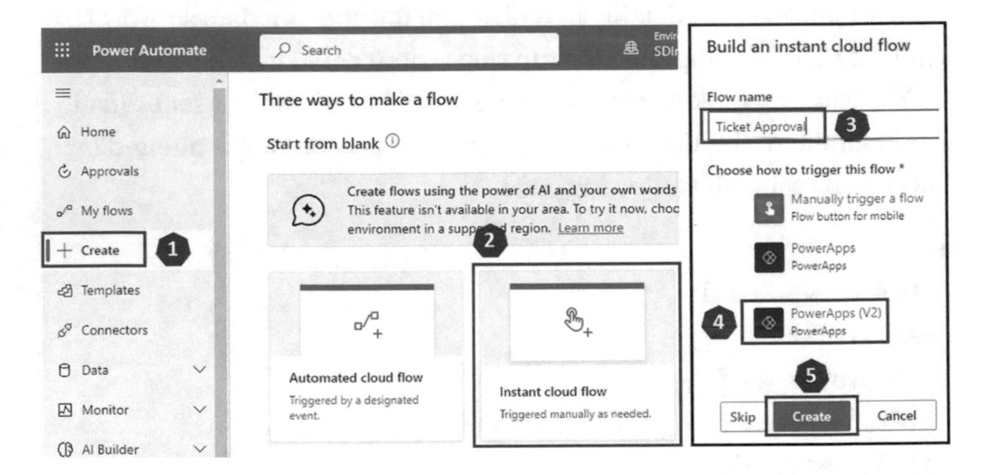

Figure 10-37. *Initiate the ticket approval flow*

The following initiates the approval, as shown in Figure 10-38.

1. Click Next step.
2. Type approval in the search box.
3. Select Start and wait for the approval.
4. Select Approve/Reject - First to respond.

5. In the approval action, select **show advanced options**.

6. Note the Title, Assigned to, Details, and Requestor fields.

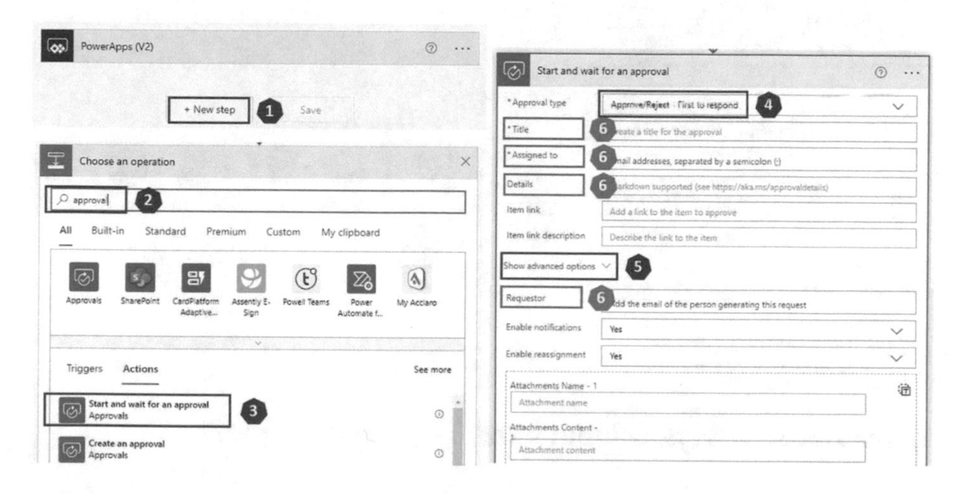

***Figure 10-38.** Initiate the approval action*

There are a few options in the approval actions in step 3. You would generally go for **Start and wait for an approval**. You soon understand why the *wait* in this option is critical to the approval flow development.

The four fields noted in step 6 come from the app, so they must be added to the Power Apps trigger as input parameters. You can add them by doing the following, as shown in Figure 10-39.

1. Select Add an input from the Power Apps (V2) action box.

2. Select Text.

3. Replace the text in the left box with **approval_title**.

4. Repeat processes 1–3 for approval_detail.

5. Repeat processes 1–3 for approver_email.

6. Repeat processes 1–3 for requestor_email.

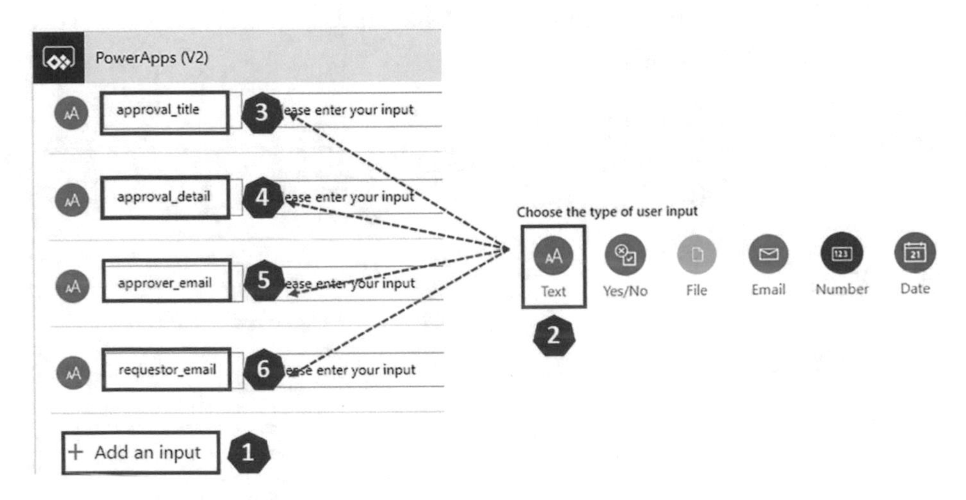

Figure 10-39. *Add flow input parameters to the trigger*

Once the input parameters have been added to the flow, you can use them later in the approval action by doing the following, as shown in Figure 10-40.

1. Click the Title input field in the Approval action box and select **approval_title** from the dynamic content selection options.

2. Repeat step 1 for the **Assigned to** field.

3. Repeat step 1 for the **Details** field.

4. Repeat step 1 for the **Requestor** field.

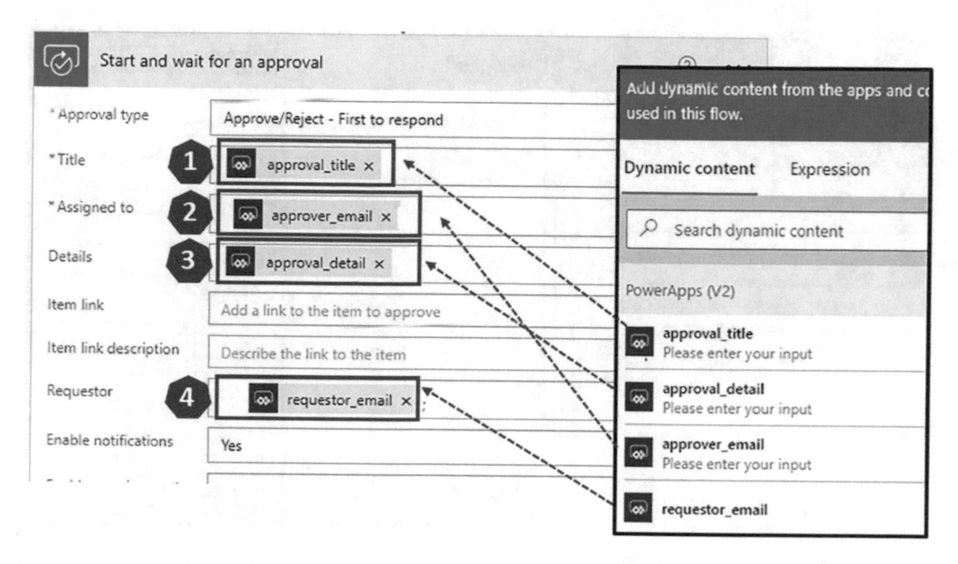

Figure 10-40. *Add the input parameter into the required fields*

If you save and manually test the flow, you must log in to the approvals and insert the required input parameters. The following steps explain how to do it, as shown in Figure 10-41.

1. Click **Test**.
2. Choose **Manually**.
3. Enter a test phrase in the approval_title, approval_detail, approval_email, and requestor_email input fields.
4. Click **Run flow**.

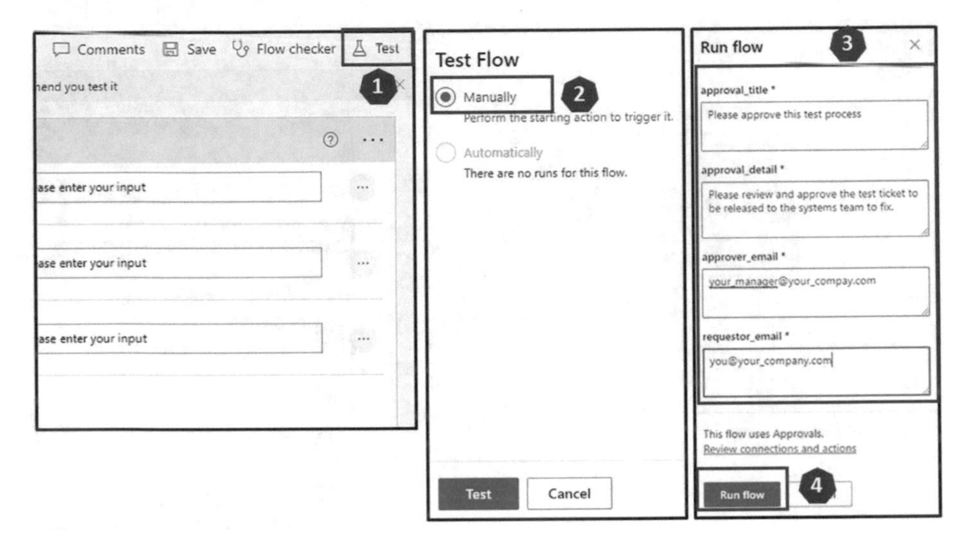

***Figure 10-41.** Testing the approval flow*

This approval flow triggers an alert in Microsoft Teams and an email in Outlook. Both communications are actionable, as shown in Figure 10-42.

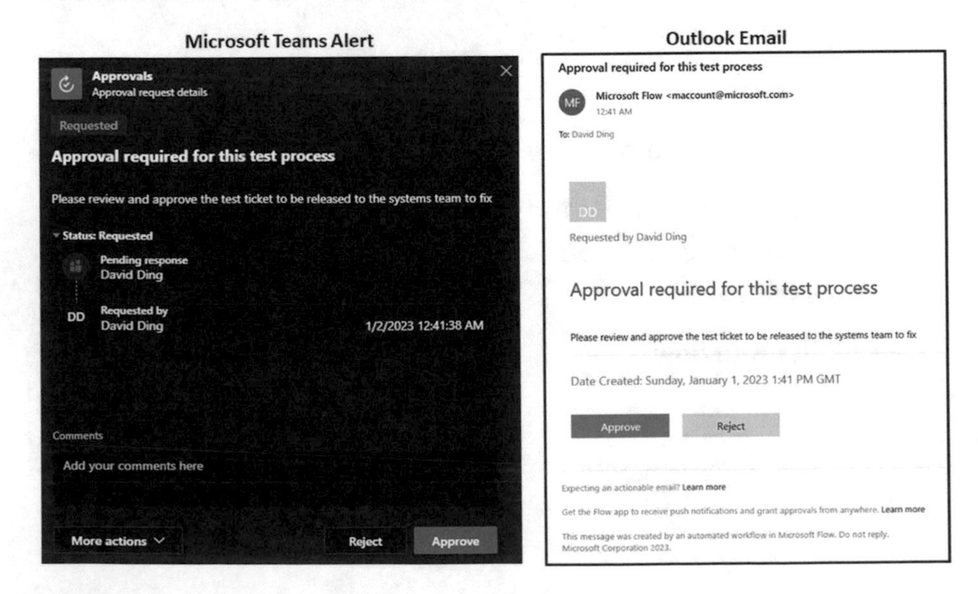

***Figure 10-42.** Approval communications*

You can record the approval decision in either Teams or Outlook by clicking the Approve or the Reject button.

According to Microsoft, the approval decisions are stored in a *dataverse* dataset. This creates some additional complexity when trying to access the decisions. Luckily, you have chosen the **Start and wait for an approval** action. **Wait** allows the flow to pause and wait for the approver to decide between continuing to the next action. The next action(s) can bc conditional based on the outcome of the approval.

This allows us to customize the actions that follow the decision. In your scenario, you first create three new approval-related columns in the SharePoint list. The following steps explain how to do this, as shown in Figure 10-43.

1. Select **Add column.**
2. Add an Approval Status text column.
3. Add an Approved By text column.
4. Add an Approval Date text column.

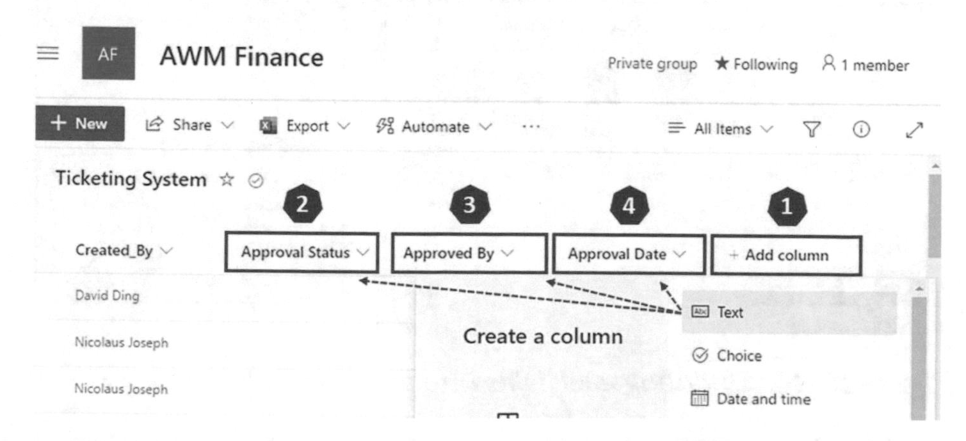

***Figure 10-43.** adding new approval columns in SharePoint list*

The next steps implement the action following the decision to populate the newly created fields in the SharePoint list, as shown in Figure 10-44.

1. Click **New step**.
2. Search for the SharePoint list.
3. Choose **Update item**.
4. Select the site from the drop-down.
5. Select the list name from the drop-down.
6. Note the mandatory fields ID and Title.

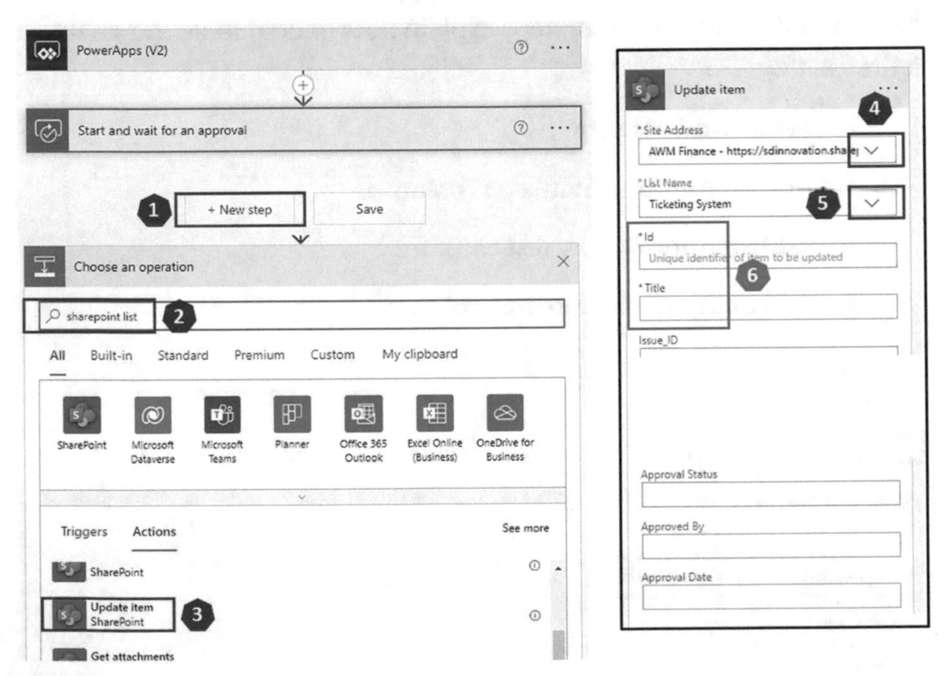

Figure 10-44. *Updating new fields into the SharePoint list*

If you think carefully about where the ID and title should come from. You may conclude that it must come from the app. The only way to get data from the app into the flow is via input parameters. This means you'll need to add two more input parameters in the Power Apps trigger, as shown in Figure 10-45.

***Figure 10-45.** Adding new input parameters*

Once they've been added, you can return to complete the **Update item** action by doing the following, as shown in Figure 10-46.

1. Drag and drop item_id from the dynamic content options into the Id field.

2. Drag and drop item_title from the options into the Title field.

3. Drag and drop the Outcome option into the Approval Status field.

4. Drag and drop the approver_email option into the Approved By field.

5. Enter the following code as an expression to get today's date in the text.

```
convertFromUTC(utcNOW(),'AUS Eastern Standard
Time','yyyy-MM-dd')
```

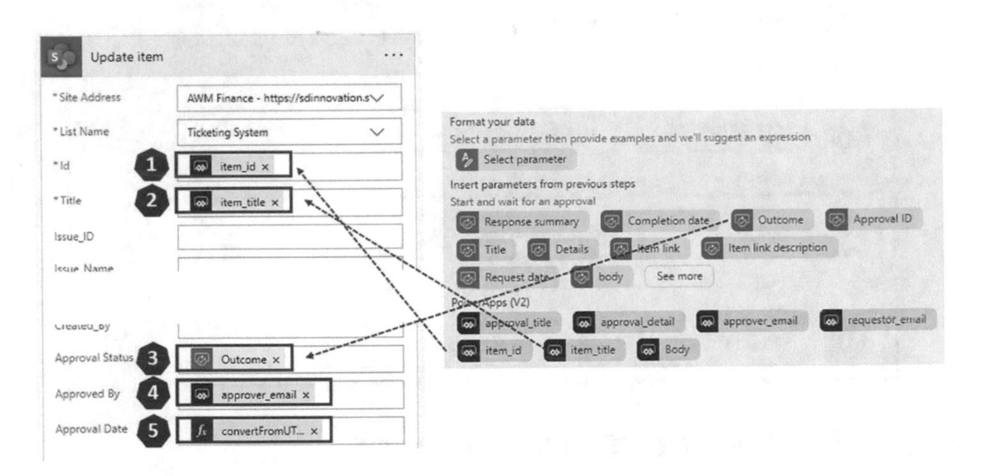

Figure 10-46. *Add the required fields in Update items action*

You can now test the updated flow by completing the following steps, as shown in Figure 10-47. For the ID and Title field, please pick a real ID and Title from your SharePoint list (e.g., 10058 and X). Note that the ID is different from the Issue_ID.

1. Select Test.

2. Choose Manually.

3. Enter the required input parameters (ID: 10058 and Title: X).

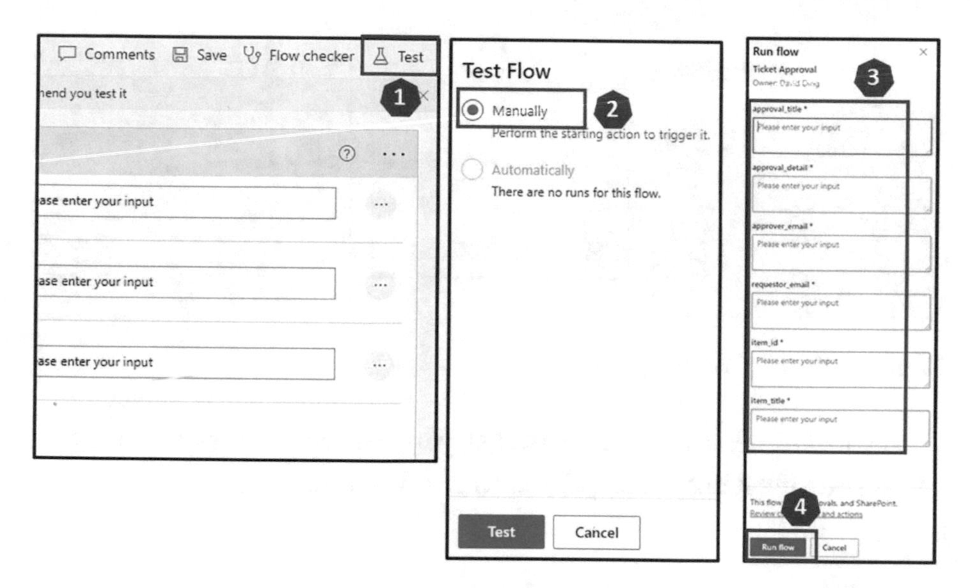

Figure 10-47. *Test the wait in approvals flow*

You can see the *waiting* in action (no pun intended), as shown in Figure 10-48.

1. After the flow has been triggered, the email and alert are issued to the approver, and the **Start and wait for an approval** action continues to run and wait for a decision.

2. Users can click the Approve button in either the email or the alert when they are ready to decide.

3. After the decision has been made, the flow completes the Update item step.

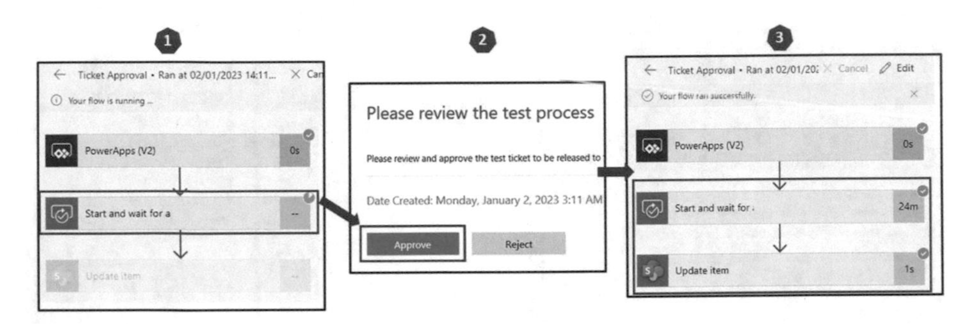

Figure 10-48. Waiting in action

If you go back to the SharePoint list, you can see the row of data with ID 10058 and Title X has been updated with new approval columns, as shown in Figure 10-49.

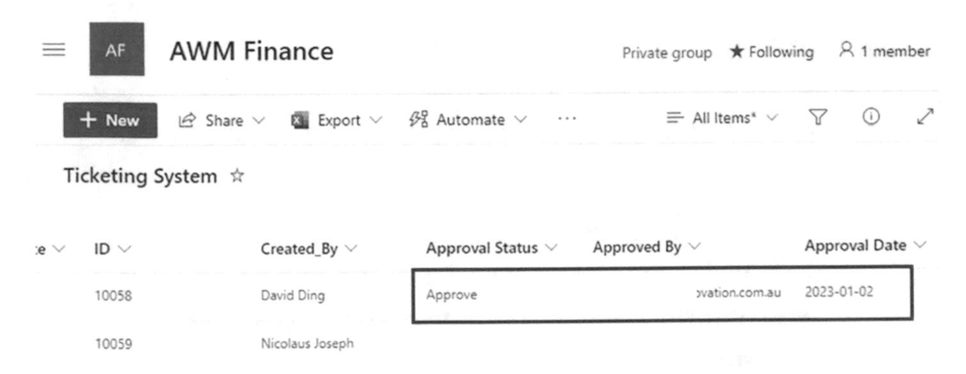

Figure 10-49. New details populated in SharePoint list

Integrate with Power Apps

With the approval flow ready, you can integrate it into Power Apps. In the scenario, this is added when a new ticket is added or changed. This means you can call the flow.Run() function from the Save button. After adding the new flow, you can use the code to add to the end of the Save button's OnSelect property.

```
TicketApproval.Run(
  // Title: change wording for New or Existing ticket
  "Please review this "
    & If(gblNewTicket, "New", "Updated")
    & " issues ticket",
  // Detail: change wording for New or Existing ticket
  "Ticket #"
    & If(
        gblNewTicket,
        gblNewID & " has been created",
        txtIssueID.Text & " has been updated"
      )
    & " in the ticketing system pending review.",
  // Approver_email
  gblApproverEmail,
  // Requestor email
  User().Email,
  // item ID
  If(gblNewTicket, gblNewID, txtIssueID.Text),
  // item Title
  "X"
);
```

Except for the gblNewID variable, the logic for the various input parameters should be self-explanatory. gblNewID is created because when a new ticket is created, the txtIssueID does not know the new ID. You can get the ID for this new item from the Patch() function using the following code.

```
Set(
  gblNewID,
  Patch(
```

```
    'Ticketing System',
    Defaults('Ticketing System'),
    {data fields ...}
  ).ID
)
```

Other than submitting information to the SharePoint list, the Patch() function can also get information back from SharePoint. When a new item is created, Patch() function can return the various fields populated by the SharePoint list back to Power Apps. This allows you to get the .ID field for the newly created item.

This completes the approval process by calling PowerAutomate flows from the app.

Approvals Customization Options

There are two additional helpful customization options for the approval flow that is useful to be used in other approval scenarios.

Custom Responses in Approval Flow

The Custom Responses approvals allow you to change the responses. For example, you can add a Defer option to Approve/Reject the decision, as shown in Figure 10-50. You can also change this to suit the different scenarios.

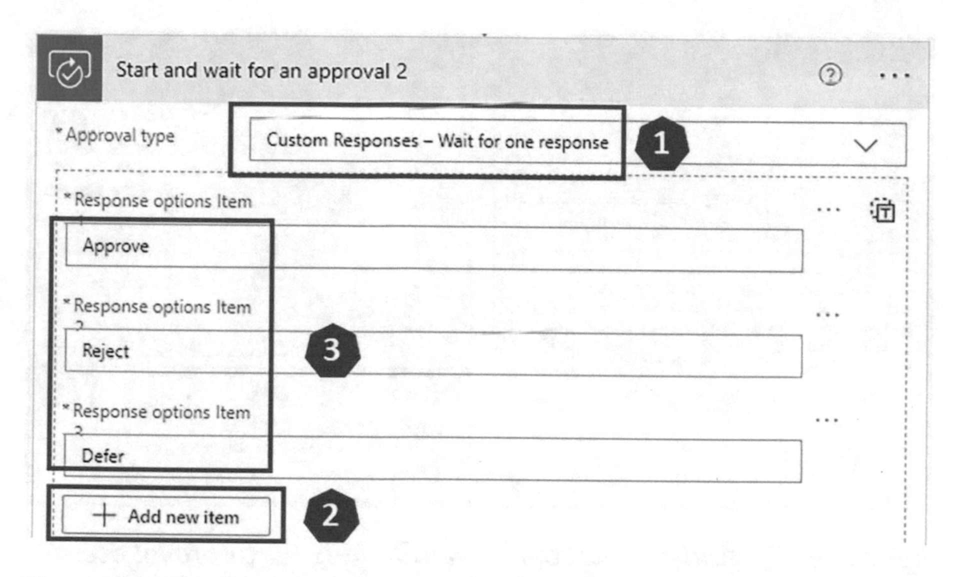

***Figure 10-50.** Custom responses for the approval flow*

Add Conditional Actions After the Approval Flow

Power Automate flow also provides a condition step that can trigger different actions based on different conditions. For example, you can trigger different actions based on the *outcome* of the decision. The following steps explain how, as shown in Figure 10-51.

1. Click **New step.**
2. Search and select Condition.
3. Select Approval Outcome in the first input box.
4. Configure the condition to **Outcome is equal to Approve**.

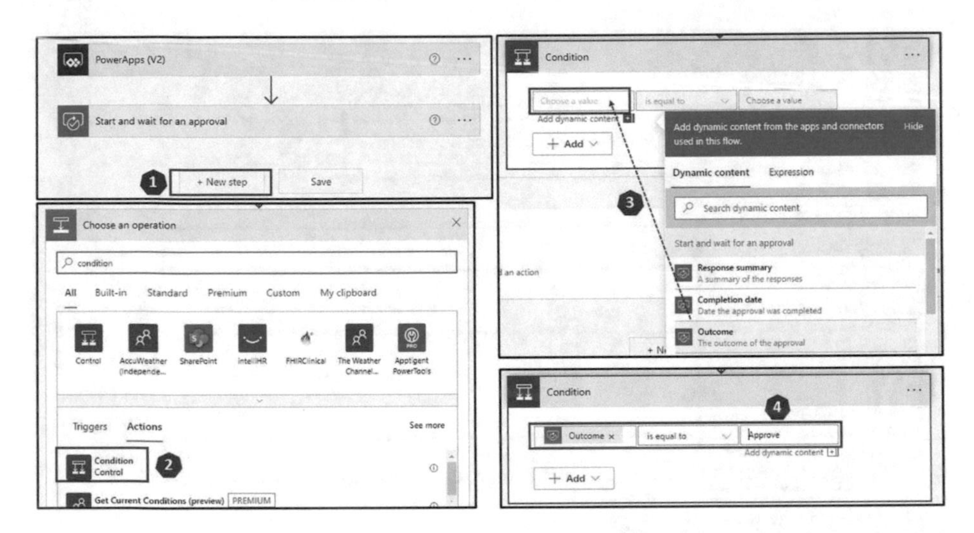

Figure 10-51. Adding new condition following the approval decision

The Outcome condition leads to two branches of action. If the request is approved or rejected. Figure 10-52 shows that you can configure flow actions separately based on the condition.

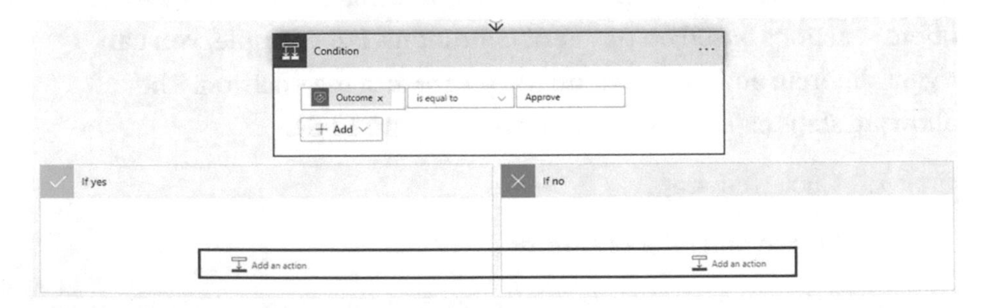

Figure 10-52. Conditional actions

Note It is not often necessary to configure separate processes. You can build some of the logic inside the app instead of in the flow. The app is preferred to flow because it is much easier to handle complex logic in Power Apps compared to Power Automate.

Conclusion

Congratulations on learning the basics of Power Automate. In this chapter, through the three examples, you have gained a practical understanding of how Power Automate works. More importantly, you now know how to develop Power Automate flows on the following three common requirements at work.

- Send out email notifications.
- Create backups.
- Streamline approval processes.

Mini-Hackathon

With your understanding of Power Automate and the business requirements, there is always room for improvement. In this chapter, you have quickly created three different flows. Once you have completed one task, you should already know one or two additional tasks to help improve the existing solutions.

Objectives

There are two objectives in this hackathon.

Objective 1

The backup flow is slow to execute. You have learned a few methods to improve the efficiency, but there might be a better way to capture the backup data.

Hint Since you're in control of the SharePoint list, the app, and the flow, you can modify any of them to capture relevant information.

Objective 2

In addition to the improvement, one benefit of creating the backup is allowing the leadership team to better understand the issue resolution process. You also need to use the backup data to create a management report in Power BI that shows at least the following measure over time.

- Resolution time
- System performance

Time Limit

Objective 1 be completed within three hours. Objective 2 be completed within five hours.

CHAPTER 11

Integrated Solutions

So far in this book, you have learned about a wide range of Power Platform solutions. In this chapter, you bring them all together to construct an amazing integrated solution that lets users know what action to take, record the action and see the live result from a report. For the past 20 years, the goal for data analytics and business intelligence is about delivering actionable insights. This integrated framework allows users to go beyond actionable insight because you can take the action or record the decision.

Business Scenario

Kim has received a new task to help with a product review process. She has some historical sales data. She needs to use this data to do the following.

- Classify different products into different opportunity groups.
- Identify the biggest opportunity products in each group.
- Provide an app with an action list to guide the production managers in reviewing.
- Check and report on review progress.

D. Ding, *Transitioning to Microsoft Power Platform*,
https://doi.org/10.1007/978-1-4842-9239-6_11

After successfully delivering the ticketing app with all the enhancements, Kim feels like she's at the top of the world. It has only been six months since she joined the company and five months since her first Power BI report. She has been recognized by the leadership team and might win the employee of the year award. What really shocked her was how little she knew about Power Platform five months ago and how easy it is to pick up on these tools. She also really enjoys collaboration with the business and other developers.

Kim really enjoyed that she learned something new every day during this period. Because this is mostly problem-driven learning, she can apply this to solve real business problems immediately. This was much better than formal training because she remembered what she learned through practice.

Kim came across another interesting functionality while researching Power Apps. That is the ability to embed Power Apps into Power BI reports. Kim has been struggling with the following restrictions of Power Apps when developing the ticketing system.

- The data sources are limited. The core table is stored in SQL Server, which requires a premium connector.
- Data transformation is complicated in Power Apps. More complex data processing better take place outside of the app. The transformed data must then be loaded into a SharePoint list, adding extra steps.
- Power Apps has limited reporting functionalities. Use a separate Power BI link to see reports.

Kim hopes the Power Apps embedding in Power BI can help to address these issues.

Power Apps Visual in Power BI

As you learned in the Power BI chapters of this book, Power BI has three key components to analyze and visualize data: Power Query, data models, and DAX. Power Apps visual can pass pre-processed data to a canvas app. The data feeding the app can also be updated according to filter actions taken in the report.

Limitation of Power Apps Visual

Before you embed the Power Apps visual, it is important to understand the key limitations of such an approach, as follows.

- Only a single table with 1,000 records can be passed into the Power Apps visual. The table can contain as many columns as you see fit, but only the first 1,000 rows of records are passed to the app.
- The app cannot pass information back to Power BI.
- Even if the Power BI report is linked to the same data source, any record created/updated in the app is not automatically picked up in the report.
- Access must be managed separately for the Power BI Report and the app.

These limitations may have changed since this book was published. Please Google *Power Apps visual for Power BI* to check the official list of limitations. The official list also provides a complete list of limitations.

Integrated Architecture

In the previous chapters of this book, you learned about Power BI, SharePoint, Power Apps, and Power Automate. This chapter introduces an integrated architecture that requires all the tools. This design combines each solution's advantages while balancing some limitations. It is also flexible in solving a wide range of business problems.

The previous architecture for Power Apps is shown in Figure 11-1. This architecture allows users to use a standard Power Apps and Power Automate connector to interact with the data source in a SharePoint list. As you have seen in previous chapters, you can develop many useful applications with this.

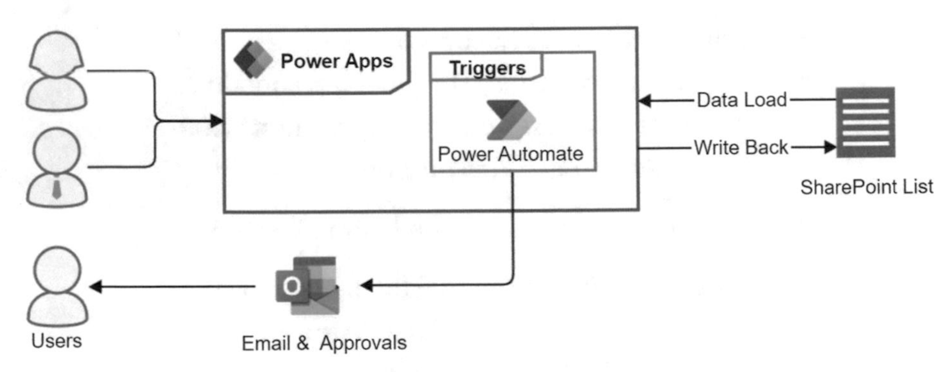

***Figure 11-1.** Power Apps solution architecture*

To overcome the restrictions mentioned at the beginning of this chapter, you can embed this architecture into a Power BI report, as illustrated in Figure 11-2.

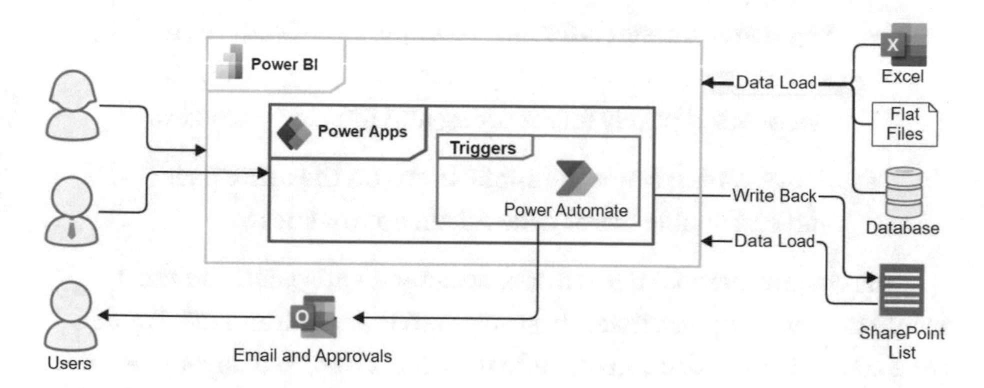

***Figure 11-2.** Embed Power Apps solution architecture*

This architecture provides three additional features compared to the Power Apps view.

- Additional data connection (read-only) without premium connector
- Combines report and app in the same Power BI link
- Allows data transformation/modeling to take place in Power BI

Power BI can pass a data table into Power Apps via the Power BI Integrated Data in this architecture. This table can transfer up to 1,000 rows of data from Power BI into Power Apps. There is also no limit on the number of columns for the table.

Application Design

There are two high-level layout options to combine Power BI and Power Apps: separated screen and same screen.

- **Separated screen** is for more complex applications. It makes sense to keep the app on its own page. Power BI can be used purely for reporting and filtering purposes.
- **Same screen** keeps the applications on the same page and can enable better data-driven approaches.

This chapter demonstrates the same screen approach. You must decide the layout option that is best suited to the requirements. For the same screen layout, you can fit it into the Power BI design layout described in Chapter 4. If the app is designed in landscape format, you can use the design shown in Figure 11-3. Otherwise, if the app is designed in a portrait format, you can use the design in Figure 11-4.

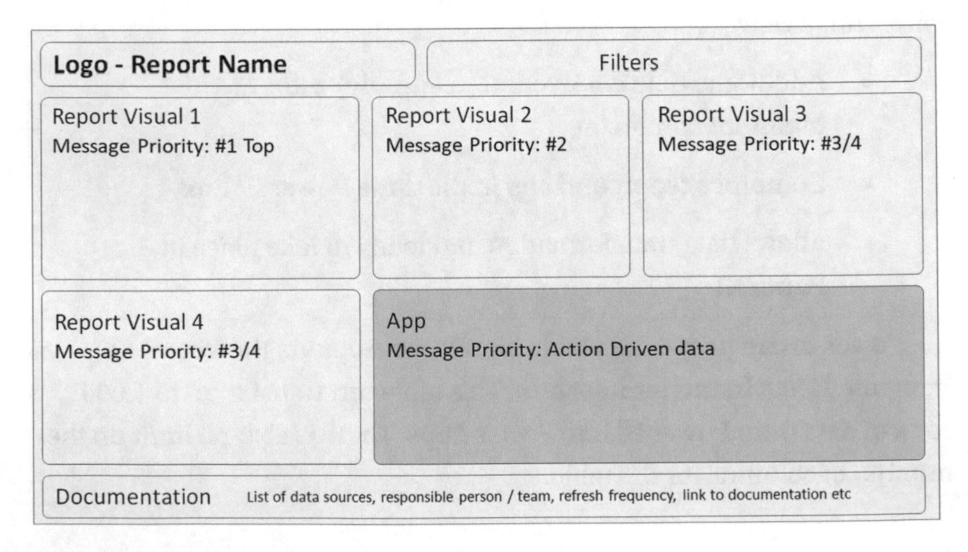

***Figure 11-3.** Landscape app in the same page layout*

The yellow portion of the app represents Power BI, and the purple portion represents Power Apps.

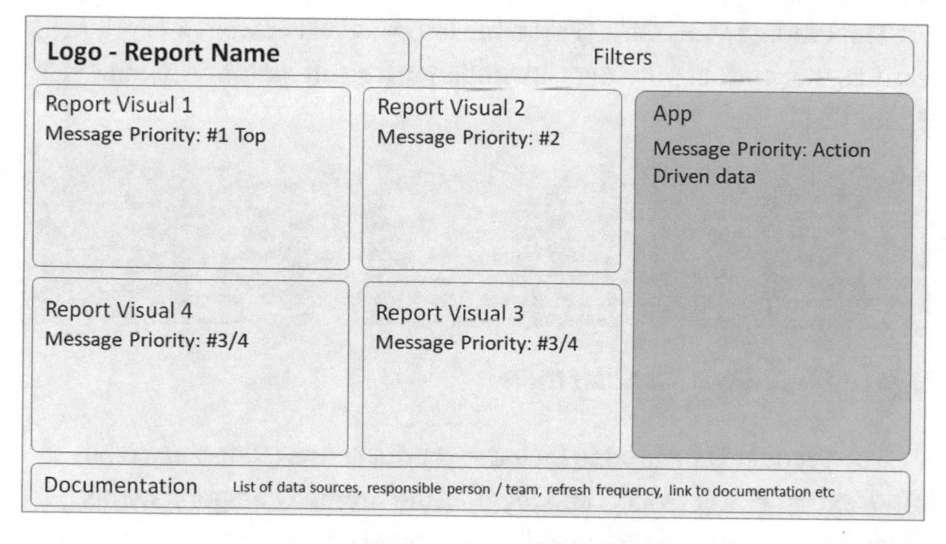

Figure 11-4. *Portrait app in the same page layout*

Both designs assume a data-driven process, with the app being the last step in the process, either as a data entry tool or to record simple actions. Because users typically start reading from the top left-hand corner, having the app at the bottom right-hand corner suits the flow as the last step in the user experience design.

Implement the Same Page Layout

You can now implement the same page design. The portrait design layout is used in this case.

Underlying Data

The fact and reference tables have been preloaded into the Power BI report 11.1 product review.pbix report. You can find this file in the Chapter 11 file folder. The following describes the fact and reference tables included.

The Product Sales table (Fact table imported via Power Query) features the product sales data, including price and quantity, as shown in Figure 11-5.

SalesOrderID	SalesYear	SalesMonth	SalesOrderDetailID	OrderQty	ProductID	UnitPrice	Sales LineTotal	OrderDate
59210	2013	2013-11	70817	1	870	4.99	5	Friday, November 1, 2013
59212	2013	2013-11	70826	1	870	4.99	5	Friday, November 1, 2013
59216	2013	2013-11	70838	1	870	4.99	5	Friday, November 1, 2013
59229	2013	2013-11	70870	1	870	4.99	5	Friday, November 1, 2013
59236	2013	2013-11	70883	1	870	4.99	5	Friday, November 1, 2013
59239	2013	2013-11	70895	1	870	4.99	5	Friday, November 1, 2013

Figure 11-5. *Product Sales table*

The Product Details table includes product costs (Reference table imported via Power Query) and features the product category, name, variations, and product cost, as shown in Figure 11-6.

ProductID	ListPrice	StandardCost	ProductCategoryName	ProductSubcategoryName	ProductDescription	Product	Variations
1	0	0	NULL	NULL	Adjustable Race	Adjustable Race	
2	0	0	NULL	NULL	Bearing Ball	Bearing Ball	
3	0	0	NULL	NULL	BB Ball Bearing	BB Ball Bearing	
4	0	0	NULL	NULL	Headset Ball Bearings	Headset Ball Bearings	
316	0	0	NULL	NULL	Blade	Blade	
317	0	0	NULL	NULL	LL Crankarm	Crankarm	LL
318	0	0	NULL	NULL	ML Crankarm	Crankarm	ML

Figure 11-6. *Product detail table*

The Calendar table (reference table created in DAX) contains the calendar grouping and display columns, as shown in Figure 11-7.

Date	Calendar Year	Month_Display1	Month_Display2	YearMonthInt	Start of Month
1/1/2013 12:00:00 AM	2013	Jan-13	Jan-2013	201301	1/1/2013 12:00:00 AM
1/2/2013 12:00:00 AM	2013	Jan-13	Jan-2013	201301	1/1/2013 12:00:00 AM
1/3/2013 12:00:00 AM	2013	Jan-13	Jan-2013	201301	1/1/2013 12:00:00 AM
1/4/2013 12:00:00 AM	2013	Jan-13	Jan-2013	201301	1/1/2013 12:00:00 AM
1/5/2013 12:00:00 AM	2013	Jan-13	Jan-2013	201301	1/1/2013 12:00:00 AM
1/6/2013 12:00:00 AM	2013	Jan-13	Jan-2013	201301	1/1/2013 12:00:00 AM

Figure 11-7. *Calendar table*

The three tables are combined into a data model, as shown in Figure 11-8. Note that the filter direction between the sales and product tables is applied in *both* directions.

***Figure 11-8.** Data model*

The _Measures table has also been created to store all measures later.

Use Scatter Plot for Strategic Categorization

A scatter plot is one of the most useful data visualization options for strategic decision-making. It can visually group products into multiple quadrants based on two value axes. For example, if the two axes are sales and quantity, you can derive the four quadrants as high value–low volume, high value–high volume, low value–low volume, and low value–high volume, as shown in Figure 11-9.

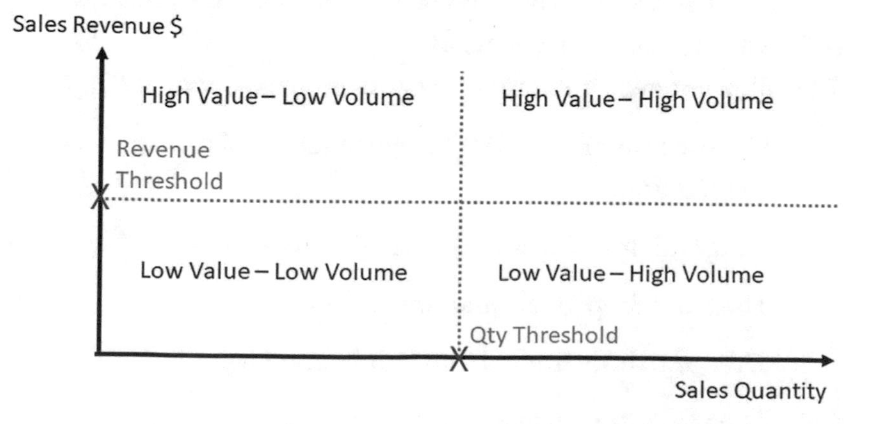

***Figure 11-9.** Four quadrants based on Value and Volume*

Let's use Gross Profit Margin and Sales Quantity for the two axes, but when you try to apply this to real situations, finding the right axis and getting the leadership team to agree to it can be challenging. You must approach this with an open mind and be ready to take criticism.

Before you can create the scatter plot, the following five measures must be created.

```
Total Quantity = SUM('monthly product sales data'[OrderQty])

Total Revenue = SUM('monthly product sales data'[Sales
LineTotal])

Total Cost =
  SUMX(
    'monthly product sales data',
    RELATED('product cost data'[StandardCost])
  )

Total Profit = [Total Revenue] - [Total Cost]

Gross Profit Margin = DIVIDE([Total Profit], [Total Revenue])
```

Gross profit margin (profit divided by revenue) is chosen over profit margin (profit divided by cost) because GPM is limited to between 0 to 1 (or –1 to 1 if a loss has occurred). While PM works in many instances, it doesn't allow the data to be cut nicely.

The following steps initialize the scatter plot, as shown in Figure 11-10.

1. Choose Scatter Plot from Visualizations and resize the plot area.
2. Drag and drop Gross Profit Margin to X-Axis.
3. Drag and drop Total Quantity to Y-Axis.
4. Drag and drop Product into Values and Legend.
5. Drag and drop Total Profit to Size.

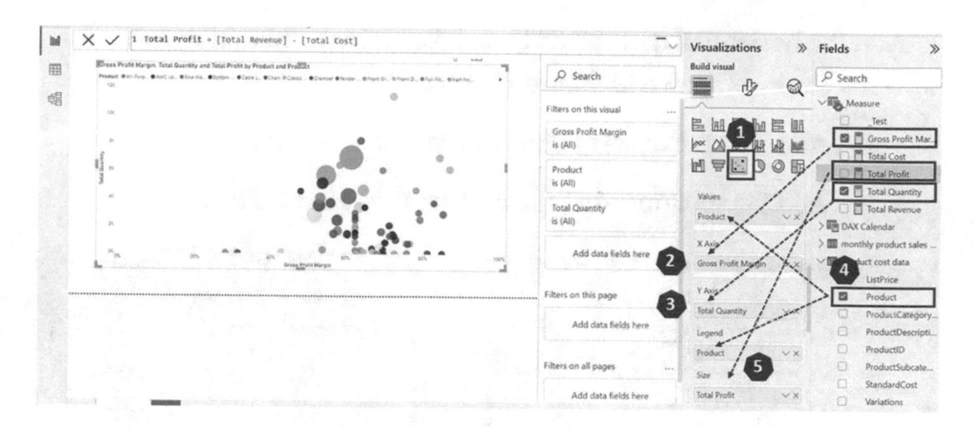

Figure 11-10. Initialize the scatter plot

Note the top left-hand side of the scatter plot. There are a lot of blank areas. You can use the logarithm option on the Y-axis to allow the circles to spread out further. Adding a range limit to the X-axis also help. You can apply this using the following steps, as shown in Figure 11-11.

1. Change the X-axis range to be between 0.2 and 0.9.
2. Switch on the Y-axis **Logarithmic scale**.

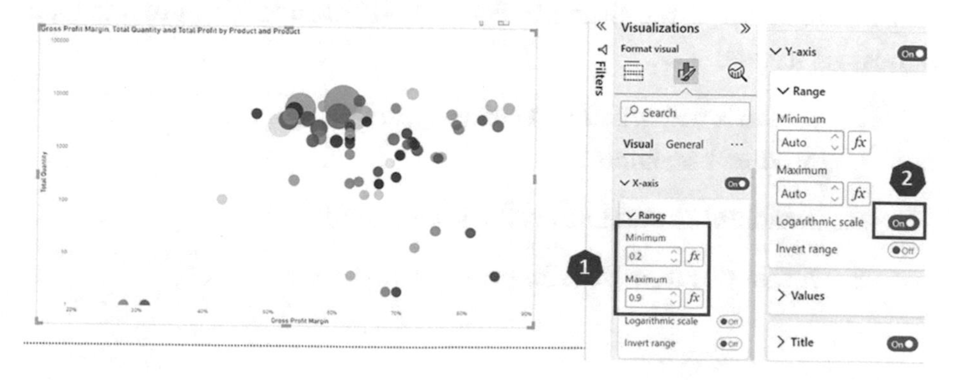

Figure 11-11. Spread out the circles on the scatter plot

Add the horizontal and vertical lines. You can do this using the following steps, as shown in Figure 11-12.

1. Add X-Axis Constant Line and set the value to 0.7.

2. Add Y-Axis Constant Line and set the value to 2000.

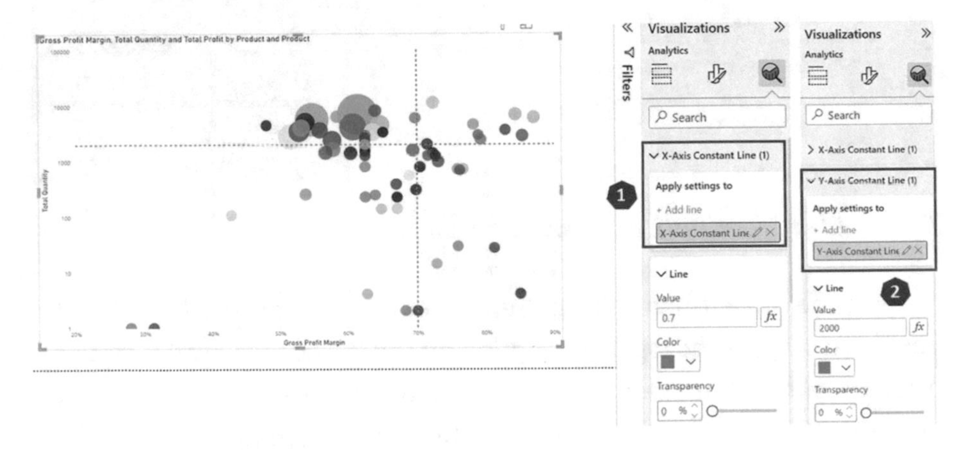

Figure 11-12. *Adding horizontal and vertical lines*

You can now add more visuals to provide additional information to the user. The following three visuals in Figure 11-13 are added to provide total values and trends.

1. A matrix with a product category, subcategory, and product breakdown

2. A monthly trend line for the total quantity

3. A monthly trend line for gross profit margin

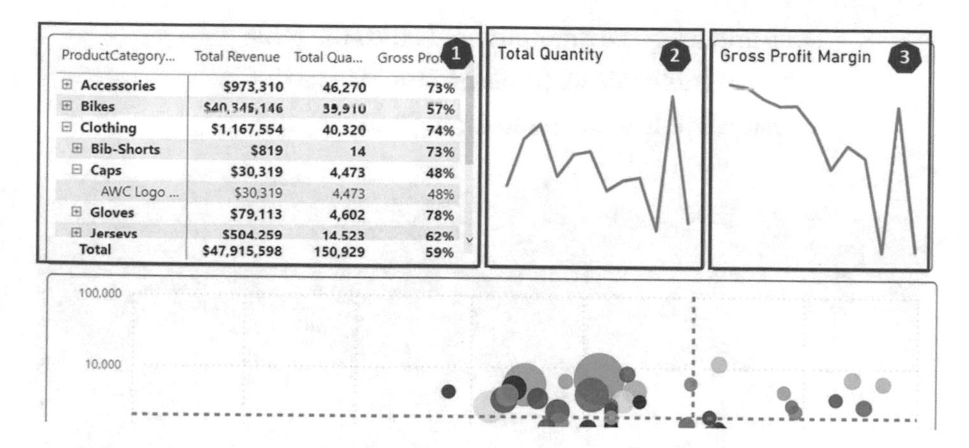

Figure 11-13. *additional visualizations to add information*

The matrix visual was selected to be used as a Summary table and for filtering.

Note For a better interactive experience, you may want to fix the value range for both axes. This allows you to know the strategic position of the selected products relative to other products.

Create an Embedded App

With the report and data ready, you can start to create the embedded app on the right-hand side of the screen. Power BI can only pass a single table into the app. This means you must consider what data to include in the table and later be passed into the app. You can do this by creating a table visual and later converting this into the app visual. The following steps explain how to do this, as shown in Figure 11-14.

1. Select the table visual.
2. Resize the table visual.

3. Drag and drop the different fields into the table columns (note this includes product ID, as the decision needs to be made for individual products).

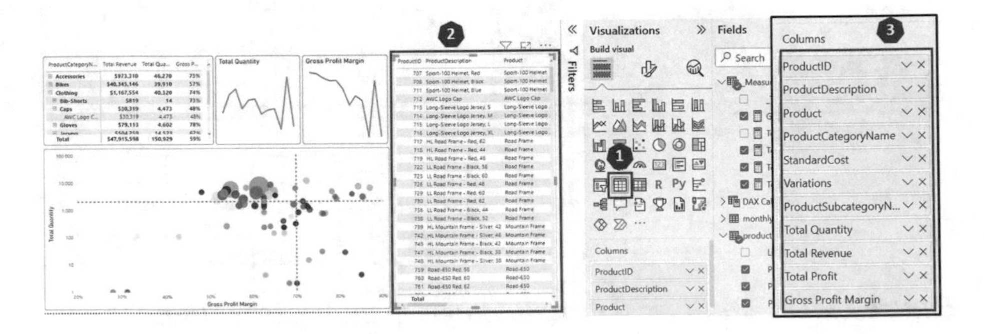

***Figure 11-14.** Creating the table for the app*

Because the embedded app has no column limit, you can drag most columns. You can decide later about which column is required in the app. You can also remove/change /add columns later.

Now you can convert this table into Power Apps using the following process, as shown in Figure 11-15.

1. Make sure to first select the table and then click the Power Apps for Power BI visual.

2. Click Create New.

3. Wait for the pop-up window to appear and select OK to open the browser.

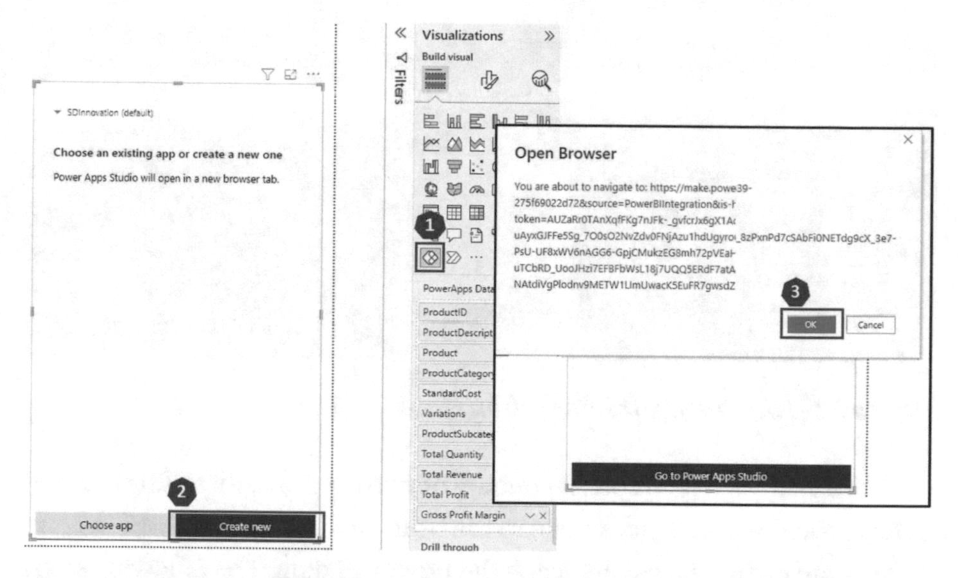

Figure 11-15. *Create a new app*

Because the app requires Power BI data input, you almost always start by creating a new app.

For now, you can follow the Power Apps online webpage link, as shown in Figure 11-16. It defaults to the Mobile layout, which fits the portrait design well. More importantly, you can now see the PowerBIIntegration icon between App and Screen1 on the menu. If you remember from the chapter on Power Apps, the order means that Power BI integration happens after app start and before Screen1 is loaded. The order is important because you can't load Power BI data until the Screen1 loads.

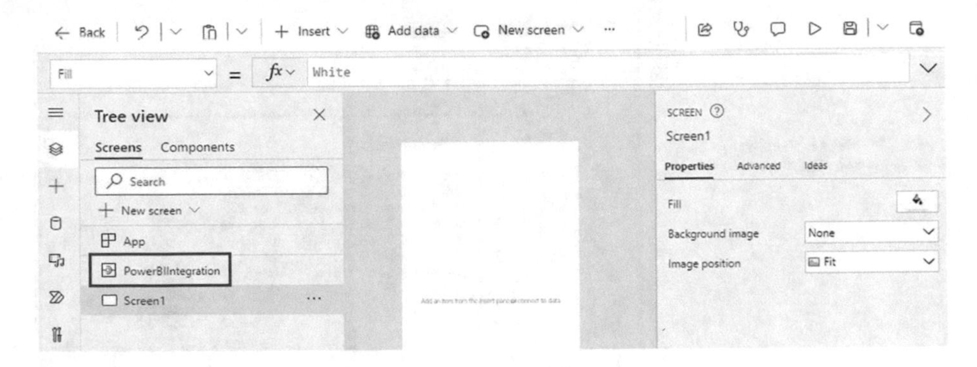

Figure 11-16. *Power BI integration data*

The app design is simple; it consists of two sessions, including a gallery in the top and a user input screen in the bottom.

You can create the gallery using the Power BI data. The following steps explain how to do this, as shown in Figure 11-17.

1. Drag and drop **Blank vertical gallery** onto the screen.

2. Resize the gallery to fit the top half of the screen.

3. Select the Item property of the gallery and enter **PowerBIIntegration.Data**. This contains the entire Power BI table data up to 1,000 records.

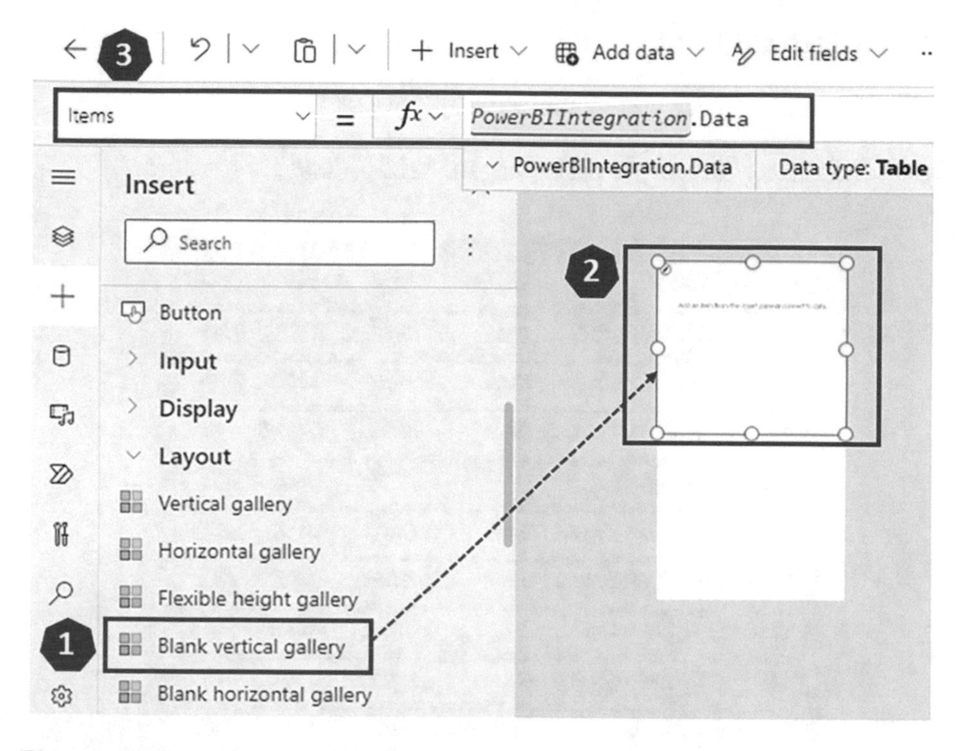

***Figure 11-17.** Create a vertical gallery with data from Power BI*

You can test the connection by dragging and dropping a text field into the first row in the gallery. You should be able to use ThisItem.Product to obtain a list of Products. If ThisItem cannot bring up any values, you may have entered the wrong column name, or the connection is not created properly. You may need to close the app without saving, go back to Power BI, and try to convert the table into Power App to create a new app again.

Once the PowerBIIntegration.Data can bring in the data fields. You can further customize the top half of the app into something similar to Figure 11-18. The following formatting codes are used for the quantity (Qty), revenue (Rev), and gross profit margin (GPM) columns. The screen also contains the hover effect using a rectangle box.

```
Text(ThisItem.'Total Quantity', "0,0")

Text(ThisItem.'Total Revenue'/1000, "$0,0K")

Text(ThisItem.'Gross Profit Margin'*100, "0%")
```

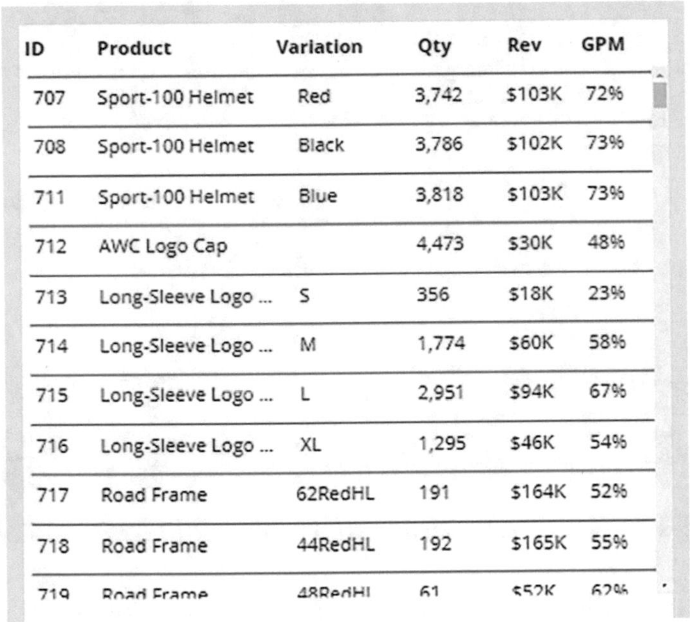

ID	Product	Variation	Qty	Rev	GPM
707	Sport-100 Helmet	Red	3,742	$103K	72%
708	Sport-100 Helmet	Black	3,786	$102K	73%
711	Sport-100 Helmet	Blue	3,818	$103K	73%
712	AWC Logo Cap		4,473	$30K	48%
713	Long-Sleeve Logo ...	S	356	$18K	23%
714	Long-Sleeve Logo ...	M	1,774	$60K	58%
715	Long-Sleeve Logo ...	L	2,951	$94K	67%
716	Long-Sleeve Logo ...	XL	1,295	$46K	54%
717	Road Frame	62RedHL	191	$164K	52%
718	Road Frame	44RedHL	192	$165K	55%
719	Road Frame	48RedHL	61	$52K	62%

Figure 11-18. *Gallery configuration*

You can also add the form to capture the decision using the following design, as shown in Figure 11-19. Note that the top text fields in the form use Gallery1.Selected.column_name to display the selected item from the gallery.

712 Clothing → Caps → AWC Logo Cap
Full Description: AWC Logo Cap
Decision: Terminate
Priority: High
Assign to: Email address
Comments: Comment about the decision
Submit

Figure 11-19. *The bottom half of decision-making form*

To avoid losing the hard work until now, you can save this app before moving on to the next task. You know that a SharePoint list is required to store this data. Knowing the required fields, you can create a SharePoint list, as shown in Figure 11-20.

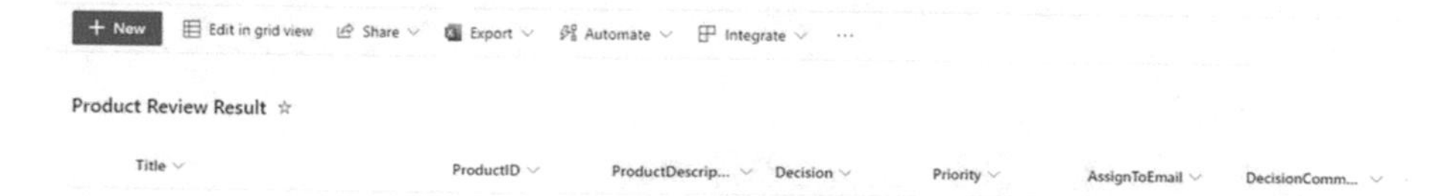

Figure 11-20. *SharePoint list columns*

You can add the new Product Review Result list to the app connection, as shown in Figure 11-21.

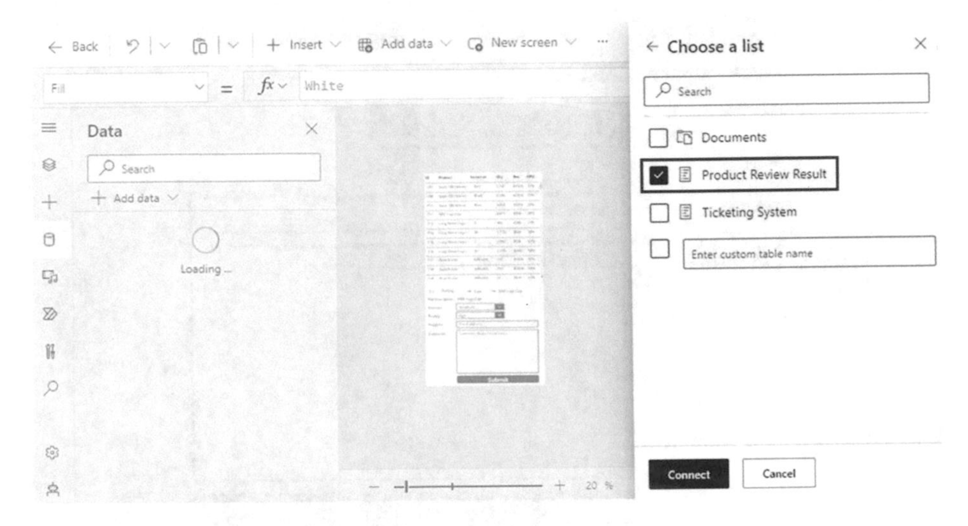

Figure 11-21. *Connect to the SharePoint list*

You can now configure the Submit button using the following code to save the detail to the SharePoint list.

```
Patch(
    'Product Review Result',
    Defaults('Product Review Result'),
    {
        Title: "X",
        ProductID: lblProductID.Text,
        ProductDescription: lblProductDesc.Text,
        Decision: drpDecision.SelectedText.Value,
        Priority: drpPriority.SelectedText.Value,
        AssignToEmail: ipAssignTo.Text,
        DecisionComments: ipComment.Text
    }
)
```

You can save and test this from the Power BI report again. It is important to publish the app once you're ready to release the changes you've made by clicking the Publish button, as shown in Figure 11-22.

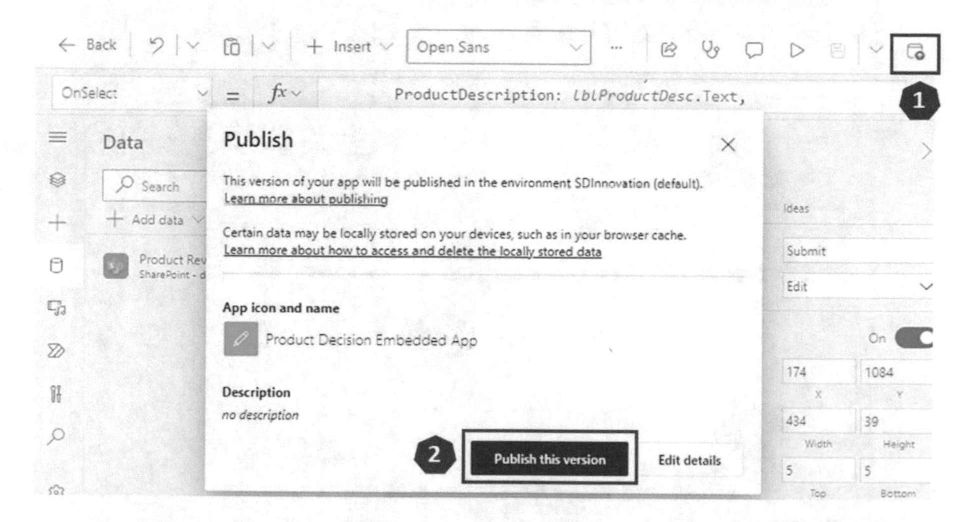

***Figure 11-22.** Publishing the changes*

First, you must ensure the app is properly embedded into Power BI. The following steps explain how to do this, as shown in Figure 11-23.

1. Select the app and convert it into a table visual.
2. Convert the table into an app again.
3. Click Choose App and select the app you have just created.

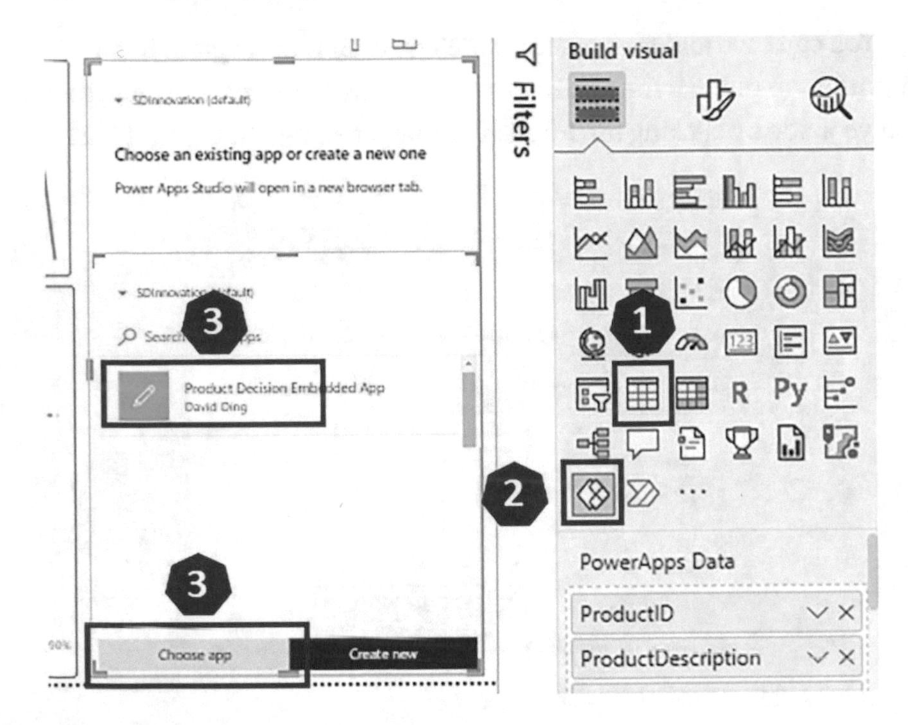

Figure 11-23. *Embed the app*

Once it's loaded, you can test it. The following steps explain how to do this, as shown in Figure 11-24.

1. Choose an outlier at the bottom left-hand corner with a low GPM and quantity.

2. Select the product.

3. Fill out the form.

4. Click Submit.

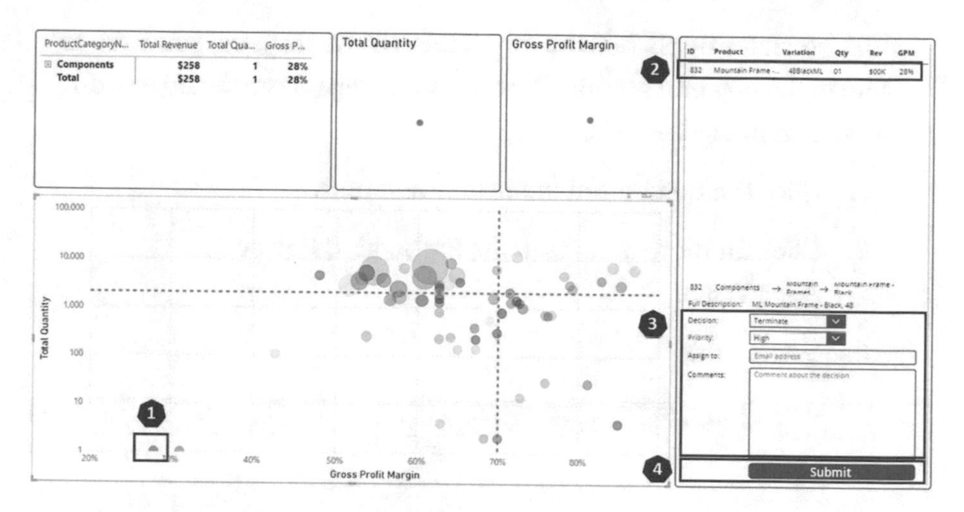

Figure 11-24. Testing the app

If the app is configured properly, you should see the SharePoint list populated as you submit the decisions, as shown in Figure 11-25.

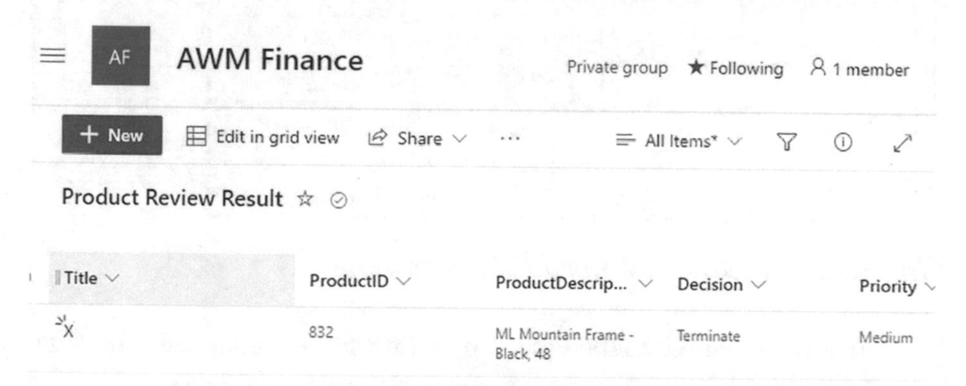

Figure 11-25. Populated SharePoint list

You can now submit about 10 to 20 more decisions, so you have enough data to go into the next step. When making further changes, you can also publish this report into the Power BI Workspace for a better integration experience between Power BI and Power Apps.

After you have published the report into the workspace, you need to navigate to the app edit screen. The following steps describe how to do this, as shown in Figure 11-26.

1. Click the Edit button in the online report.
2. Click for more options in the Embedded Power Apps Visual.
3. Select Edit.

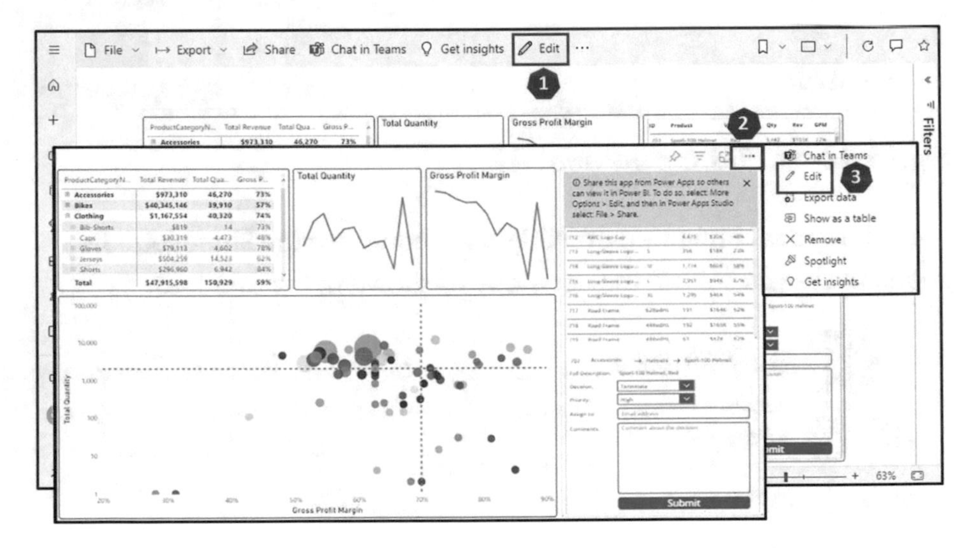

Figure 11-26. *Edit the Embedded Power Apps*

You now be taken into the Power Apps Edit screen again with the data from Power BI ready and loaded into PowerBIIntegration.Data.

Include Decisions Back into the App

To avoid creating multiple conflicting decisions on the same product ID, you need to determine if the product already exists in the list in the app. You can check this using the filter function in the Decision drop-down, as shown in Figure 11-27. The code in the Default property displays the Decision field if the ProductID matches any existing records in the SharePoint list.

```
First(
  Filter(
    'Product Review Result',
    ProductID = lblProductID.Text
  )
).Decision
```

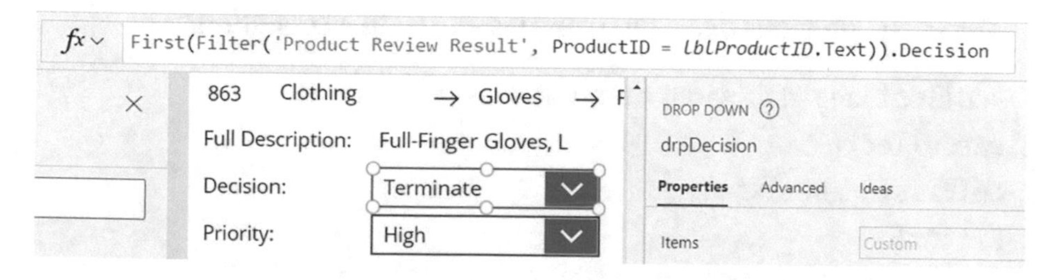

Figure 11-27. *Decision drop-down formula example*

You can repeat this for Priority, Assign to, and Comments. You can also add another two text labels that show the last decision date, as shown in Figure 11-28. If no prior decision has been made on the product, the date field does not display any value.

```
Text(
  First(
    Filter(
      'Product Review Result',
      ProductID = lblProductID.Text
    )
  ).Modified,
  "yyyy-mm-dd"
)
```

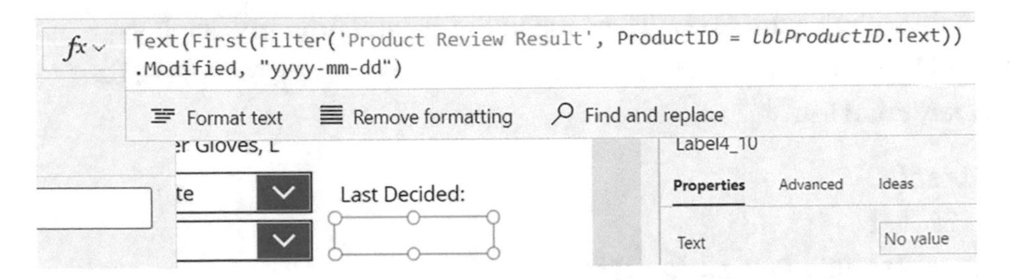

Figure 11-28. Display the last decision date

The OnSelect property of the Submit button is implemented in two steps.

First, the previous decision (if any) is stored in a collection.

```
// collect any decision that matches ProductID
ClearCollect(
  colExistingDecision,
  Filter(
    'Product Review Result',
    ProductID = lblProductID.Text
  )
);
```

Second, an IF logic is implemented to update existing decisions or create a new decision record.

```
If(
  // check if the prior decision exist
  CountRows(colExistingDecision) > 0,
  // if prior decision exist, update existing decision
  Patch(
    'Product Review Result',
    First(
```

```
        Filter(
          'Product Review Result',
          ProductID = lblProductID.Text
        )
      ),
      {
        Title: "X",
        ProductID: lblProductID.Text,
        ProductDescription: lblProductDesc.Text,
        Decision: drpDecision.SelectedText.Value,
        Priority: drpPriority.SelectedText.Value,
        AssignToEmail: ipAssignTo.Text,
        DecisionComments: ipComment.Text
      }
    ),

    // if no prior decision exist, create new record
    Patch(
      'Product Review Result',
      Defaults('Product Review Result'),
      {
        Title: "X",
        ProductID: lblProductID.Text,
        ProductDescription: lblProductDesc.Text,
        Decision: drpDecision.SelectedText.Value,
        Priority: drpPriority.SelectedText.Value,
        AssignToEmail: ipAssignTo.Text,
        DecisionComments: ipComment.Text
      }
    )
);
```

You can now save and publish this app. If you run the test in Power BI, you see the previous decision displayed, as shown in Figure 11-29.

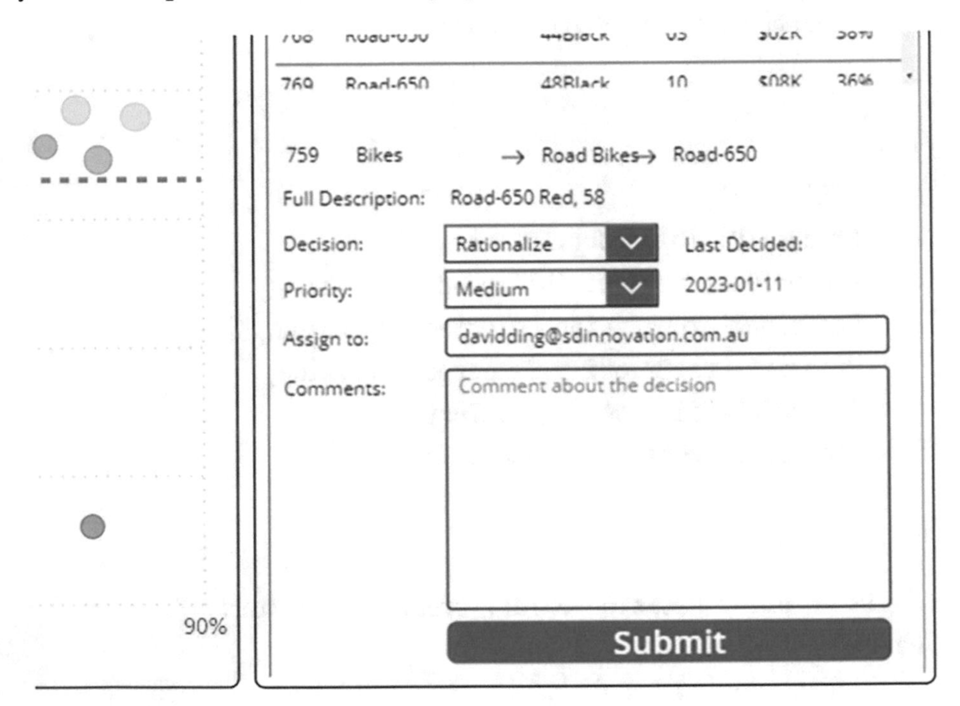

Figure 11-29. *Display previous decision*

Include Decisions Back into the Report

It is also useful to integrate the decision data back into the report. This help decision-makers track the product review progress. To do this, open the saved Power BI report .pbix file from Power BI Desktop.

Add the SharePoint list by following these steps, as shown in Figure 11-30.

1. Choose **Get data**.
2. Select **More** options.

3. Search SharePoint and choose SharePoint Online List.

4. Enter the SharePoint site URL.

5. Select 2.0.

6. Choose Default and click OK.

7. Select the list and click Load (next screen, not in this figure).

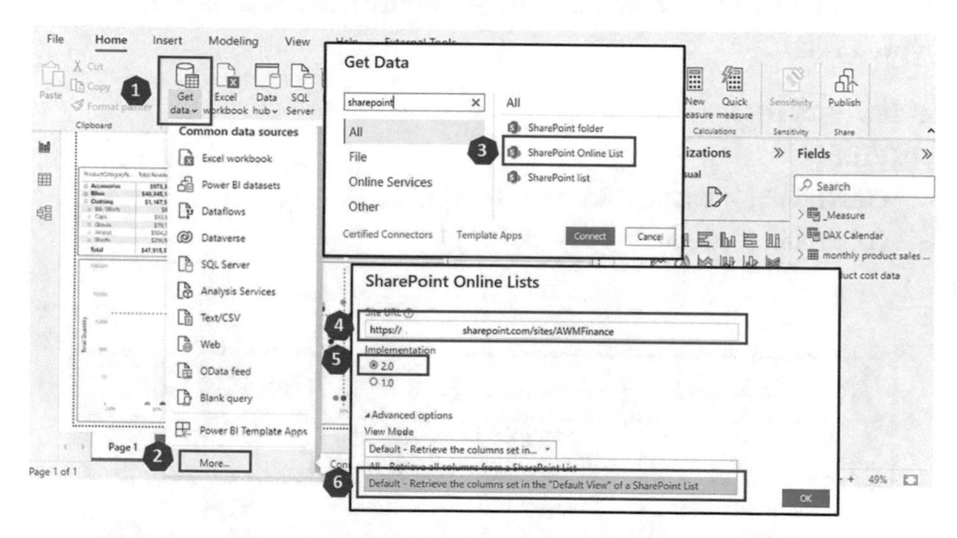

***Figure 11-30.** Adding SharePoint list*

You can add the SharePoint list using the data model shown in Figure 11-31. The decision table is connected to the product cost data table via ProductID.

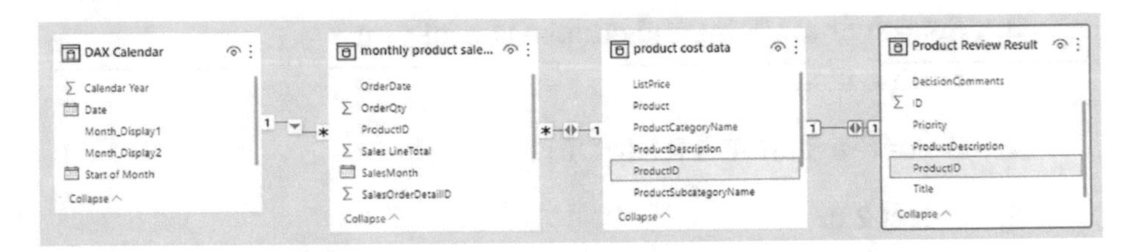

***Figure 11-31.** Data model with decision data table*

You can now create a measure called *review progress* using the following DAX code and add it to the Summary table as shown in Figure 11-32.

```
Review Progress =
  DIVIDE(
    COUNTROWS('Product Review Result'),
    COUNTROWS('product cost data')
  )
```

ProductCategoryN...	Total Revenue	Total Qua...	Gross P...	Revie...
⊞ **Accessories**	**$973,310**	**46,270**	**73%**	**3%**
⊟ **Bikes**	**$40,345,146**	**39,910**	**57%**	**11%**
⊞ Mountain Bikes	$13,831,962	12,245	61%	
⊞ Road Bikes	$13,882,412	14,520	55%	26%
⊞ Touring Bikes	$12,630,772	13,145	55%	
⊞ **Clothing**	**$1,167,554**	**40,320**	**74%**	**3%**
⊞ **Components**	**$5,429,588**	**24,429**	**64%**	**1%**
Total	**$47,915,598**	**150,929**	**59%**	**5%**

***Figure 11-32.** Add review progress*

Once you publish this report to the Power BI cloud, schedule the report to be refreshed eight times daily. The following steps explain how to do this, as shown in Figure 11-33.

1. Select dataset schedule refresh from the workspace.
2. Turn on **Scheduled refresh**.
3. Select the frequency and time zone.
4. Add the times.

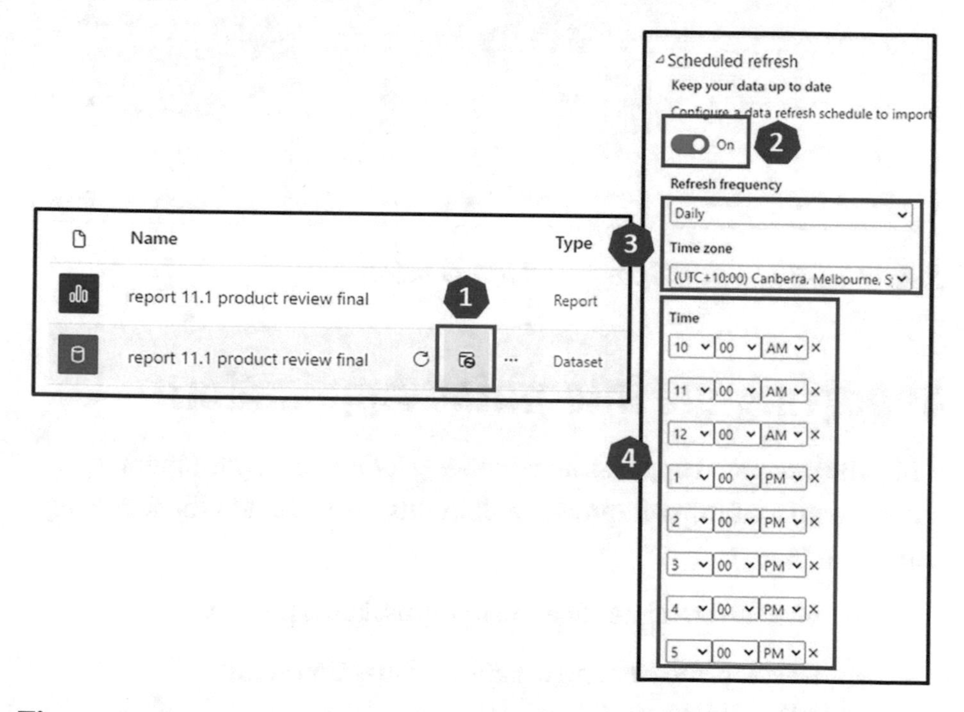

Figure 11-33. *Setup schedule refresh*

Congratulations on developing your first integrated application, as shown in Figure 11-34. This application can be powerful in driving effective decisions. It can break traditional barriers in large organizations to improve delivery outcomes and efficiency when used effectively in management and operational meetings.

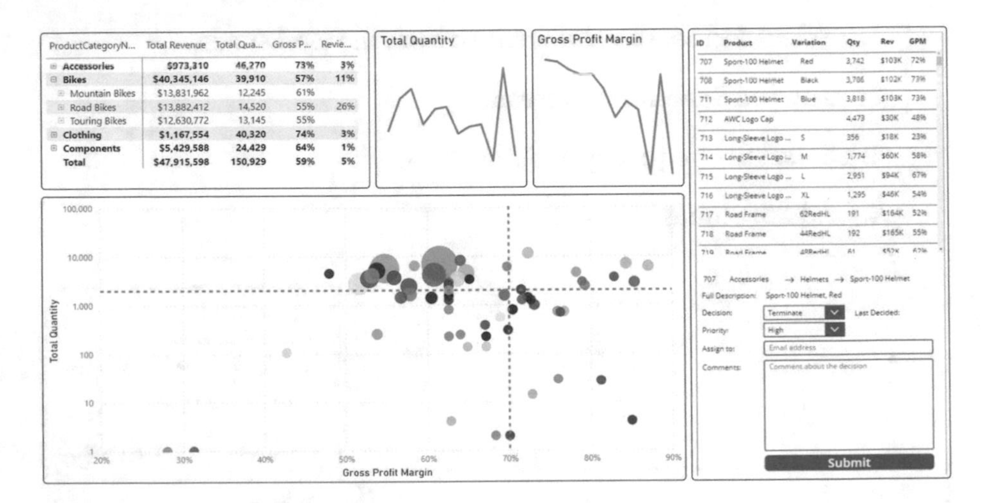

Figure 11-34. *Initial integrated application*

Improving the Integrated Application

While the integrated application is working for the most part, there is always room for further improvement. In this case, you add the following features.

- Send an email notification to the assigned person.
- Develop another app to capture the action status following the decision.

Add Email Notification

You can do this via Power Automate. If you have implemented the email notification process in the previous chapter, you can re-use the same flow by calling it from the Submit OnSelect property. The following steps explain how to do this, as shown in Figure 11-35.

1. Go to the Flow pane in Power Apps and click **Add flow**.

2. Select the email flow from the previous chapter.

3. Select the Submit button and go to the OnSelect property.

4. At the end of the property, insert the following code to populate the required fields in the email flow.

```
// send email notification
Chpt10TestEmail.Run(
  ipAssignTo.Text,
  "A new product reivew task has been assigned to you",
  "Product ID: " & lblProductID.Text
    & " has been flagged for " & drpDecision.
    SelectedText.Value
    & ". Please review."
)
```

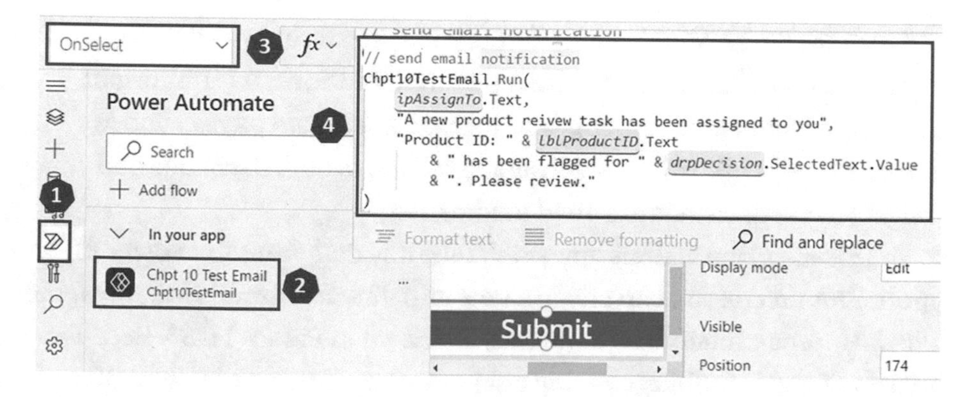

***Figure 11-35.** Adding email notification flow*

When the button is pressed, an email is sent to the assigned email address to notify the person about the decision.

Add Action Status Tracker

After the assignee has been notified about the decision, he/she needs to act upon it. This process can also be incorporated into the integrated application via an additional app.

Sometimes, one decision can lead to multiple actions and even across multiple teams. It is possible to have a separate SharePoint list to create a one-to-many relationship between the decision and the actions. For now, you can assume the same decision SharePoint list can be modified to incorporate this new requirement by adding the following new columns.

- **Action status** tracks if a decision has been made.
- **Action date** records the action date (an automatic date).
- **Action detail** records more detail about the action taken.
- **Action comments** records any additional comments related to the action.

Once the SharePoint list is updated, you can continue to develop the app to record the actions. With integrated solutions, it is important to consider where to fit this app. You include a separate page called My Actions for this app. This allows users to see the overall decisions and navigate to the decisions assigned to him/her.

At this stage, the actions app won't take any data from the Power BI report. This means you can create a new app directly from Power Apps; the following steps explain how to do this, as shown in Figure 11-36. Because the default app is mobile, creating a separate app outside of Power BI provides more design flexibility.

1. Go to make.powerapps.com and select Create.
2. Click **Blank app**.

3. Select **Blank canvas app**.

4. Name the app.

5. Choose Tablet.

6. Click Create.

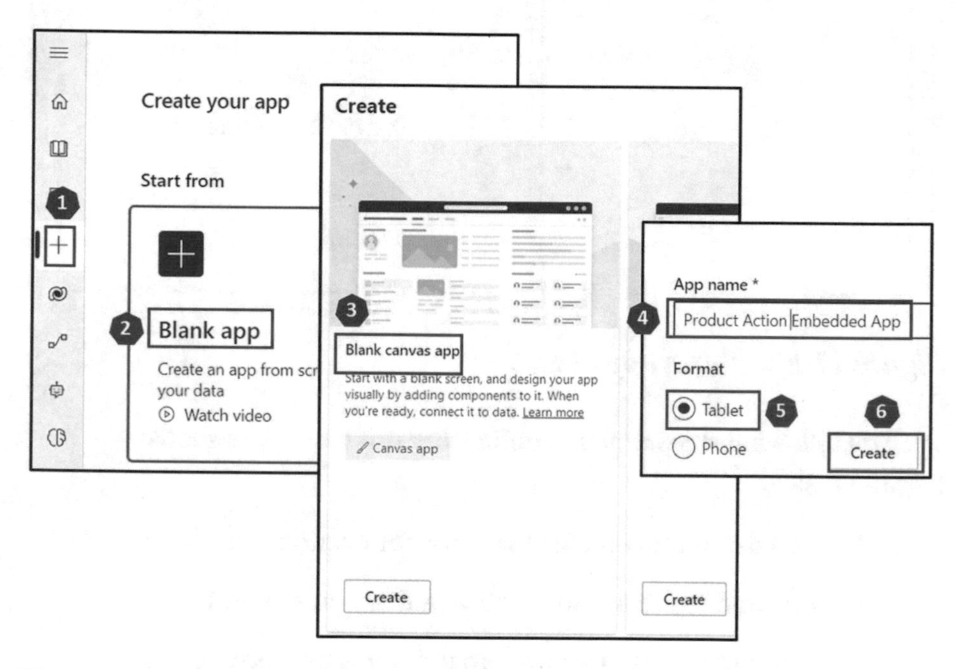

Figure 11-36. *Creating a new blank canvas tablet app*

Once the blank app is created, you can follow the tablet app design in Chapter 9 to have a gallery on the left and a data input form on the right, as shown in Figure 11-37.

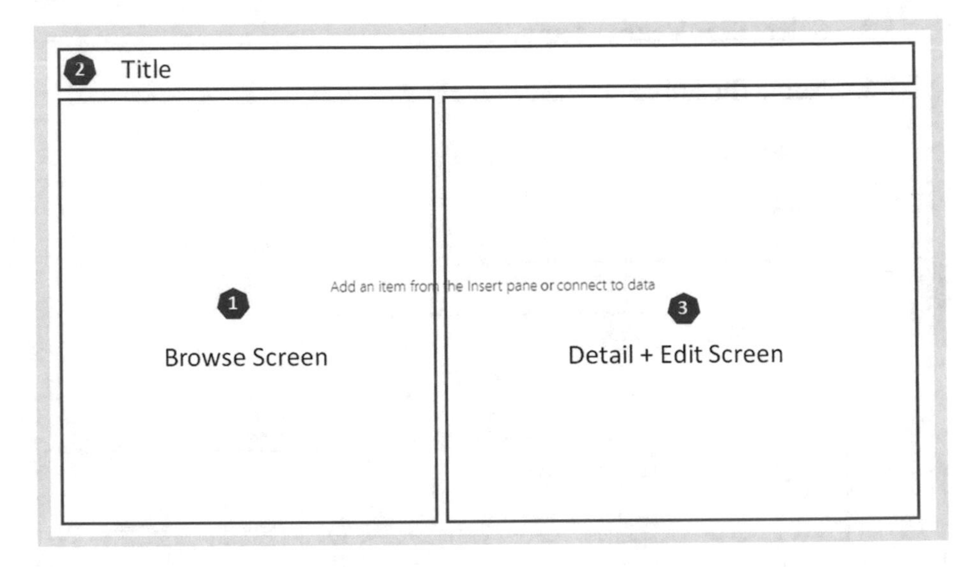

Figure 11-37. *Tablet app design*

The following steps explain configuring this gallery, as shown in Figure 11-38.

1. Add the SharePoint list to a data connection.
2. Add and resize a blank vertical gallery on screen 1.
3. Insert the Filter function into the Items property of the gallery.
4. Add the title labels on top of the gallery.
5. Configure the first row of the gallery.

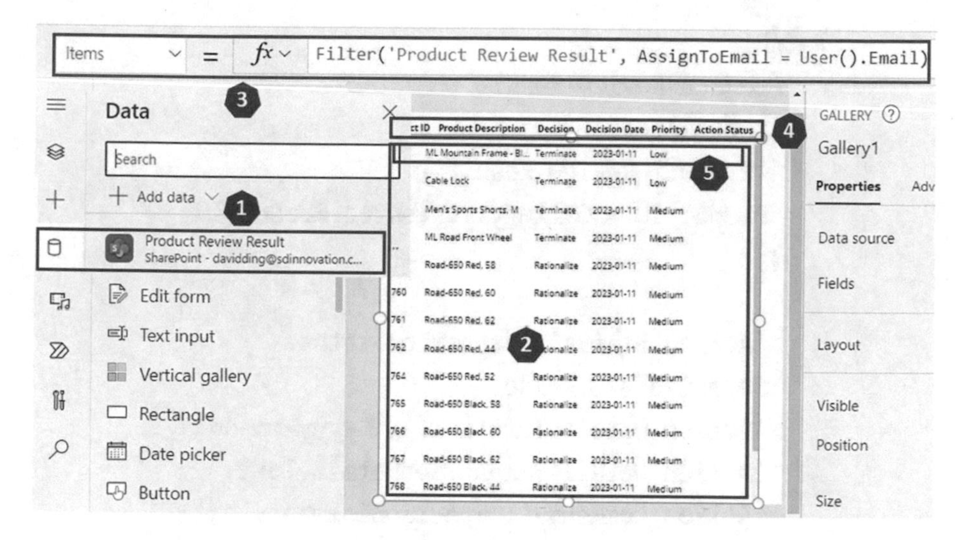

***Figure 11-38.** Create and configure the gallery*

You can create the user input fields. The following steps explain how to do this, as shown in Figure 11-39.

1. Bring existing decision fields into the screen.
2. Create an Action Status drop-down list.

   ```
   ["Backlog", "In Progress", "Testing", "Completed",
   "Cancelled"]
   ```

3. Create the Action Details and Comments input fields.
4. Create a Submit button.
5. In the OnSelect property of the Submit button, enter the patch function code to update the record.

```
Patch(
  'Product Review Result',
  First(Filter(
    'Product Review Result',
    ProductID = Gallery1.Selected.ProductID
  )),
  {
    'Action Status': drpActionStatus.
    SelectedText.Value,
    'Action Date': Text(Now(), "yyyy-mm-dd"),
    'Action Detail': ipActionDetail.Text,
    'Action Comments': ipActionComment.Text
  }
)
```

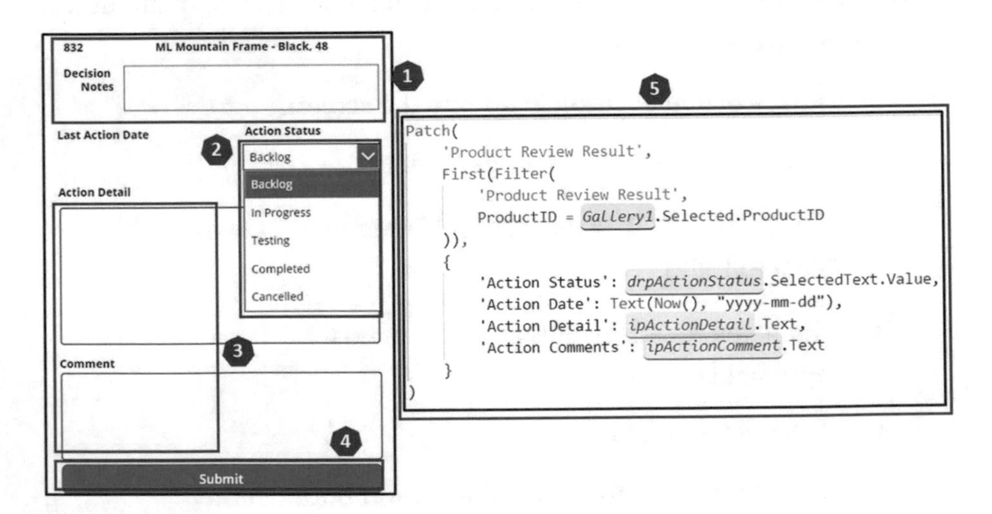

Figure 11-39. *Configure the user input fields*

Try to run a few tests on this app to see if it is working, as shown in Figure 11-40.

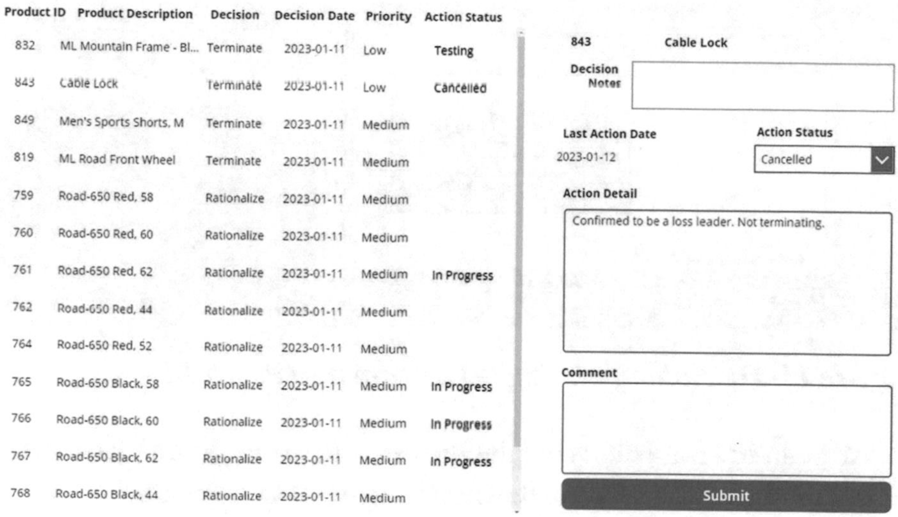
Product ID
Product Description
Decision
Decision Date
Priority
Action Status
832 ML Mountain Frame - Bl... Terminate 2023-01-11 Low Testing
843 Cable Lock Terminate 2023-01-11 Low Cancelled
849 Men's Sports Shorts, M Terminate 2023-01-11 Medium
819 ML Road Front Wheel Terminate 2023-01-11 Medium
759 Road-650 Red, 58 Rationalize 2023-01-11 Medium
760 Road-650 Red, 60 Rationalize 2023-01-11 Medium
761 Road-650 Red, 62 Rationalize 2023-01-11 Medium In Progress
762 Road-650 Red, 44 Rationalize 2023-01-11 Medium
764 Road-650 Red, 52 Rationalize 2023-01-11 Medium
765 Road-650 Black, 58 Rationalize 2023-01-11 Medium In Progress
766 Road-650 Black, 60 Rationalize 2023-01-11 Medium In Progress
767 Road-650 Black, 62 Rationalize 2023-01-11 Medium In Progress
768 Road-650 Black, 44 Rationalize 2023-01-11 Medium
843
Cable Lock
Decision Notes
Last Action Date
2023-01-12
Action Status
Cancelled
Action Detail
Confirmed to be a loss leader. Not terminating.
Comment
Submit

Figure 11-40. *Testing the Actions app*

Add this app to the Power BI report. The following steps explain how to do this, as shown in Figure 11-41. Embedded Power Apps must take some data from the Power BI report.

1. Create a new report page.
2. Add a Power Apps visual and resize it to fit the screen.
3. Add ProductID to the Power Apps data field.
4. Select **Choose app**.
5. Select the Actions app and add it.

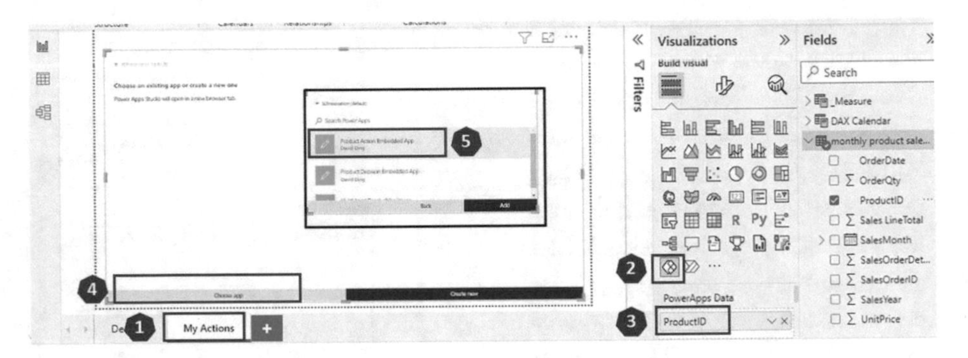

Figure 11-41. *Adding existing action app into Power BI*

You can add the action completion measure to the Summary table, as shown in Figure 11-42, by creating the following DAX measures.

```
// count completed actions, status = Completed or Cancelled
Actions Completed =
  CALCULATE(
    COUNTROWS('Product Review Result'),
    (
      'Product Review Result'[Action Status] = "Cancelled"
      || 'Product Review Result'[Action Status] = "Completed"
    )
  )

// calculate completion %
Action Completion =
  DIVIDE(
    [Actions Completed],
    COUNTROWS('product cost data')
  )
```

ProductCategoryN...	Total Revenue	Total Qua...	Gross P...	Review Pro...	Action Comp...
⊞ **Accessories**	**$973,310**	**46,270**	**73%**	**3%**	**3%**
⊟ **Bikes**	**$40,345,146**	**39,910**	**57%**	**11%**	
⊞ Mountain Bikes	$13,831,962	12,245	61%		
⊞ Road Bikes	$13,882,412	14,520	55%	26%	
⊞ Touring Bikes	$12,630,772	13,145	55%		
⊞ **Clothing**	**$1,167,554**	**40,320**	**74%**	**3%**	
⊞ **Components**	**$5,429,588**	**24,429**	**64%**	**1%**	
Total	**$47,915,598**	**150,929**	**59%**	**5%**	**0%**

Figure 11-42. *Adding the action completion % measure to the Summary table*

Once you publish this into the Power BI cloud, the users can navigate to the My Action page to perform the assigned actions, as shown in Figure 11-43.

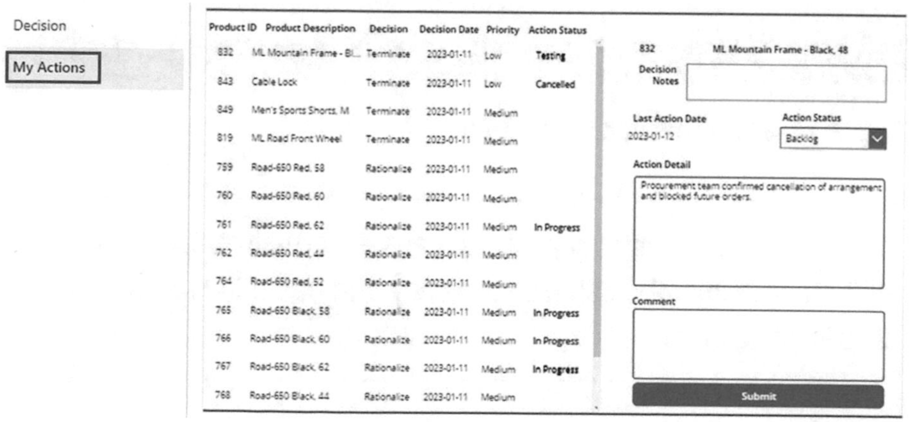

Decision

My Actions

Product ID	Product Description	Decision	Decision Date	Priority	Action Status
832	ML Mountain Frame - Bl...	Terminate	2023-01-11	Low	Testing
843	Cable Lock	Terminate	2023-01-11	Low	Cancelled
849	Men's Sports Shorts, M	Terminate	2023-01-11	Medium	
819	ML Road Front Wheel	Terminate	2023-01-11	Medium	
759	Road-650 Red, 58	Rationalize	2023-01-11	Medium	
760	Road-650 Red, 60	Rationalize	2023-01-11	Medium	
761	Road-650 Red, 62	Rationalize	2023-01-11	Medium	In Progress
762	Road-650 Red, 44	Rationalize	2023-01-11	Medium	
764	Road-650 Red, 52	Rationalize	2023-01-11	Medium	
765	Road-650 Black, 58	Rationalize	2023-01-11	Medium	In Progress
766	Road-650 Black, 60	Rationalize	2023-01-11	Medium	In Progress
767	Road-650 Black, 62	Rationalize	2023-01-11	Medium	In Progress
768	Road-650 Black, 44	Rationalize	2023-01-11	Medium	

832 ML Mountain Frame - Black, 48

Decision Notes

Last Action Date 2023-01-12

Action Status Backlog

Action Detail

Procurement team confirmed cancellation of arrangement and blocked future orders.

Comment

Submit

Figure 11-43. *My Actions page*

Access Management

Previously, you have looked at access management separately for Power BI, Power Apps, and SharePoint lists. Unfortunately, there is no simple way to manage access across the systems at the time of writing. This is mainly because SharePoint lists and Power Apps do not accept email distribution lists. The only option to centrally manage this would be via Active Directory Security Groups. The creation and management of Security Groups are normally managed through IT, and the process can be complicated.

When granting user access, you must usually add the users to three separate places. This can become very messy very quickly. You are strongly advised to use an Excel spreadsheet (or other methods) to manage user access for each integrated solution. An example User Access Register is shown in Figure 11-44.

User Email	Approved By	Access Date	Power Apps	SharePoint List	Power BI	PBI RLS
you@company.com	your manager@company.com	1/10/2022	Yes	Yes	Yes	NA

Figure 11-44. *User Access Register*

Tip The user's company email address is normally used across all three systems. You can also use copy and paste to create initial access using email.

Conclusion

Congratulations, this is the final chapter of the book. You learned to build an integrated application using integrated solution architecture. The integrated application incorporates everything you have learned in this book.

Now everyone is happy with Kim's work in AWM. The integrated solution is so powerful that it cuts through the organization's red tape and functional boundaries. The near real-time reporting solution also greatly improved transparency and allowed the leaders to understand the bottleneck in the review and implementation process. Kim's solution is estimated to be half the cost of similar product review programs.

Mini-Hackathon

You are an investment analyst. You are tasked to identify countries with investment opportunities using International Monetary Fund (IMF) datasets (`www.imf.org/en/Data`).

Objective

Build an integrated application using IMF datasets to help fund managers to review opportunities by country based on analysis of historical data by country. The integrated application must include the following features.

- Detailed analysis of the relative performance of each country over time
- High-level risk and opportunity analysis
- The ability for fund managers to record decisions in the application and assign tasks
- The ability for leaders to monitor and review progress daily

Time Limit

The objective should be completed within seven hours.

Final Thoughts

The objective of this book is to empower you. If you feel empowered, this feeling should come from acquiring new knowledge to solve real business problems. If you enjoy this feeling, you may consider the following actions to sustain it over a long period of time.

- I recommend another book on Power BI: *The Definitive Guide to DAX* by A. Ferrari and M. Russo (Microsoft Press, 2019).
- Help other people with their Power Platform questions. While establishing your credibility, you gain much wider exposure to problems outside your work/team/organization. If you don't find many problems within your immediate circle, Stack Overflow (`https://stackoverflow.com/questions/tagged/powerbi`) could be a good source.
- Learn Python. Python is seen as the second-best programing language for everything. You need to learn the basics of Python to open the door to a whole new world of possibilities.

This book also incorporated a practical and agile approach to problem-solving MVP (a minimal viable product). You can learn more about this from the book *The Lean Startup* by Eric Ries (Currency, 2011).

At last, your mindset is critical to your future. I also suggest *Extreme Ownership* by Jocko Willink and Leif Babin (St. Martin's Press, 2017).

I hope you enjoyed the book. If you have any feedback, questions, or business opportunities, feel free to contact me via my LinkedIn profile at `www.linkedin.com/in/david-ding-38442721/`. I look forward to hearing from you.

Index

A

B

D. Ding, *Transitioning to Microsoft Power Platform*,
https://doi.org/10.1007/978-1-4842-9239-6

C

D, E, F

G

H

I, J

K, L

M, N

U

V, W, X, Y, Z

Printed in the United States
by Baker & Taylor Publisher Services